Genius: the natural history of creativit[illegible]
creativity, based on the personality [illegible]
geniuses. Starting with the fact that [illegible]
psychopathology, it uses modern resea[illegible] the causes of cognitive over-inclusiveness to suggest possible applications of these theories to creativity. Professor Eysenck reports experimental research to support these theories in their application to creativity, as well as considering the role of intelligence, social status, gender and many other factors that have been linked with genius and creativity. The theory traces creativity from DNA through personality to special cognitive processes to genius.

Problems in the Behavioural Sciences

Genius

Problems in the Behavioural Sciences

1. Contemporary animal learning theory
   A. DICKINSON
2. Thirst
   B. J. ROLLS & E. T. ROLLS
3. Hunger
   J. LE MAGNEN
4. Motivational systems
   F. TOATES
5. The psychology of fear and stress
   J. A. GRAY
6. Computer models of mind
   M. A. BODEN
7. Human organic memory disorders
   A. R. MAYES
8. Biology and emotion
   N. MCNAUGHTON
9. Latent inhibition and conditioned attention theory
   R. E. LUBOW
10. Psychobiology of personality
    M. ZUCKERMAN
11. Frustration theory
    A. AMSEL
12. Genius
    H. J. EYSENCK

Genius

The natural history of creativity

H. J. Eysenck
Institute of Psychiatry, University of London

Published by the Press Syndicate of the University of Cambridge
The Pitt Building, Trumpington Street, Cambridge CB2 1RP
40 West 20th Street, New York, NY 10011-4211, USA
10 Stamford Road, Oakleigh, Melbourne 3166, Australia

First published 1995

Printed in Great Britain at the University Press, Cambridge

A catalogue record for this book is available from the British Library

Library of Congress cataloging in publication data

Eysenck, H. J. (Hans Jurgen), 1916–
Genius : The natural history of creativity / H. J. Eysenck.
p. cm. – (Problems in the behavioural sciences : 12)
Includes bibliographical references and index.
ISBN 0 521 48014 0. – ISBN 0 521 48508 8 (pbk.)
1. Genius. 2. Creative ability. I. Title. II. Series.
BF412.E97 1995
153.9′8 – dc20 94-32136 CIP

ISBN 0 521 48014 0 hardback
ISBN 0 521 48508 8 paperback

SE

To Sybil,
who gave flesh and blood to the
abstract concept of 'psychoticism'
which may hold the key to a better
understanding of genius and creativity.

Contents

Introduction

The important thing in science is not so much to obtain new facts as to discover new ways of thinking about them.

W. L. Bragg

It is just over 50 years ago that I wrote my first article on the topic of intelligence (Eysenck, 1939). I was reviewing Thurstone's famous monograph (Thurstone, 1938) in which he criticized Spearman's (1927) theory of intelligence as a single entity or concept, spreading over all cognitive activities; he postulated instead a number of 'primary abilities', separate and uncorrelated. Reanalysing his very extensive data, I found evidence *both* for a general factor of intelligence, very much as postulated by Spearman, and also for a number of primaries, very much as postulated by Thurstone. Both Spearman (Spearman and Jones, 1950) and Thurstone (Thurstone and Thurstone, 1941) finally agreed that a hierarchical system of description of the cognitive space incorporating both a general factor of intelligence and various group factors or special talents, was most in line with the available facts, and there is now considerable consensus on some such system (Vernon, 1979; Eysenck, 1979; Brody, 1992; Carroll, 1993).

While quite happy with such an account at the descriptive level, I felt that if there was a strong genetic basis for IQ, as there undoubtedly is (Woodworth, 1941; Eysenck, 1979; Plomin, De Fries and McClearn, 1990), there must be some physiological or hormonal intermediaries between DNA and behaviour – it is impossible for DNA to influence behaviour directly. I have tried to discover some of the threads in question, and I believe that some important connections at least have been unearthed (Eysenck, 1982, 1983a, 1986a,b, 1987b, 1991b; Eysenck and Barrett, 1985, 1993b; Barrett, Daum and Eysenck, 1990; Barrett and Eysenck, 1993, 1994; Deary and Caryl, 1993).

This book, however, is not concerned with the biological basis of intelligence. Consider the three major conceptions of intelligence, as presented in Figure 0.1. We see that psychometric intelligence (IQ) is shown in the middle, determined in part by biological and genetic factors on the left, and itself partly responsible for social or practical intelligence, on the right (Eysenck, 1988). This latter concept of intelligence deals with the use or application of intelligence in everyday life, in business, education, the armed forces, com-

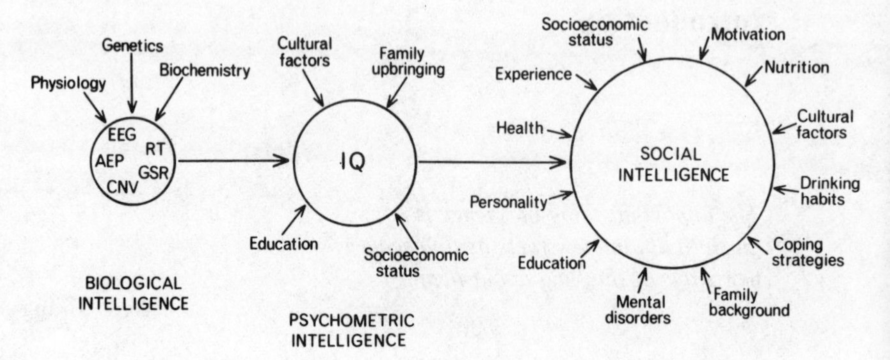

Fig. 0.1 The relations between biological intelligence and psychometric intelligence, and social or practical intelligence.

monsense, art, science, industry and so on. Clearly complexity increases as we go from left to right, and the scientific approach becomes more and more difficult and questionable as we approach such intricate and complicated matters.

This book is concerned with one particularly interesting part of this third aspect of intelligence, namely creativity and genius. From the beginning of my becoming acquainted with the literature on intelligence, I had been convinced that it was possible to study this concept scientifically. Critics tended to point out that there were important aspects of cognition that seemed to be beyond the scope of science, and creativity and genius were among those most frequently mentioned. Having many other research areas to interest me and take up time and energy, I tried not to think about these topics, but finally temptation became too strong, and I succumbed (Eysenck, 1983a, 1989a) and started to theorize about creativity, and by extension genius. I published my major conclusions in a kind of monograph (Eysenck, 1993a), but it became obvious that the whole subject was too complex to be dealt with in anything short of book form. Hence the present volume.

There are many books on these topics, and perhaps an explanation is due for inflicting another one on the general public. This book does not aspire to textbook status; there are many aspects of genius that are not treated in detail but only mentioned in passing. In essence what I have done is to try and answer a number of questions relevant to the study of genius and creativity from the psychological point of view (and of course there are many others). Among these questions are the following: Can genius be defined and measured? Can creativity be defined and measured? What role does intelligence play in the development of either? What is the contribution of personality? Is there any relation between genius and 'madness', and if so what is it? Can we formulate a cognitive theory to account for creativity, and describe the workings of the creative mind? Can we define and measure intuition, as one of the alleged characteristics of the creative person? What is the role of the unconscious, if any? These and many other, related questions are dealt with in this book; they seemed enough to fill its pages, without going into the large number of other questions that go to make up the many textbooks and handbooks on these subjects.

In this discussion of genius I have included both science and art, but I have left out a third aspect included for instance by Simonton (1984a,b) in his book on *Genius, Creativity and Leadership*. This whole topic of leadership is different in many ways from eminence in science or the arts, and there is considerable evidence (Brody and Page, 1975; Simonton, 1984a) suggesting that personal qualities, while very important in scientific and artistic achievement play a much smaller part in all the varieties of social leadership. As Simonton points out (Simonton, 1984b) his inquiry was partly inspired by a casual observation made by Sorokin (1925, 1926) to the effect that the most illustrious rulers tend to be the longest lived. If eminence is founded on a profusion of historical events, then this may be due to the following sequence of events. (1) Long life in monarchs increases the probability of their having a long reign. (2) During a long reign, more events of historical importance are likely to happen. (3) Distinction is a function of such procession of historical events. (4) The personal attributes of monarchs play little part in this chain.

Simonton (1984a,b) carried out a detailed historiometric investigation of this hypothesis. When he broke down the composite index of historical events into good and bad, he found that long tenure in itself contributed to a person's prominence as a leader regardless of outcome!

> Those monarchs whose reigns were damned with more battlefield defeats, territorial losses, economic difficulties, famines, massacres, and executions, earn eponymous credit as much as those monarchs whose reigns were blessed with more battlefield victories, territorial gains, economic improvements, treaties, and laws (p. 138).

Personal involvement did not seem to matter, either; it made no difference whether the ruler led his armies into battle personally, or let his officers do it. There was no difference in the influence exerted by events for which the ruler had immediate responsibility, or by those for which no responsibility could be attributed to him.

Simonton (1984a,b) published a path analysis of the many influences determining a ruler's eminence, using the historical achievements of 342 European rulers. The combined influence of intelligence, morality and leadership amounted to less than 10% of the total, situational factors appeared to be vastly more important. Simonton concludes that 'greatness as a leader seems to be less a matter of being the right person and more a matter of being in the right place at the right time' (p. 139). There may be some quibbles about such a conclusion, but they do not change the final verdict. For instance, the relationship between leader morality and historical eminence appears to be U-shaped; those who are neither particularly virtuous or vicious claim neither much action during their reign nor much fame posthumously. (A similar finding applies to presidential leadership.) Overall then, if our interest is in the psychological factors which promote or are related to genius, rulers and leaders, whether elected or by divine right, are not a very useful source of information. It is clearly preferable to concentrate on scientists and artists, and this I propose to do in this book. Some of the conclusions arrived at may also be relevant to leaders (particularly to military leaders like Napoleon, Alexander the Great, Hannibal, or Frederick the Great), but the argument is not based on them.

But can art and science be grouped together? Flaubert argued that 'poetry is as exact a science as physics', and Martindale (1990) has given some substance to this aphorism. We do not have to agree with Flaubert's comment to try and see whether the concept of 'creativity' does in fact cover science and art; let experiment decide!

There are two main ways to tackle the subjects of genius and creativity. As Fechner pointed out in his *Vorschule der Aesthetik* (Fechner, 1897), topics of this kind can be attacked 'von oben' (from above) or 'von unten' (from below), i.e. in a purely abstract, philosophical way, or from an empirical, experimental point of view. Experimentalists like myself tend to read with a complete lack of understanding philosophical books like *Genius: The History of an Idea* (Murray, 1989a). Words like 'imagination', 'insight' and so forth are sprayed about without definition, without possibility of measurement, without forming a testable theory, or even usually a comprehensible one. As Lord Kelvin said: 'One's knowledge of science begins when he can measure what he is speaking about, and expresses it in numbers.'

Philosophers seldom take kindly to such views, and point to the obvious imperfections of scientific measurement, particularly in its early stages. They look with amusement at our attempts to measure creativity, or intuition; this is not what they mean by these words. Yet what they mean is by no means clear, or referable to any kind of reality; each philosopher uses terms such as these in his own manner, and there clearly is no way in which we can decide between them. Philosophers have debated many of the issues involved for at least 2000 years, but what has been the outcome? There clearly is no agreed knowledge, no certain conclusion, no proper theory. The experimenter may not do any better, but he can certainly do no worse, and the attempt seems worthwhile.

Bertrand Russell once indicated very clearly the difference between the philosopher and the scientist when he pointed out that 'Aristotle mentioned that women have fewer teeth than men; although he was twice married it never occurred to him to verify this statement by examining his wives' mouths.' (Actually Russell exemplifies this characteristic of philosophers to write without examining the evidence better than Aristotle! As Dunbar (1993) has pointed out, Aristotle was one of the greatest observers who ever lived; he created the scientific study of biology, and laid the basis for modern work in that field.) But the situation is more complicated than would be covered by saying that both sides criticize each other, and the truth probably lies in the middle. Like the ski resort full of girls hunting for husbands, and husbands hunting for girls, the situation is not as symmetrical as it might seem. Science attempts proof, and can be judged on rational grounds; philosophy cannot in the nature of things prove or disprove its assertions.

Of course science often leaves the straight and narrow path of righteousness; frequently psychologists in particular use statistics as a drunken man uses a lamp-post – for support rather than illumination. Again, psychologists in particular often search for 'facts' without stating a prior theory. But as Poincaré has pointed out: 'Les faits ne parlent pas'. Science, he goes on to say, 'is built up with facts, as a house is with stones. But a collection of facts is no more science than a heap of stones is a house.'

I have tried to escape the obvious dangers of the experimental approach, without imagining for any moment that any firm conclusion could be obtained with the primitive tools at our disposal. But if and when we go wrong, science has the antidote: criticism, replication, improvement. Our efforts spiral towards greater accuracy, better theories, improved understanding, errors rejected, genuine insight retained. Philosophers often assert that measurement is not possible of complex human behaviours such as those involved in creativity and intuition. I would rather side with Thorndike in his famous saying that 'everything that exists exists in some quantity and can therefore be measured'. There is no doubt that such measurement is usually difficult, often complex, and never perfect. Measurement, even in the hard sciences, involves inevitable error; our effort has to be directed to reducing such error (which can never be completely eliminated, of course!). But even poor measurement is better than no measurement; it can be improved and built upon, in a way that philosophical or psychoanalytic speculation cannot.

One important source of such improvement is the correction of factual and interpretive errors. The scientific study of genius began with Galton's findings of familial aggregation of high eminence – talent seemed to be concentrated in certain families, and Galton interpreted this in terms of genetic endowment. Modern research and re-analysis has thrown doubt on the alleged facts, and advances in behavioural genetics have suggested that the argument is essentially false – although the conclusion is essentially correct! Concepts like epistasis and emergenesis, unknown to Galton and his contemporaries, have altered and advanced our understanding, and genetic studies of creativity in identical and fraternal twins, separated at birth, have thrown much needed light on problems of this kind.

Laymen often misunderstand the role of measurement, and imagine that nothing in psychology can be as accurate as measurement in physics; they often conclude that consequently it must be worthless. But remember that when Galileo laid the foundations of modern science, studying the behaviour of balls rolled down an inclined plane, he had to measure the time elapsed. Having no watches or clocks remotely accurate enough for such measurement, he used to feel his pulse, or else sing a rhythmic song, counting the beats! Shockingly inaccurate by our modern standards, but sufficient to elucidate the laws of movement, and lay the basis of modern science. This is the sort of measurement modern psychology can equal; hopefully it may allow us to glimpse the beginnings of a proper theory of creativity.

How does this book differ from the many excellent textbooks and other volumes that have been written on the topics of creativity and genius in recent years? These have been mainly of two kinds. We have what may be called the anecdotal–historical variety, exemplified by Arthur Koestler's (1964) book, or Ghiselin's (1954). It is difficult to know what to make of the historical accounts and introspections detailed therein. On the one hand we would be unwise to disregard what genius has to say about himself; who would know better? But on the other hand if there is one thing psychology teaches us it is to distrust unaided memory and self-justifying introspections. Genius is no exception to human frailty, and no more to be believed than the notorious man on the Clapham omnibus. We have all heard of Kekule's story of dreaming of snakes biting their own tails, which suggested to him that the molecules of certain organic compounds are not open structures, but closed chains or 'rings' – 'like the snake swallowing its tail', says Koestler (p. 118). But of course there were no dreams – 'I dozed', says Kekule. And there were no snakes – he states that his 'mental eye' could distinguish 'larger structures, of manifold conformation, long rows, sometimes more closely fitted together; all twining and twisting in snakelike motion.' We know how quickly such dream-like events vanish from memory, unless immediately recorded – we will never know what Kekule actually experienced. Much the same is true of the other stories so assiduously collected by many historians; their evidential value is low unless we can fit them into a testable theory.

Again, alleged theoretical conceptions to 'explain' creativity, like Koestler's 'bisociation' or Rothenberg's (1979) 'janusian' and 'homospatial' thinking only give the problem another name. Explanation means referral to a better known, experimentally studied process which would predict and encompass the behaviour to be explained; this cannot be done along these lines. The observations may be apposite, but they do not benefit by renaming.

Quite a different approach is the psychometric one, i.e. the art of producing alleged tests of creativity, and analysing their correlations with each other, and with certain criteria. Here we approach a scientific way of looking at the phenomena, but the approach is somewhat arid – describing rather than explaining. The description is now more precise, being in principle mathematical, even when dealing with historical phenomena, as in historiometry (Simonton, 1984b). There is no doubt that this approach has given us much valuable information, and the rest of this book bears witness to the debt I owe to all those who have laboured in this field.

Why then am I not happy, and why do the editors of the recent *Handbook of Creativity* (Glover, Ronning and Reynolds, 1989) feel that it is a 'degenerating' research programme (Brown, 1989)? Cronbach, in his 1957 presidential address to the American Psychological Association, argued that there were two scientific disciplines in psychology, the experimental and the correlational, and that their failure to get together prevented the appearance of a genuinely scientific, united psychology. I have always agreed with this view, and have attempted to further such co-operation. What to my mind has been lacking in the field of creativity research has been a proper reference to the storehouse of knowledge accumulated by experimental psychologists. It is such a unification that I have tried to bring about, with what success it is impossible to say at present.

Danziger (1990) argues that the history of psychology exhibits three models of research; the experimental (Wundt), the psychometric (Galton), and the clinical (Kraepelin). To these I would add the psychophysiological–genetic approach (Helmholtz), which is needed to fill in the picture. I would extend Cronbach's argument to include the clinical and the biological approaches and research paradigms, and insist that our understanding of creativity and genius must be based on all four, and that our present dissatisfaction with research in this area is due to restricting empirical and theoretical work, largely to one area – the psychometric study of individual differences. Hence my attempt to draw on many different kinds of knowledge. The attempt may be abortive, but it seemed worth while trying to make the field more inclusive.

This approach may be contrasted with that adopted by psychologists who adopt the idiographic approach, such as Wallace and Gruber (1989). This approach favours a kind of hermeneutical point of view in which each creative person and his environment is looked upon as a unique configuration of characteristics that cannot be 'decontextualized' into measurable variables. According to this approach, the constituents of creativity aggregate in systems and interact dynamically. What this means is apparently that this process of interacting within an evolving system may bring about changes in the constituent characteristics. This of course makes the system untestable; it thus shares the major fault of all idiographic theories. If a person is unique we cannot study him scientifically, because we cannot measure his unique aspects, or compare him with others. We cannot even prove that personality is unique, because that would involve measurement, which is explicitly condemned as disregarding uniqueness! Indeed, we would have to abandon all psychological terms and concepts which allow us to compare individuals; strictly speaking, we cannot even speak about *creative* people because this implies a non-unique continuum from creative to non-creative! This whole approach is thus based on an oxymoron (non-unique uniqueness), and leads to a completely non-scientific approach to the study of creativity and personality.

At this point it may be useful to state briefly and concisely the theory which this book is meant to explicate. I start with the assumption that genius, defined as supreme creative achievement, socially recognized over the centuries, is the product of many different components acting synergistically, i.e. multiplying with each other, rather than simply adding one to the other. Among these components are high intelligence, persistence, and creativity, regarded as a *trait*. Trait creativity may or may not issue in creative achievement, depending on the presence of the many other qualifications and situational conditions. Prominent among these additional qualifications are certain *personality traits*, such as ego-strength, i.e. the inner strength to function autonomously, to resist popular pressure, and to persist in endeavour in spite of negative reinforcement.

Creativity can be measured and shown to be a necessary but not sufficient condition for great achievement. It is postulated to co-exist, and can be causally related to, psychopathological personality traits often found, to a debilitating effect, in functional psychotics (schizophrenics and manic-depressives). Geniuses and creative persons generally are thus postulated to have high scores on the personality dimension of psychoticism, i.e. a temperamental dispositional trait rendering a person more likely to succumb to a functional psychosis, given sufficient stress, and showing cognitive features similar to those of psychotic patients.

Chief among these cognitive features is a tendency to *overinclusiveness*, i.e. an inclination not to limit one's associations to *relevant* ideas, memories, images, etc. In psychotics this overinclusiveness is rampant and uncontrolled, leading to the incomprehensible word salad of the chronic schizophrenic. Less extreme, and properly controlled, this tendency to have a shallow rather than a steep associative gradient, i.e. to admit more unusual associations, ideas, memories and images than is customary, is the mark of the creative person. The positive relation between ego-strength and psychopathy is unusual in the general population, but it is characteristic of the creative person and the genius.

Overinclusiveness can be measured, and shown to be characteristic of psychopathology and of creativity; we now have theories, with good empirical backing, which anchor such cognitive peculiarities in psychophysiological features of the organism, such as *latent inhibition* (or rather its absence). Such theories, also implicating neurotransmitters like dopamine and serotonin, form the basis of modern theories of schizophrenia, and are, as I shall argue, equally fundamental in creativity. Thus the theory proposed ranges from DNA (implicated because of the strong genetic component in creativity and genius), through psychophysiological and hormonal mechanisms to psychoticism, and hence to creativity and genius. There are many steps in the argument, and each is carefully documented and discussed. For some steps there is sufficient evidence to consider the point proven; for others the evidence is largely circumstantial and for some, meagre. There is no hint that the theory is more than a suggestion of how many disparate facts and hypotheses can be pulled together into a causal chain, explaining this apparent miracle of human creativity, and the apogee of human endeavour – genius. If the theory has one point in its favour it is that every step can be tested experimentally, and that many steps have already received positive support from such testing. Possibly the publication of the theory will prove productive in generating further search for experimental support (or disproof!) of the various parts of the theory proposed.

I am not suggesting that the various parts of the theory are all original. Experts will recognize many familiar features, and obviously I have relied very strongly on the outstanding work of others, carefully described and acknowledged. The original parts are inevitably the weakest, most in need of further support; this is of course inevitable in proposing a new theory. If the argument appears overinclusive (as I am afraid it will to many readers), my excuse must be that overinclusiveness is a causal aspect of creativity!

There are probably many objections that will be made to the very idea of treating genius and creativity as proper subjects for a natural science approach. One is the 'fuzziness' of the concepts; as with intelligence, it may be asked if genius and creativity actually exist. The obvious answer, of course, is that neither 'exist' in the sense that pigs or rocks can exist; both are scientific concepts, like gravitation or electricity, evolution or metals, and are to be judged by their usefulness or uselessness. What goes for genius and creativity goes for black holes, the Higgs Boson, the gravitano, the photino, the gluon, the axion, the top quark (or indeed any quarks!), the magnetic monopole, cosmic strings, or any number of concepts in physics for which the evidence is nothing like half as good as that for genius or creativity. For the time being, these are useful concepts, and worthy of study; whether they will be supplanted by better ones in due course is a matter for the future.

When I started writing, I had in mind an audience of intelligent and educated people with an interest in the topic, but without special information. As I went on it became increasingly obvious that on many points a more detailed, academic approach was indicated. When you are putting forward new ideas, unfamiliar models, original conceptions, then it is mandatory that you must explain these in sufficient detail for experts to assess their value, and you must give a list of references to justify your views. Hence the present text will lie uneasily between two extremes, falling between two stools as any such attempt to please two different masters must. The expert will have to excuse my explaining points he is only too familiar with; the layman must be willing to accept certain conclusions without necessarily understanding all of the scientific justification offered. I have done my best to accommodate both types of readers, but am not too happy about the outcome. This paragraph will have to do for an apology!

What, then, is new in this book? What is new is intimately linked with the difficulty I have just described in making the book readable for the average reader, interested in the topic but without academic training in experimental psychology. Let me begin with a definition of what a scientist is trying to discover. He looks for general laws, usually of the form: $a=(f)\ b$, i.e. a is a function of b. The inverse square law of gravitation is a good example; Boyle's law is another: $E = MC^2$ is now perhaps the most famous of all such laws. No such law will be found in text-books of creativity (Ochse, 1990; Glover *et al.*, 1989); all that we find is correlations. There is also a dearth of theories which might lead to the discovery of such laws; clearly without the elaboration of a useful theory, no functional laws are likely to be found.

I have tried to put forward such a theory, and suggest lawful functions capable of being tested experimentally. But of course such laws must initially be stated in terms of concepts which are familiar to the expert, but not to the casual reader (and often not to the psychologist not expert in that particular area of psychology!). To say that creativity is an attribute of psychoticism, and is mediated by the absence of latent inhibition is not meaningful to readers who have never come across psychoticism, and who have never heard of latent inhibition. Further to be told that latent inhibition is itself a function of the dopamine receptors of a given person is to add insult to injury if the meaning of dopamine, and its relation to schizophrenia, is unfamiliar. It is not a question of using scientific jargon to blind the reader with science; these are technical terms which carry a lot of meaning, and which are absolutely essential for any understanding of the theory here offered. But they require explanation and elaboration, and at least some discussion of the background experimentation that allows us to postulate certain functions for these concepts. Inevitably this makes for hard and difficult reading; I have done my best to put over the essential meaning of concepts such as these, but am only too aware of my inadequacy. Even terms like 'unconscious', which are apparently familiar to most readers, carry a lot of superfluous baggage, and require to be carefully defined and purified, and if possible related to their experimental referents, before they can play a useful part in the procession.

I am indebted to many friends and colleagues who have made important contributions. I was fortunate enough to know Spearman, who as we shall see originated work on 'divergent' tests of creativity, and Guilford who did so much to improve them. I was lucky enough to get to know many of the members of the Institute of Personality Assessment and Research (IPAR) at Berkeley when I spent some time there as Visiting Professor; among them Donald McKinnon, Frank Barron, Harrison Gough, Richard Crutchfield and George Welsh, all pioneers in the empirical study of creativity. D. K. Simonton, Colin Martindale and Mark Runco have provided many important data and analyses. Many eminent research workers were invited to discuss my original monograph (Eysenck, 1993a), and to all of them I am indebted for valuable comments and criticisms. Among my own research colleagues I would single out Gordon Claridge, Chris Frith, David Hemsley, K. O. and Karin Götz; my successor at the Institute of Psychiatry, Jeff Gray, and above all my wife, Sybil, who has been exceptionally creative in developing the concept of 'psychoticism', which is fundamental to the theory here developed.

1 The nature of genius

When a true genius appears in the world,
you may know him by this sign, that the
dunces are all in confederacy against him.

Jonathan Swift

Popular concepts of genius

There are many books on the topic of genius; since the days of Aristotle and Plato, philosophers, artists, teachers, scientists, psychiatrists and lately psychologists have combined to tell us what genius is, how it is produced, how it relates to madness, how it can be cultivated. These contributions consist of writings which rely on common-sense, historical anecdotes, and descriptive passages extolling the wonders of genius. Thus 'genius' is depicted as the possessor of a mystical gift that cannot be explained by the ordinary laws of human nature – a conclusion that would immediately make impossible the realization of the research project on which this book is based. It is possible to bring together popular notions of genius by citing a number of definitions. Some of these make genius seem remarkably commonplace. Thus for Buffon, genius was 'but a great aptitude for patience'. Frederick the Great thought it was a 'transcendent capacity for taking trouble'. Edison considered it 'one per cent inspiration and ninety-nine per cent perspiration', while Disraeli agreed with Buffon – 'patience is a necessary ingredient of genius'. Thus we learn that genius means hard work, true but not very revealing. It may highlight the absurdity of modern educational methods which stress the alleged natural creativity of children, but refuse to impart the necessary knowledge without which creativity cannot function.

More interesting are quotations telling us about the *creativity* of genius. As Schumann said: 'Talent works; genius creates', and Lincoln maintained that 'towering genius disdains a beaten path. It seeks regions hitherto unexplored'. Hazlitt declared that 'rules and models destroy genius and art'. Others point out that genius may be opposed. J. S. Mill declared that 'Genius can only breathe freely in an atmosphere of freedom', and Burckhardt pointed out that 'mighty governments have a revulsion against genius'. Tacitus believed that 'the persecution of genius fosters its influence', and O. Wilde declared that 'the public is wonderfully tolerant. It forgives everything except genius'. For Kierkegaard, 'genius is like a thunderstorm: it rushes against the wind, frightens people and cleans the air. The status quo has in defence invented the lightning-conductor'.

Most agreement, however, can be found on the proposition that, as Dryden said, 'great wits are sure to madness near alli'd, and thin partitions do their bounds divide.' Aristotle was one of the first to declare that 'no great genius has ever been without some madness', and G. B. White expressed the same thought slightly differently by saying that 'genius is more often found in a cracked pot than in a whole one'. For John Adams 'genius is sorrow's child', and Browning added: 'Since when has genius been found respectable?'

Genius is different from talent. A. Meredith said that 'Genius does what it must, and Talent does what it can'. Doyle thought that 'mediocrity knows nothing higher than itself, but talent instantly recognizes genius'. Goethe declared that 'talent learns everything, genius knows everything'. But for practical purposes genius may be less useful than talent; as Schopenhauer said, 'for everyday life, genius is as useful as a giant telescope in the theatre'.

Some writers have issued warnings against taking the notion of genius too seriously. Ortega y Gasset said: 'Better beware of notions like genius and inspiration; they are a sort of magic wand and should be used sparingly by anybody who wants to see things clearly.' This is well said. Appeals to genius and inspiration explain nothing; they simply label the problem. Common wisdom may describe, more or less accurately, some aspects of supreme achievement, but it does nothing to understand it. Common-sense sets the stage: genius is highly creative, easily misunderstood, often odd or even near mad; genius needs hard work to achieve anything, and will almost certainly be opposed by the great mediocre majority. This at least is the received wisdom, with only a few disclaimers, like Ezra Pound, who declared: 'The concept of genius as akin to madness has been carefully fostered by the inferiority complex of the public'. But then he *was* mad!

The growth of a concept

These tentative generalizations are based on concepts of 'genius' that have been dominant since the Renaissance; previous to that *sanctity* was the major concern of humanity for over 2000 years. Even now, as Gustave le Bon has pointed out:

> At the bidding of a Peter the hermit millions of men hurled themselves against the East; the words of an hallucinated enthusiast such as Mahomet created a force capable of triumphing over the Graeco-Roman world; an obscure monk like Luther bathed Europe in blood. The voice of Galileo or a Newton will never have the least echo among the masses. The inventions of genius hasten the march of civilization. The fanatics and the hallucinated create history.

This 2000 year-long dream (or nightmare) followed upon an early flowering of genius, and concern with its nature and causes, in ancient Athens. Many of our modern ideas and thoughts about genius were already voiced then in an embryonic form.

The word 'genius' is of course derived from Latin, but had quite a different meaning in ancient Rome. It first appeared in Plautus, around the third century BC, denoting a kind of tutelary spirit embodied in each man, not entirely identical with him, but intimately connected with his personality – perhaps like the Greek *Daiman*. What *we* mean by *genius* derives rather from the Latin *ingenium*, meaning both *natural disposition* and *innate ability*, an early forerunner of Spearman's notion of general ability or intelligence. (Cicero's 'intelligentia' was more concerned with acquired knowledge.)

Even then, views about the nature of genius were common which are still widespread. Thus Plutarch describes Archimedes, a man who 'possessed so lofty a spirit, so profound a soul, and such a wealth of scientific inquiry that ... he had acquired through his inventions a name and reputation for divine rather than human intelligence'. Plutarch also discussed the reasons for Archimedes' exceptional ability in geometry.

> Some attribute this to the natural endowments of the man, others think it was the result of exceeding labour that everything done by him appeared to have been done without labour and with ease. For although by his own efforts no-one could discover the proof, yet as soon as he learns it, he takes credit that he could have discovered it. . . . For these reasons there is no need to disbelieve the stories told about him – how, continually bewitched by some familiar siren dwelling with him, he forgot his food and neglected the care of his body, and how, when he was dragged by force, as often happened, to the place for bathing and anointing, he would draw geometrical figures in the hearths, and draw lines with his finger in the oil with which his body was anointed, being overcome by great pleasure and in truth inspired of the Muses.

This picture contains all the major aspects of genius later emphasized by other writers; natural endowment (intelligence), hard work, divine inspiration (creativity) and a personality which indulges in behaviour which is distinctly unusual, to use a rather neutral term; in anyone lacking in the achievements of Archimedes we might be tempted to call it mad.

Genius and heredity

These ideas were common in ancient Greece, although the Greeks did not have a word for either genius or creativity. Murray (1989a,b) has told the tale of the way these ideas developed in considerable detail; inspiration (creativity) and natural talent (intelligence; special abilities) already find description in Homer, and are made explicit by Plato and Pindar. In her chapter, Murray discusses briefly the question of heredity; as she says: 'There is no indication ... that poetic talent was thought to be inherited or passed down from generation to generation.' (p. 13). She thus makes the same error as Galton (1869) in his book on *Hereditary Genius*, which marks the beginning of scientific interest in the topic of genius. He argued for the importance of hereditary factors because there *was* such generation-to-generation sequence of eminence; she argued against it because there was not. For a geneticist, however, such generation-to-generation sequence is ambiguous; the sequence could be due either to genetic or to environmental factors, so that nothing can be deduced from it. But the emergence of genius from poor soil is difficult to account for except in terms of the segregation of genes, and hence speaks powerfully for heredity. Newton, Gauss, Faraday all came from poor, undistinguished families; there is little environmental background for their genius. We can argue about Bach (Terry, 1929) and Mozart, where clearly nature and nurture co-operated, or the mathematically gifted Bernouillis (Behr-Pinnow, 1933). But what other than heredity can account for the achievement of Berlioz, whose family was not musical, and opposed his choice of profession vehemently; who grew up in a small provincial town with hardly any musical culture?

The point is too important to leave. Galton in his book dealt with reasonably eminent people, but in spite of the title of his book there was hardly any genius among them. What would happen if we looked at the families of true geniuses? Consider the 28 most famous mathematicians of all time, each one a genius in the opinion of qualified mathematicians (Bell, 1939). I have taken all the names in Bell's book, except where no information was given, or where, as in the case of Zeno, Endoxus and Archimedes, the passage of the centuries made any comparison with modern times difficult. What do we find when we look at the fathers of these men, or their close relatives? There is hardly a trace of any mathematical ability. Descartes's father came from the minor nobility. Fermat's was a leather merchant; Pascal's a civil servant who forbade his son to look at mathematics books. Newton's, Laplace's and Gauss's fathers were peasants or small farmers. Leibnitz had a philosopher for a father, Euler a priest, Lagrange a treasurer, Monge a peddler and knife-grinder, Fourier a tailor, Cauchy a lawyer, Lobatchevski a civil servant, Abel a pastor, Jacobi a banker, Hamilton a solicitor. The father of Galois was mayor of a village, Cailey's father was a merchant, Weierstrass's a customs officer, Boole's a shopkeeper, Hermite's a cloth merchant, Kronecker's a businessman, Riemann's a pastor. Kummer had a physician for a father, Dedekind a professor of law, Poincaré a physician, and Cantor a merchant.

There is here no evidence of the clustering of ability which Galton discussed. The Bernouillis, a family which produced in three generations eight mathematicians, three of them outstanding, but only one a 'genius', are the exception which proves the rule. We may add that Euler's father was an accomplished mathematician, though certainly not outstanding or original, and that Jacobi had a brother who was a highly thought of mathematician in his day, though not regarded highly now; that is the sum-total of mathematicians clustering around these famous geniuses. What appears is the sudden emergence of genius from apparently barren ground – barren, that is to say, of any mathematical talent. If Galton had looked at talent, rather than genius, he would have found some evidence for his thesis. When we look at talent, measured by psychological tests, we do find ample evidence of the clustering Galton noted. There is evidence in relation to musical ability (Mjoen, 1925), drawing ability (Krause, 1932; Haecker and Ziehen, 1931), ability at sport (Weiss, 1977) and particularly mathematical ability (Weiss, 1982). Weiss found in his analysis of specifically gifted winners in mathematical talent competitions held in the former East Germany, that the families of these mathematically outstanding probands nearly always contained mathematically active persons, in contrast to probands not showing such promise. How can we explain these differences between talent and genius? Anticipating later more detailed explanations, I would suggest that mathematical ability constitutes a genetically transmitted talent which forms a *necessary but not sufficient* condition for mathematical genius. To find such talent in families is not unexpected, but the combination of all the elements required to constitute genius demands a very unusual segregation of genes which would be so unlikely to occur that we would look in vain for anything resembling it in the family of the genius.

In fact, Galton made another error by disregarding regression, i.e. the fact that any mental or physical trait which has a heritability less than unity will show *regression to the mean* (Eysenck, 1989b). Taking IQ as an example. Fig. 1.1 shows what most people imagine to be the case when told that IQ is largely inherited. According to this notion, very dull parents have very dull children, dull parents have dull children, average parents have average children, and very bright parents have very bright children, thus presaging an endless caste society. But in reality things resemble the diagrammatic picture of Fig. 1.2. Children of very dull or very bright parents *regress* to the mean, as do rather less strongly the children of dull and bright parents. Variation is maintained by the children of average parents, some of whom are bright or dull, with a few very bright or very dull. This regression to the mean is largely responsible for the social mobility which is so characteristic of our society (Eysenck, 1979).

Actually it was Galton himself who discovered this vitally important law of genetics (Galton, 1877), although he first talked about 'reversion' rather than regression (Stigler, 1986). It will be obvious that this law would invalidate any genetic tendency for genius to cluster as Galton thought; not only intelligence, but also all the other genetic variables contributing to genius or eminence, such as creativity, motivation and persistence would regress to the mean, leaving a much less distinguished progeny, on the average. There is no record in history of a genius begetting another genius; all history records is regression to the mean. This regression is of course also aided by the fact that heredity implicates both parents; regression goes from the *mean* of mother and father, and in spite of assortive mating (bright men marrying bright girls, on the whole), the wives of geniuses seldom if ever attained similar heights of intellect,

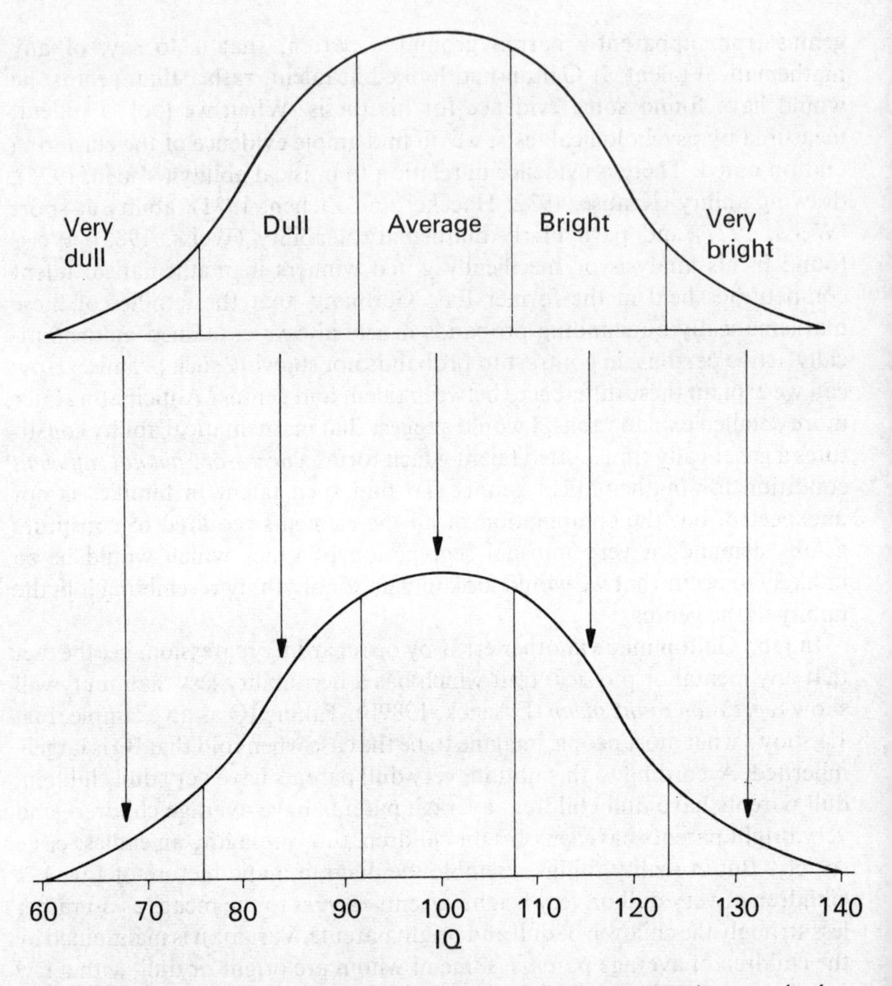

Fig. 1.1 Relation between IQ and parents and children as it is often erroneously assumed to be.

creativity, motivation, etc. On these grounds we would expect the clustering Galton discovered to be due to environmental rather than purely genetic causes.

However that may be, the battle about the importance of *physis* (natural heredity) and *techne* (art, environment) continued for centuries, very much like the debate about nature and nurture in our own century – except that it inevitably remained philosophical and argumentative, in the absence of scientific knowledge of the principles of heredity, which has enabled us to call a truce; the evidence compels us to admit that both play a part, the relative importance varying from time to time, from place to place, and from trait to trait.

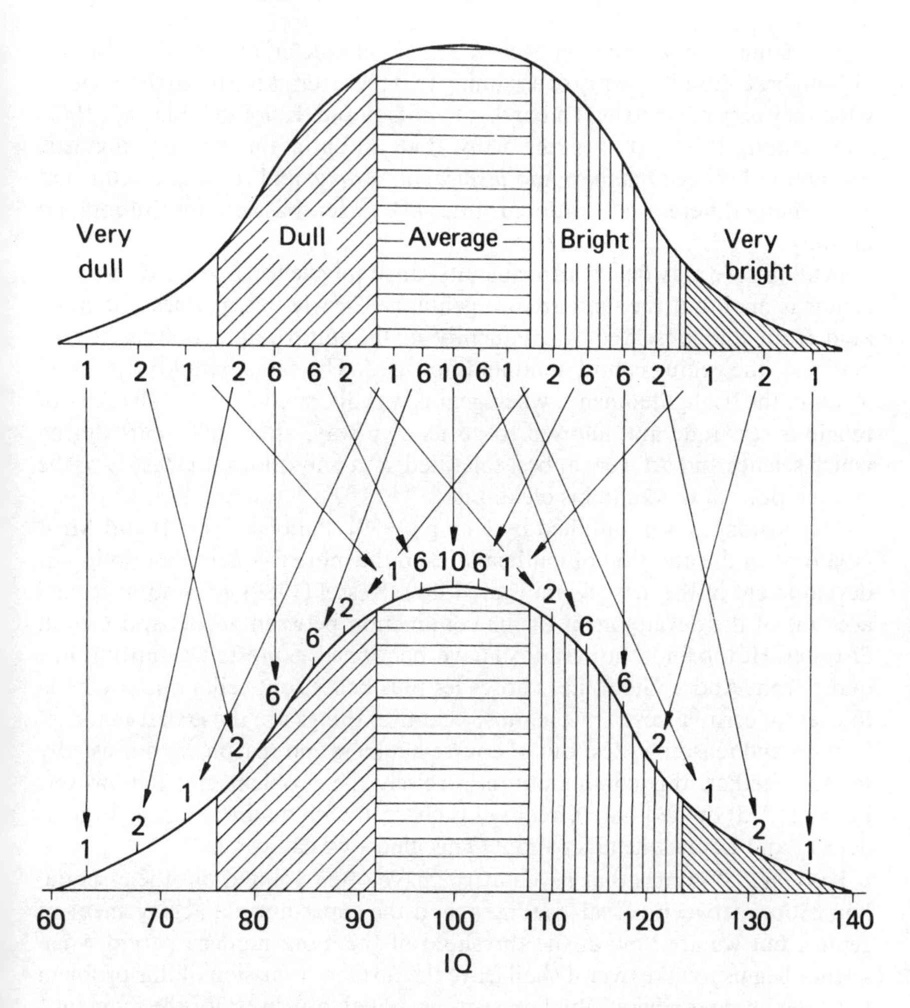

Fig. 1.2 Relation between IQ and parents and children as it actually appears in nature.

Genius and madness

The notion of the *madness of genius* was taken up by many writers. Thus Plato, having discussed Apollonian (prophetic) and Dionysian (ritual) madness, anticipates Aristotle and goes on to say this about *furor poeticus*:

> Third is the possession and madness which comes from the Muses. It takes hold of the tender and untouched, rousing it up and exciting it to frenzy in lyric and other kinds of poetry . . . But whoever comes to the gates of poetry without the Muse's madness, persuaded that art will make him a good poet, is ineffectual himself, and the poetry of the sane man is eclipsed by that of the mad.

Both mania and depression were cited as characteristic of great artists, although we should be wary of assuming that these terms meant to the ancients what they now mean to us (Klibarsky, Panofsky and Saxl, 1964; Flashar, 1966; and Simon, 1978). It is noteworthy that in Latin there is no linguistic distinction between *madness* and *inspiration*. *Mania* and *furor* are terms that cover many different non-rational states like anger, passion, inspiration and insanity.

In all this we may detect an ambiguity, an ambivalence, and a paradox; the genius is mad, but the mad are not geniuses. Clearly this madness differs in kind from that observed so frequently in the unfortunate victims of true madness; the genius is both mad and not-mad. The same paradox appeared again in the 'Enlightenment', when genius was liberated from the shackles of religious servitude and allowed to go its own way, after 2000 years during which science and art were at best tolerated, and only allowed to testify to the greater glory of God and his creations.

This history is well outlined by Kemp (1989), Panofsky (1962) and Most (1989), with the addition of mathematics to the more frequent description of development in the arts (Kilminster, 1989). Kessel (1989) gives an historical account of the development of the conjunction between genius and mental disorder. He too finds the paradox I have mentioned, expressed beautifully in a quote from Andre Gide: 'Les choses les plus belles sont celles que souffle la folie et qu'ecrit la raison' – the most beautiful things are those that madness inspires and reason writes. But of course madness and reason are not usually found together; this conjunction is precisely the problem, not the answer. Pascal (1925) opined that 'Great wit is charged with madness just as is great defect', and Lamartine talks about 'This illness we call genius'.

Kessel (1989) notes that psychiatrists have often pointed out the essential opposition between actual 'madness' and the super-normal achievement of genius, but we are now at the threshold of the more modern period when science begins to take over. I shall leave the further discussion of this problem to a later chapter where I shall present one possible answer; for the moment I shall continue our discussion of the nature of genius by looking at various more modern writers, and the ideas they have put forward, usually based on common-sense.

Common-sense notions, based on experience, may have an intuitive appeal, but they need winnowing, testing, and validating before they can be accepted. Hence in the last hundred years science has taken a hand, and we now have a number of articles and books concerned with the scientific study of genius. Most of these are devoted to what the Germans call 'Geisteswissenschaft' – history, sociology, philosophy, literary criticism, and other purely descriptive areas of study, differing from common-sense by subjecting themselves to some form of academic discipline. The difference from common-sense is one of degree, rather than of kind, but it is nevertheless noticeable, and should not be under-estimated. *Geisteswissenschaft* should be contrasted with *Naturwissensachaft*, i.e. natural sciences, obeying the dictates of the hypothetico-deductive method, using quantitative models, controlled observation, and experimental investigations. A good example of the best that Geisteswissenschaft has to offer is Lange-Eichbaum's (1956) *Genie, Irrsinn und Ruhm*, with its 2860 references which cover an astonishing variety of subjects, all concerned with the genius and madness problem.

Genius as eminence

Naturwissenschaft at the beginning of its reign is best represented by Sir Francis Galton's (1869) *Hereditary Genius*, republished in 1978, and more recently the works of Simonton (1984a,b; 1988a,b). Many other works in this group will be referred to in later chapters, so no more will be said here about them. All these approaches assume, of course, that the concept of 'genius' has some assignable meaning on which there is agreement, and the possibility of empirical testing.

It was Sir Francis Galton (1869), who first defined genius in terms of achieved distinction, or *reputation*, which in turn he defined as 'the opinion of contemporaries, revised by posterity . . . the reputation of a leader of opinion, of an originator, of a man to whom the world deliberately acknowledges itself largely indebted' (p. 33). This notion has proved seminal, but it has some obvious difficulties. One point to be made is that the title of 'genius' is bestowed by a group, a group which contains not only experts but is much wider, and may use standards and criteria which are strictly irrelevant. Lange-Eichbaum in particular has documented this point; as he says: 'When human beings bestow fame in someone, they usually consider not only purely intellectually some cognitive achievement, but they follow the dictate of their own emotional drives' (p. 135). The fundamental point insisted on by Lange-Eichbaum is simply that there are two sides to the equation; supreme achievement on the one, recognition on the other, and they may not balance out. Supreme achievement may not be recognized – Mendel is an obvious example, as is Lobachevski, who invented non-Euclidean geometry, or Semmelweis, who discovered a method for avoiding puerperal fever. Recognition may be given to unworthy pseudo-genius.

Another problem is change over time – someone regarded as a genius now may not be so regarded a century hence, or conversely, someone not considered a genius now may be so considered in the future. Examples are numerous – Mendel, Lobachevski, Semmelweis. Lange-Eichbaum discusses the point in detail, and gives many other examples. The reception of Newton in France, or that of Shakespeare on the continent generally, are too well known to require discussion. If, as Emerson says, 'genius always finds itself a century too early', then clearly renown and acceptance may have to await the elapse of the century! Indeed, it is well recognized in the arts as well as in science that novel ideas are strongly resisted by the great majority representing orthodoxy. Darwin, in his autobiography, wrote: 'What a good thing it would be if every scientific man was to die when 60 years old, as afterwards he would be sure to oppose all new doctrines'. And Max Planck (1949), in his autobiography, had this to say: 'The way in which a new scientific truth usually becomes accepted is not that its opponents are persuaded and declare themselves enlightened, but rather, that its opponents gradually die off and the following generation grows up accepting the truth from the start.'

When we add to these problems those of nationalist rivalry, religious and political fanaticism, and differences in language, social organization and freedom of expression, we can see that the definition of genius becomes decidedly fuzzy; so fuzzy indeed that the term may lose all scientific respectability. That, it would seem, is to regard very real difficulties as insuperable. If there is disagreement, we can quantify it, and try to see whether sufficient agreement remains to distinguish genius from talent. Few if any musicians would deny that Mozart or Beethoven was a genius, Salieri or Telemann merely talented; centuries after a person has died there may be sufficient agreement on his achievement.

However that may be, Galton argued (and gave some proof) that excellence in performance was continuous, and roughly distributed in the form of a normal (Gaussian) curve. He argued that some 250 men in a million became 'eminent', and that there were 400 idiots and imbeciles in a million, with the rest, varying in 'excellence', in between. He originated a system of grading that has become so fundamental to modern discussions of genius that it is quoted at some length.

> The number of grades into which we may divide ability is purely a matter of option. We may consult our convenience by sorting Englishmen into a few large classes or into many small ones. I will select a system of classification that shall be easily comparable with the numbers of eminent men. We have seen that 250 men per million become eminent; accordingly, I have so contrived the classes in the following table that the two highest, F and G, together with X (which includes all cases beyond G, and which are unclassed), shall amount to about that number – namely, to 248 per million.

This is shown in Table 1.1.

> It will, I trust, be clearly understood that the numbers of men in the several classes in my table depend on no uncertain hypothesis. They are determined by the assured law of deviations from an average. It is an absolute fact that if we pick out of each million the one man who is naturally the ablest, and also the one man who is the most stupid, and divide the remaining 999998 men into 14 classes, the average ability in each being separated from that of its neighbours by *equal grades*, then the numbers in each of those classes will, on the average of many millions, be as is stated in the table. The table may be applied to special, just as truly as to general ability. It would be true for every

Table 1.1 *Classification of men according to their natural gifts (Galton, 1869)*

Grades of natural ability, separated by equal intervals		Numbers of men comprised in the several grades of natural ability, whether in respect to their general powers, or to special aptitudes							
		Propor-tionate, viz.	In each million	In total male population of the United Kingdom, viz. 15 millions, of the undermentioned ages					
Below average	Above average	one in	of the same age	20–30	30–40	40–50	50–60	60–70	70–80
a	A	4	256 791	651 000	495 000	391 000	268 000	171 000	77 000
b	B	6	162 279	409 000	312 000	246 000	168 000	107 000	48 000
c	C	16	63 563	161 000	123 000	97 000	66 000	42 000	19 000
d	D	64	15 696	39 800	30 300	23 900	16 400	10 000	4700
e	E	413	2423	6100	4700	3700	2520	1600	729
f	F	4300	233	590	450	355	243	155	70
g	G	79 000	14	35	27	21	15	9	4
x all grades below g	X all grades above G	1 000 000	1	3	2	2	2	—	—
On either side of average . .			500 000	1 268 000	964 000	761 000	521 000	332 000	149 000
Total, both sides			1 000 000	2 536 000	1 928 000	1 522 000	1 042 000	664 000	298 000

examination that brought out natural gifts, whether held in painting, in music, or in statesmenship. The proportion between the different classes would be identical in all these cases, although the classes would be made up of different individuals, according as the examination differed in its purport.

It will be seen that more than half of each million is contained in the two mediocre classes a and A; the four mediocre classes a, b, A, B, contain more than four-fifths, and the six mediocre classes more than nineteen-twentieths of the entire population. Thus, the rarity of commanding ability, and the vast abundance of mediocrity, is no accident, but follows of necessity, from the very nature of these things.

The meaning of the words 'mediocrity' admits of little doubt. It defines the standard of intellectual power found in most provincial gatherings, because the attractions of a more stirring life in the metropolis and elsewhere, are apt to draw away the abler classes of men, and the silly and imbecile do not take a part in the gatherings. Hence, the residuum that forms the bulk of the general society of small provincial places, is commonly very pure in its mediocrity.

The class C possesses abilities a trifle higher than those commonly possessed by the foreman of an ordinary jury. D includes the mass of men who obtain the ordinary prizes of life. E is a stage higher. Then we reach F, the lowest of those yet superior classes of intellect.

On descending the scale, we find by the time we have reached f, that we are already among the idiots aud imbeciles. We have seen that there are 400 idiots and imbeciles, to every million of persons living in this country; but that 30 per cent of their number, appear to be light cases, to whom the name of idiot is inappropriate. There will remain 280 true idiots and imbeciles, to every million of our population. This ratio coincides very closely with the requirements of class f. No doubt a certain proportion of them are idiotic owing to some fortuitous cause, which may interfere with the working of a naturally good brain, much as a bit of dirt may cause a first-rate chronometer to keep worse time than an ordinary watch. But, I presume, from the usual smallness of head and absence of disease among these persons, that the proportion of accidental idiots cannot be very large.

Hence we arrive at the undeniable, but unexpected conclusion, that eminently gifted men are raised as much above mediocrity as idiots are depressed below it; a fact that is calculated to considerably enlarge our ideas of the enormous differences of intellectual gifts between man and man.

Galton's view of genius is defined by two key notions which recur again and again in later discussions. Essentially genius is defined in terms of a person's posthumous *reputation*, making it a social and interpersonal construct. But Galton also recognized and indeed emphasized natural ability as the major source of genius, and hence of reputation. By 'natural ability' he meant 'those qualities of intellect and disposition, which urge and qualify a man to perform acts that lead to reputation, I do not mean capacity without zeal, nor zeal without capacity, nor even a combination of both of them, without an adequate power of doing a great deal of very laborious work. But I mean a nature which, when left to itself, will, urged by an internal stimulus, climb the path that leads to eminence, and has a strength to reach the summit – one which, if hindered or thwarted, will fret and strive until the hindrance is overcome . . .' (Galton, 1869, p. 33).

We may incorporate Galton's view in a diagram:

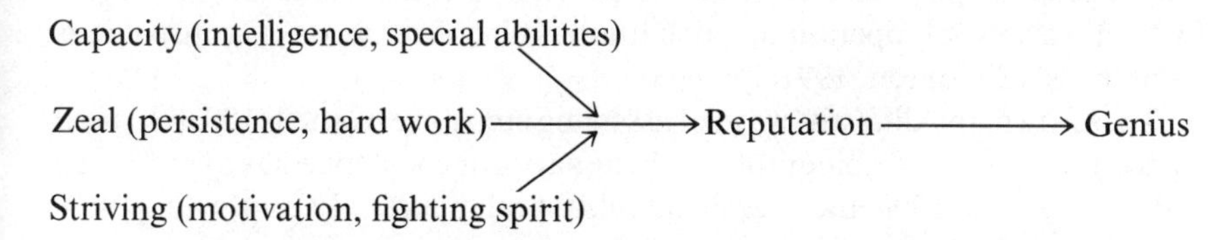

This combination of qualities, which Galton apparently regarded as synergistic, i.e. multiplicative rather than additive, almost infallibly leads to eminence: 'It is almost a contradiction in terms, to doubt that such men will generally become eminent' (p. 33), for 'the men who achieve eminence, and those who are naturally capable are, to a large extent, identical' (p. 24).

Such a hypothesis is testable, if only in part. We can try to discover whether men of genius, as defined by reputation, do indeed have great capacity, zeal or striving, and the results will be shown to bear Galton out; we can never prove that there are not some men of great capacity, zeal and striving who fail to reach eminence – the 'mute inglorious Miltons' and 'village Hampdens' of Gray's Elegy. Carlyle put forward the view that 'fame, we may understand, is no sure test of merit, but only a probability of such'. Simonton (1991a) believes that Galton would not have dissented, and if indeed the relation between the psychology of the individual and the eminence he achieves is as the theory suggests, the correlation would be what statisticians call heteroscedastic – all geniuses have the right psychological attributes, but not all who have these attributes are or become geniuses!

The empirical quest

We may divide the empirical quest for evidence regarding this whole question into two parts. The first question is whether Galton was right in defining genius in terms of reputation; the second, and perhaps much more difficult and important, whether he was right in linking eminence with the personal qualities of the 'genius'. Simonton (1991a) has closely investigated the first problem, and I shall follow his documentation here. As he points out, if we use *reputation* as our measure of genius, we must apply the usual psychometric criteria of reliability and validity to these measures, and see if Galton's theory of reputation as a general concept transcending time and method of measurement stands up to scrutiny. The first question, then, is whether there is indeed any consensus between experts in science and art on the degree of eminence of painters, composers, astronomers, physicians, writers, poets, and even psychologists? Simonton (1990, chapter 4) has reviewed the evidence and has found internal consistency reliability of .86 for artistic distinction, .94 for philosophical fame, .94 for scientific eminence, and so forth. 'So long as the sample of historical figures is sufficiently inclusive as to incorporate a heterogeneous pool of claimants for posterity's praise, multiple indicators with a variety of operational definitions all converge on a conspicuous consensus' (Simonton, 1991a, p. 608).

There are many different ways of assessing eminence. Cattell (1903a,b) used space allotments in six biographical dictionaries or encylopaedias to produce a ranking of the 1000 most eminent historical figures. Farnsworth (1969) obtained rankings for 100 classical composers by questioning members of the American Musical Society. Sorokin (1937) had experts rate 2000 philosophers on a 12-point scale. 'Still other investigators have used measures that tap frequency of citation, quotation, or performance, or tap inclusion in chronologies, histories, biographical dictionaries, or anthologies' (Simonton, 1991a, p. 608; Simonton, 1984b). It is indeed remarkable that there is such great congruence between such diverse methods of measurement.

These results refer to *internal* reliability; is there test–retest reliability, i.e. reliability over time? Apparently there is considerable such consistency for musicians (Farnsworth, 1969), literary figures (Rosengren, 1985), scientists (Simonton, 1991b), general historical figures (Simonton 1990), and psychologists. Indeed, consistency has been demonstrated even over the separate products on which reputation must be founded; Simonton (1989) found that the comparative merits of Shakespeare's 154 sonnets have been consistent for at least a century of literary appreciation.

Having established the reliability of the measures used to test reputation and eminence, Simonton (1991a) goes on to point out that Galton's hypothesis requires an additional step. He postulates four models, only one of which would be in agreement with Galton's theory, although all would predict the observed consistency over time of different modes of measurement. Model 1, the Galtonian model, postulates a *general factor* to account for the observed correlations between different measures of eminence; in addition, of course, it posits error terms associated with each type of measurement. Thus the formula for a given type of measurement (m_1), with a given error (e_1) of Galton's general factor of genius (G) would be $V_G = V_{m_1} + \mathrm{V}_{e_1}$, i.e. the variance of G is made up by the added variance of m_1 and e_1. Simonton's second model offers a slight elaboration of model 1 which is still in the spirit of Galton. It permits the errors of two or more measures to be correlated, and it permits two different measures to be correlated over and above the correlation due to G. If two different measures share a particular national background – say m_1 and m_2 are both derived from French sources – then this introduced correlated errors. Again, if m_1 precedes m_2 in time, the authors of m_2 may be influenced by the author of m_1, thus producing a correlation not due to the influence of G.

The other two models are not compatible with Galton's theory. Thus 'any consensus among a series of measures may reflect merely a continual borrowing of judgments in a first-order auto-regressive process' (Simonton, 1991a, p. 609). Thus the fame of philosophers may be determined by the first historian to evaluate their contributions; later critics follow his lead, with independent or correlated error producing variability around their consensus derived not from G but from sheep-like herd instinct making each one follow tradition. This model is perhaps implausible because new philosophers are constantly introduced into the field who were not evaluated by the first source, but the model can accommodate such an influx by making subsequent critics follow the estimation made by the first critic to evaluate the newcomer in relation to his predecessors. Similarly, Farnsworth's (1969) demonstration that the eminence of 92 classical composers was consistent over consecutive decades merely means that each generation follows its predecessors when expressing its artistic judgments; having heard much Mozart, but little Salieri or Telemann in its youth the younger generation models its taste on that which governed the taste of its fathers.

Model d, finally, combines models b and c, by introducing a series of latent variables which determine contemporaries by the observables $m_1, m_2, m_3, m_4 \ldots m_n$; it also contains error elements $e_1, e_2, e_3 \ldots e_n$. The model is easier to fit to reality, because it contains more variables, but it is also far less parsimonious. However that might be, all models produce testable consequences, and are hence scientifically acceptable. The actual testing of the models is a complex statistical process, and it would not be appropriate to enter into these intricate matters here; Simonton (1991a) gives a very thorough account of his method of testing the models. I will merely indicate very roughly the sort of argument which can be used to decide whether any more complex models than a or b are needed to account for the observed results. Spearman (1904) originally suggested that postulating a general factor G to account for the observed correlations between tests required for its confirmation that if we take any four tests a, b, p and q, their correlation must fit the *tetrad* equation: $r_{ap} \times r_{bq} - r_{bp} \times r_{aq} = 0$, and the same for any set of correlations. If the tetrad criterion holds, i.e. if the matrix of correlation is of rank one, the hypothesis of a general factor being responsible for the observed correlation is supported. Does this rule hold for Galton's G factor?

Let us consider a sample of 2012 philosophers. Simonton (1976b) used a factor composite made up, of 10 indicators spanning 37 years, comprising ratings; number of pages mentioning the thinker; mention in a history of civilization; status in the Great Books anthology; status in an Encyclopaedia of Philosophy; etc. Factor loadings were found to range from .605 to .897, and Simonton (1991a) concluded that 'the variance structure is dominated by Galton's *G*' (p. 613). (He also found an instance of correlated errors, due to the fact that two artifacts corresponding to the 'difficulty factors' sometimes found in the analysis of ability tests.) Simonton precedes his detailed discussion of five such data sets by saying: 'One generalization should be expressed at once: Models c and d are simply untenable. This negative inference follows mathematically from the many implied partial or tetrad equations that are plainly invalid either statistically or substantially.' (p. 612). With this conclusion it would be difficult to disagree; Galton's *G* emerges with unexpected clarity from the evidence. (I say unexpected because the many possible sources of errors one could easily think up might have submerged even a substantial contribution by *G*, and clearly this has not happened.)

Agreement on who is a genius

Galton's theory demands that people should agree on who is to be called a genius; unless there is such agreement, the concept has little value. I have cited some overall evidence in the preceding section to show that there is a good deal of agreement among experts; in this section I want to put some flesh on these statistical bones, and present some data on geniuses (and near-geniuses) known to most readers. These data will also enable us to form a judgment on another important aspect of Galton's theory, namely that genius is *continuous* with near-genius, and that with lesser degrees of eminence. Galton's view contrasts quite dramatically with the traditional notion of genius as something apart from the sphere of our sorrow, quite unlike anything else, completely *sui generis*. This is an essential part of Galton's theory, and requires substantiation.

The possibility of proper quantification of 'agreement' has been explored by Simonton (1981). He used a previous study by Maranell (1970), who had 571 American historians rate 33 presidents on seven dimensions of leadership, and condensed them statistically into a single dimension of *greatness*. The greatest presidents are those who score high on general prestige, strength of action, presidential activeness, and administrative accomplishments. The final score bore close relation to the other indications of greatness, such as having monuments built in their honour in the nation's capital, having county seats or a state capital named after them, to appear on low-denomination paper currency, and so on. There was considerable agreement with F. D. Roosevelt (9.04), Lincoln (8.26), Washington (5.69), Jefferson (5.98), T. Roosevelt (6.01) and Wilson (5.94) among the highest placed, and Pierce (−6.92) and Coolidge (−5.10) among the lowest. Not only is there close agreement among experts, but their verdict reflects closely popular esteem. There clearly is a good deal of validity to these judgments.

A second example will illustrate some of the strengths and weaknesses of this approach to genius by expert consensus. At best, only one or two of the American presidents would be considered a 'genius'; what would happen if we took 17 of the best-known classical composers, added two modern popular composers (Victor Herbert and Edward MacDowell, to define a bottom line), and had them rated by members of the New York Philharmonic, and the Boston, Minneapolis and Philadelphia Symphony Orchestras? Would there be agreement on who was the greatest, down to who was the least possessor of musical genius? Folgmann (1933) used Thurstone's scaling methods for the

Table 1.2 *Ranking of 19 composers by members of 4 famous orchestras (Folgmann, 1933)*

	Phila. Orch.	N.Y. Phil.	Bost. Symph.	Minn. Symph.	All Orch.
Bach	3	5	3	6	5
Beethoven	1	1	1	1	1
Berlioz	15	11	10	12	12
Brahms	2	3	5	2	2
Chopin	13	13	15	15	14
Debussy	8	9	6	10	8
MacDowell	18	19	18	19	18
Cesar Franck	14	14	11	13	13
Grieg	17	16	17	14	17
Haydn	7	7	8	11	7
Victor Herbert	19	18	19	18	19
Mendelssohn	11	10	12	8	10
Mozart	4	2	4	4	3
Schubert	6	6	7	5	6
Schumann	9	8	9	9	9
Stravinsky	12	17	16	16	16
Tchaikovsky	10	15	13	7	11
Verdi	16	12	14	17	15
Wagner	5	4	2	3	4

purpose, each judge being asked to compare each of the 19 composers with each of the rest, indicating his preference. Here I will give only the main results, ranking composers in order, and disregarding details about the distances between orders involved; these details are interesting but irrelevant to the purpose. Table 1.2 shows the results, for the 4 orchestras separately and combined.

Clearly there is considerable agreement. Beethoven is ranked first in every case, and Victor Herbert and MacDowell last. Bach, Brahms, Mozart, Wagner and Schubert are all ranked high; Grieg, Cesar Franck, Verdi and Stravinsky low. Agreement is not perfect, but of course the range is extremely limited; these 17 famous composers are the best-known of all who wrote music over 300 years, and such restriction of range must severely limit the degree of intercorrelation. On the other hand, the rankings are not made by individuals, but by groups; this would eliminate very extreme subjective judgments. If we were to use scaled scores, using Beethoven as our anchor, we would get the results shown in Table 1.3; here disagreement is more obvious. An analysis of

Table 1.3. *Scaled scores of 19 composers as in Table 1.1 (Folgmann, 1933)*

	Phila. Orch	N.Y. Phil.	Boston Symph.	Minn. Symph.	Total
Bach	−0.0532	−0.9130	−0.8161	−1.0741	−0.7448
Beethoven	0.0000	0.0000	0.0000	0.0000	0.0000
Berlioz	−2.4628	−2.3229	−1.8619	−2.2159	−2.1814
Brahms	−0.0266	−0.6497	−0.9382	−0.1714	−0.4268
Chopin	−2.3601	−2.5528	−2.3513	−2.3484	−2.3392
Debussy	−1.2390	−1.6731	−1.4525	−1.6681	−1.4870
MacDowell	−3.6856	−3.5556	−3.6188	−3.2244	−3.5482
Cesar Franck	−2.3905	−2.6573	−2.0078	−2.2487	−2.2908
Grieg	−3.1320	−2.8089	−2.6877	−2.3261	−2.6667
Haydn	−1.2085	−1.1446	−1.5113	−1.7180	−1.3962
Victor Herbert	−3.7933	−3.5471	−3.6620	−3.1153	−3.5840
Mendelssohn	−2.1018	−1.8177	−2.1956	−1.5915	−1.8606
Mozart	−0.2812	−0.5940	−0.8213	−0.7097	−0.5133
Schubert	−1.0446	−1.0016	−1.4927	−1.0433	−1.0434
Schumann	−1.7102	−1.4573	−1.6438	−1.6220	−1.6634
Stravinsky	−2.1777	−3.1326	−2.3543	−2.5010	−2.5856
Tchaikovsky	−2.0318	−2.7272	−2.2050	−1.5654	−2.0854
Verdi	−2.4762	−2.4412	−2.3416	−2.6134	−2.4747
Wagner	−0.5196	−0.7071	−0.5005	−0.3977	−0.5591

the *reasons* for such disagreement as there is, may throw some light on the ways in which we may avoid artifacts.

Let us consider the position in the scale of three composers: Bach, Debussy and Stravinsky. The Philadelphia Symphony Orchestra ranks them higher than any of the other orchestras. Why? It is well known that Leopold Stokowski, the then conductor of the Philadelphia Symphony Orchestra, was a great admirer of these three composers, particularly of Bach; Stokowski has orchestrated many of Bach's compositions, and these arrangements, still in manuscript at the time Folgmann was writing, were played only by the Philadelphia Symphony Orchestra. Clearly, the conductor's enthusiasm has had an impact on the orchestra players.

Next, consider the low position of Tchaikovsky in the views of the members of the New York Philharmonic Orchestra. Arturo Toscanini, its conductor at the time, is well known to have had an aversion to Tchaikovsky, whose works were seldom performed under his leadership. It is possible, as Folgmann suggests, that Tchaikovsky's low position is a reflection 'of the attitude of its forceful leader', as is the very high scale position of Beethoven *vis-à-vis* the second-ranked composer, knowing that Toscanini was a great interpreter of Beethoven.

The known preferences of the conductor clearly play a part in influencing the voting of musicians; others mentioned by Folgmann are the nationalities represented in this orchestra membership, environmental factors such as the location of the orchestra, extra-musical activities of its members, and the different instruments played. Folgmann has shown that the highest percentage of French musicians is in the Boston Symphony Orchestra (27.3%, as opposed to 3.2%, 6.7% and 6.9% in the others), and Berlioz, Debussy and Franck, are placed unusually high in the scale for this orchestra. There is an unusually high number of musicians of Scandinavian extraction in the Minneapolis Symphony Orchestra, and this may be responsible for Grieg's unusually high position in the scale for this orchestra.

How reliable is this kind of judgment? Farnsworth (1969) obtained rankings of 100 classical composers by sending questionnaires to musicologists, and obtained rankings very similar to Folgmann, after a time interval of 36 years. Farnsworth's composers, in order, are given in Table 1.4; the agreement with Folgmann is obvious, although of course there are slight divergences. Bach is first, rather than fifth; Beethoven second rather than first; Haydn fourth rather than seventh; Brahms fifth rather than second. But remembering that different composers entered the two rankings, the agreement is close.

Farnsworth (1969) also used nominations by members of the American Musicological Society, who were asked for the names of 10 composers born in the 1870s or later whom they considered most likely to form part of our musical heritage. Replies were randomly placed in one of two piles, with 350 names in each, and rankings constructed of the 'eminence' of the nominated composers. The top 14 composers are given below; over all the 35 top names submitted, the correlation between the two piles was .97, i.e. close to identity. The top 14 composers were, in order: Stravinsky, Bartok, Hindemith, Schoenberg, Berg, Prokofiev, Ravel, Webern, Copeland, Britten, Vaughan Williams, Ives, Shostakovich and Milhaud. A similar experiment for composers born before 1870 produced two piles of 425 ballots giving a rank correlation of .98. Replication over the years produced very similar correlations, showing consistency of judgments over time. College students and various groups of children also showed agreement with the musicologists, and so did Dutch attenders at symphony concerts, demonstrating that agreement was not confined to citizens of one country. Farnsworth also looked at 'enjoyment' ratings for composers, their appearance in programmes of music played by the Boston Symphony Orchestra, other symphony orchestras, frequency of broadcast music, number of recordings, space allocations in histories and encyclopaedias, all of which show good agreement. Farnsworth concluded: 'Taste is lawful' (p. 133); and it is difficult to disagree.

An alternative way of looking at rankings is to count relative performance frequencies of the works of 250 classical composers, as Moles (1968) has done,

Table 1.4. *Ranking of 100 composers by musicologists (Farnsworth, 1969)*

1.	J. S. Bach	26.	Machaut	50.5	Fauré	77.	Praetorius
2.	Beethoven	27.	Schütz	52.	Dowland	77.	Borodin
3.	Mozart	28.	Liszt	53.	C. P. E. Bach	77.	Gounod
4.	J. Haydn	29.	Mussorgsky	54.	Rimsky-Korsakov	79.	M. Haydn
5.	Brahms	30.	Corelli	55.	Perotinus	80.5	Sousa
6.	Handel	31.	D. Scariatti	56.	Wolf	80.5	Sullivan
7.	Debussy	32.	Gabrielli	57.	Bartok	82.5	Bellini
8.	Schubert	33.	Couperin	58.	Grieg	82.5	Janacek
9.	Wagner	34.	Gluck	59.	Weber	85.	Donizetti
10.	Chopin	35.	Puccini	60.	Gibbons	85.	Webern
11.	Monteverdi	36.	Franck	61.	Sweetlinck	85.	Willaert
12.	Palestrina	37.	Dvorak	62.	Schoenberg	87.	Offenbach
13.	Verdi	38.	Buxtehude	63.	J. Strauss, Jr.	88.5	Ravel
14.	Schumann	39.	Bruckner	64.	Saint-Saens	88.5	Delius
15.	des Pres	40.	Sibelius	65.5	Telemann	91.	Elgar
16.	de Lassus	41.	Rameau	65.5	Lulli	91.	Hindemith
17.5	Purcell	42.	Frescobaldi	67.	Landino	91.	Satie
17.5	Berlioz	43.	Okeghem	68.	MacDowell	93.5	Cherubini
19.	R. Strauss	44.	Stravinsky	69.	J. C. Bach	93.5	Foster
20.	Mendelssohn	45.	A. Scarlatti	70.	Leoninus	95.	de Rore
21.	Tchaikovsky	46.	Dunstable	71.	A Gabrielli	96.5	Boccherini
22.	Vivaldi	47.	Bizet	72.5	Carissmi	96.5	Franco of Cologne
23.	Mahler	48.	Gesualdo	72.5	Pergolesi	98.5	Clementi
24.	Byrd	49.	Rossini	74.	Marenzio	98.5	Tartini
25.	Dufay	50.5	de Victoria	75.	Smetana		(The next 4 are tied)

Notes:

with results shown in Table 1.5. Again we find a good deal of agreement, with Mozart, Beethoven and Bach heading the list. Simonton (1988a,b) has shown that there is similar agreement in other fields of art, science and leadership, and there is little doubt that Galton's criterion of *renown* possesses sufficient reliability to be a reasonable guide for the recognition of genius.

Genius and the problems of continuity

Of course such rankings do not possess the quality of absolute truth. Is Monteverdi really a greater composer than Schumann; de Lassus than Berlioz, Byrd than Liszt, Dufay than Franck, Machant than Dvorak, Schuetz than Sibelius, Corelli than Puccini, Okeghau than Bartok, Buxtehude than Stravinsky, Rameau than Rossini? To ask such questions is to realize that the rankings do not aim to answer such detailed questions. Different composers wrote music for different purposes; we cannot easily compare Puccini with Vivaldi, Bizet with Hindemith, Rimsky-Korsakov with Purcell without taking into account the time when they flourished, whether they wrote operas or symphonies, and which country they came from. What we can say is that their work has stood the test of time, and that on Galton's criterion all have won, and all must have prizes!

But is there a cut-off point which singles out the genius from the also-rans? When we look at Nobel Prize winners, it has become very clear that among the also-rans there are at least some who in the opinion of their peers would have had just as much right, possibly more, to selection; accidents of timing, judgment and chance are seen to play an important part. The question of whether there is a definite point where 'genius' and 'talent' diverge is probably insolvable, but we may be able to apply a statistical technique known as the 'scree' test. This is here adapted from its prior use in factor analysis, where it sorts out significant factors from mere artifacts. Consider Table 1.2, the last column. This lists the scale separation of the composers, as ranked, but given as interval scales; in other words, the intervals between the composers indicate how far they are separated in the opinion of the judges. (A simple ranking as in Table 1.1, is an *ordinal* scale, where all differences are equal, namely just one rank, and hence have no separate meaning.) Fig. 1.3 lists the composers in order of ranking from first to last, and gives (with only two digits after the decimal point) the actual ordinal scale positions of the composers, with Beethoven having the highest score (0.00) and all others deviating from it in the downward direction (hence the minus signs).

Scree is the name given to an accumulation of rock fragments at the foot of a cliff or hillside, often forming a sloping heap. The scree test in statistics uses such a change in slope to detect a change in quality. Fig. 1.3 shows quite clearly that there is a regular decline from Beethoven to Grieg, but a precipitous change in rate of decline from Grieg to MacDowell and Victor Herbert. The

Table 1.5. *Relative performance frequencies of 250 composers (Moles, 1968)*

6.1	Mozart	1.0	Purcell	0.45	Milhaud	0.25	Glinka
5.9	Beethoven	1.0	Puccini	0.4	Bartók	0.25	Granados
5.9	J. S. Bach	0.95	Grieg	0.4	Borodin	0.25	Gretchaninoff
4.2	Wagner	0.95	Weber	0.4	Bruckner	0.25	Khatchaturian
4.1	Brahms	0.95	Prokofiev	0.4	Vivaldi	0.25	Hindemith
3.6	Schubert	0.95	Berlioz	0.4	Elgar	0.25	Lalo
2.8	Handel	0.95	Rossini	0.4	Mascagni	0.25	Leoncavallo
2.8	Tchaikovsky	0.95	Ravel	0.35	Offenbach	0.25	des Pres
2.5	Verdi	0.85	Rimsky-Korsakov	0.35	Palestrina	0.25	Poulenc
2.3	Haydn	0.85	D. Scarlatti	0.35	Monteverdi	0.25	de Lassus
2.1	Schumann	0.7	Franck	0.35	Shostakovich	0.25	Boccherini
2.1	Chopin	0.7	Gounod	0.35	Schoenberg	0.25	Bellini
1.75	Liszt	0.7	Vaughan-Williams	0.35	Walton	0.2	Telemann
1.75	Mendelssohn	0.65	Bizet	0.35	Honeggar	0.2	Pergolesi
1.7	Debussy	0.65	Couperin	0.3	Albéniz	0.2	Enesco
1.65	Wolf	0.6	Mahler	0.3	Buxtehude	0.2	J. C. Bach
1.6	Sibelius	0.6	Rameau	0.3	Chabrier	0.2	C. P. E. Bach
1.4	R. Strauss	0.6	Saint-Saëns	0.3	Delius	0.2	Berg
1.3	Mussorgsky	0.6	Massenet	0.3	Gershwin	0.2	Bruch
1.3	Dvorak	0.55	Donizetti	0.3	Lully	0.2	Britten
1.3	Stravinsky	0.45	Falla	0.3	Suppé	0.2	Corelli
1.2	Fauré	0.45	Scriabin	0.3	A. Thomas	0.2	Busoni
1.2	J. Strauss	0.45	Meyerbeer	0.25	Bloch	0.2	Dukas
1.1	Smetana	0.45	Gluck	0.25	Delibes	0.2	Ponchielli
1.0	Rachmaninoff	0.45	Paganini	0.25	Giazunov	0.2	Tartini
							150 others (1 each)

Source: from Moles (1968), with permission.

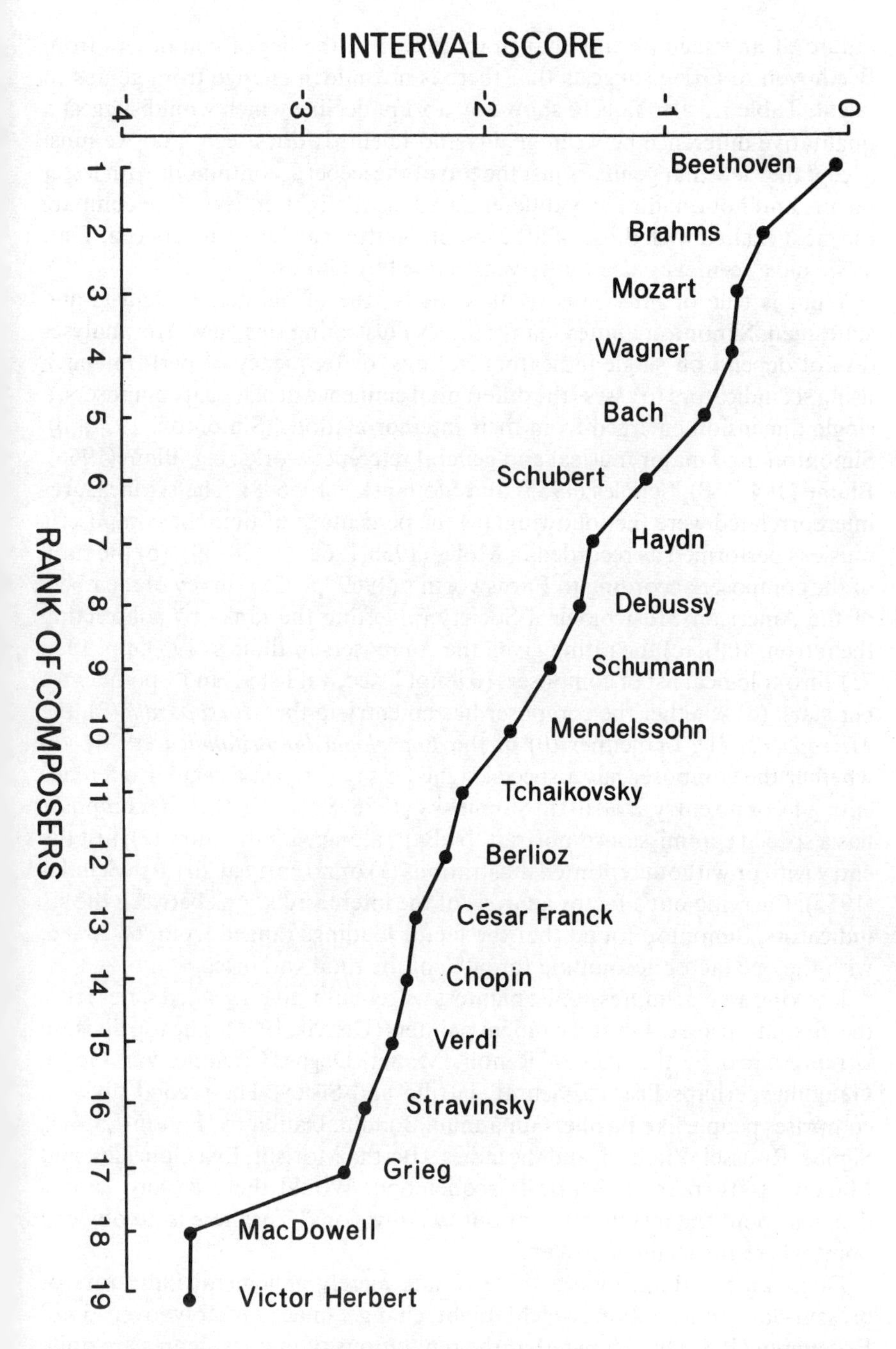

Fig. 1.3 Rated eminence of composers showing gradual decline for 17 'geniuses', and sudden decline when reaching minor modern composers.

failure of any such abrupt change to appear in the list of composers from Beethoven to Grieg suggests that there is no sudden change from genius to talent. Table 1.5 also fails to show any abrupt decline which would suggest a qualitative difference between genius and talent. It does seem that we must accept the view that genius is just the top of the iceberg, continuous with lesser talents and not qualitatively differentiated, as it might appear if we compare the most exalted with those of little talent, omitting all those in between. This, as we have seen, was also the view expressed by Galton.

What is true of musicians seems equally true of painters, scientists and statesmen; Simonton quotes many studies illustrating this view. His analyses do not depend on single indicators (ratings, or frequency of performance); using six indicators to assess the differential eminence of classical composers, a single dimension emerged from their intercorrelations (Simonton, 1977a,b). Simonton used major musical and general reference works, e.g. Blom (1966), Blume (1949–68), Scholes (1955) and Slominsky (1956–8). The six measures intercorrelated were the following: (a) the percentage of time the composer's music is performed as recorded in Moles (1958/1968, pp. 28–39); (b) the rank of the composer according to Farnsworth's (1969, p. 228) survey of members of the American Musicological Society (inverting the ranks by subtracting them from 100); (c) the rating given the composers in Illing's (1963, pp. 133–75) chronological list of composers (0 if not listed, 1 if listed, and 1 point extra per star); (d) whether the composer has an entry in the *Macropaedia* (2), the *Micropaedia* (1), or neither (0) in the *Encyclopaedia Britannica* (1974); (e) whether the composer has a special signed essay (2), just a regular reference entry (1), or no entry at all (0) in Slominsky (1956–8); (f) whether the composer has a special commissioned portrait 'by Batt' along with the entry (2), just the entry with or without reprinted illustrations (1) or no entry at all (0) in Scholes (1955). Carrying out a factor analysis of the intercorrelations between the six indicators, Simonton found that the factor loadings ranged from .68 to .86, with the one factor accounting for 60% of the total variance.

Looking at, e.g. impressionist painters will equally give a graded series from the 'first 11' to lesser but still eminent painters (Denvir, 1991). The top division is represented by the likes of Renoir, Monet, Degas, Cezanne, van Gogh, Gauguin, perhaps Pissaro, Seurat, Utrillo and Sisley. The second division comprises people like Bazille, Guilaumin, Bondin, Daubigny, Fautin-Latour, Signac, Roussel, Vuillard, and the ladies – Bertha Morisot, Eva Gonzales, and Mary Cassatt, perhaps Marie Bracquemond. Would there be any serious debates about the membership of our two divisions? Yet there is no obvious point where the chain is broken.

Does fame endure, or are these results merely ephemeral indicators of present-day appreciation which might change unaccountably over time? Rosengren (1985) has shown that the reputations of literary figures are quite durable, with a reliability of .61, a result not dissimilar to findings reported by Farnsworth (1969) for music, and Over (1982) for psychologists. There are apparent exceptions – Telemann once was rated above Bach, while now he is hardly remembered. Contemporary judgment is more fallible than that of posterity, as Galton's definition suggests.

It is sometimes objected that expert opinion, whether in science or in the arts, is only subjective and hence not worthy of scientific attention (Endler, Rushton and Roediger, 1978). Cattell (1903a,b) argued to the contrary:

> There is no other criterion of a man's work than the estimation in which it is held by those most competent to judge . . . I am somewhat sceptical as to merit not represented by performance, or as to performance unrecognized by the best contemporary judgment. There are doubtless individual exceptions, but, by and large, men do what they are able to do and find their proper level in the estimation of their colleagues.

Over (1982) was able to demonstrate that expert rankings of psychologists and number of citations in the *Citation Index* correlated remarkably highly together. It seems a contradiction in terms to think of someone as a 'genius' who is not highly regarded by his contemporaries or his successors, and whose work is only very infrequently cited by his peers!

Altogether the notion of 'subjectivity' is far from obvious. Is temperature subjective? Until the discovery of the thermometer (Middleton, 1966) subjective feelings of hot and cold were the only evidence of temperature differences; yet these feelings are certainly affected by other factors than measured temperature. The wind-chill factor is well known, of course; exercise, food and drink intake, and fever are examples of irrelevant factors which cause our estimates of temperature to deviate from the truth. Yet 'subjective' estimates are remarkably accurate on the whole, and if the temperature readings on the thermometer did not largely correspond with our 'feelings' of hot and cold, we could hardly accept them as objective indices of temperature.

As Simonton (1988a,b) has shown, statistical methods can be used to indicate objective reality underlying subjective judgments; factor analysis, path analysis and multidimensional scaling are only some of the techniques that can be used to test hypotheses regarding the objective bases for our subjective judgments. In due course, no doubt, we will approach ever nearer to complete objectivity, but our failure to do so is no reason to reject well-researched methods of peer rankings and ratings. All scientific measurement of variables such as weight (Kirsch, 1965) and mass, length and time (Feather, 1959) have gone through such a process of increasing objectification, and there is no reason to expect psychology to be the exception. Our measures of intelligence, creativity and genius are no doubt primitive when compared with long-established physical measures, but when we look at more recent estimates, such as the value of the Hubble constant (Overbye, 1991), we find even greater discord among experts in astronomy than is usual in psychology! The age of the cosmos, which is of course a function of Hubble's constant, is constantly changing; recent discoveries are doubling the age of the universe from previously accepted values (Gribbin, 1992). The value of the Hubble constant has changed from 500, when I was a student, to between 50 and 80, the difference being a battle-ground between rival factions. What price objectivity!

Creativity as trait and as achievement

We may conclude that the first part of Galton's hypothesis receives considerable support. Eminence, as judged by reputation, is a good guide to genius in all the disciplines and specialities where tests have been carried out. We are thus encouraged to pursue the second part of Galton's hypothesis a much more difficult pursuit, but one which may help us test his general theory as expressed in the diagram drawn to illustrate his views; this will be done in subsequent chapters. Here I will discuss a few points that may be relevant to the conception of 'genius'.

I have tried in this chapter to define 'genius' along the lines of Galton and McKeen Cattell in terms of creativity shown in the production of great works in science or the arts, judged and accepted as such by peers, and achieving great acclaim by contemporaries and later generations. Later chapters will be devoted to the investigation of factors (historical, social, personal, cultural) which help or hinder the development of genius, and in particular to the elucidation of the true meaning of the concept of 'creativity' which is here used simply in its everyday sense of 'being characterized by originality of thought, having or showing imagination' (*Collins Dictionary*, p. 229). The interaction of all these factors, including intelligence and various special abilities, will of course also be discussed, as will the experimental study of such concepts as 'creativity' and 'intuition'. Above all, I shall search for a solution to the problem of the paradox that creativity as a psychological trait is apparent in all people, to varying degrees, while creativity in the sense of great achievement seems confined to singularly few. Our search will be in particular for a *tertium quid* linking together these two concepts.

In order to avoid the use of the term 'creativity' in two rather different senses (trait and achievement), it might be useful to use the term 'originality', instead of 'creativity as a trait'. It is obviously possible to be original, i.e. present unusual solutions, associations, etc. without being creative in the achievement sense. Creativity implies that the original responses are *relevant*, and the production of creative objects requires a lengthy process of constructive work, defence against critics, etc. As Newton wrote to Oldenburg, 8th November, 1676: 'I see a man must either resolve to put out nothing new or to become a slave to defend it.' Originality by itself is not enough to be considered *creative*; much more is required in addition. A psychotic person's responses are original, in the sense of unusual, but they are hardly ever creative; they lack relevance.

This point seems so obvious that it has been taken for granted, and repeated by many authors, yet there are many problems with it. When Shakespeare broke the rule of the constancies (of time and place) in his plays he was certainly original, but was his originality relevant? What would relevance mean in this context? Who would decide? He was certainly condemned for his breach of convention by continental critics, until Lessing argued in his favour. When Lobachevsky invented non-Euclidian geometry, in what sense was that relevant? Relevant to what? Contemporary mathematicians certainly did not think so; posterity has decided otherwise. Judgments of relevance are decided by *orthodox* judges for whom the concept of relevance is itself defined in terms of orthodoxy; as Kuhn has pointed out so often, a revolution in prevailing paradigms rules out agreement on relevance. Lack of traditional relevance may be the defining feature of originality!

Runco and Charles (1993) have shown that much the same considerations apply in the experimental measurement of creativity in terms of divergent ideation. They showed that originality ratings predicted creativity much better than appropriateness or relevance ratings. Their results suggested that, although it is not necessary for an original item to be appropriate to be viewed as creative, original ideas are not valued less for being appropriate. However, with unoriginal ideas, appropriateness may inhibit judgments of creativity. I shall return to this complex but important concept of appropriateness or relevance in a later chapter.

Despite Galton's hypothesis of 'eminence' being normally distributed, the evidence from creativity as achievement shows it to be very abnormally distributed. This evidence comes from the highest realm of artistic or scientific achievement, but also from lesser but still notable successes. Concentrating on scientific achievement, where judgments are perhaps more objective than in art, it is well known that a small proportion of active scientists is responsible for the major number of creative works. Thus Dennis (1955) found that the top 10% most productive contributors in a variety of scientific disciplines were responsible for about half of the total works published whereas the bottom 50% were less productive and contributed only about 15% of the total output (Bloom, 1963; Davis, 1987; Shockley, 1957). In psychology, for instance, the most prolific author can claim more contributions than can 80 colleagues in the lower half of the distribution (Dennis, 1954). These data in fact underestimate the difference because they only include those who have made at least one contribution, thus leaving out of consideration all those never making any contribution at all! As Walberg *et al.* (1984) argue, the concept of normal distribution cannot be applied to exceptional performance.

Lotka (1926) and Price (1963) have attempted to formulate quantitative laws to encapsulate these and similar findings. There is agreement on the general shape of the distribution – monotonically decreasing at a decelerating rate. According to Lotka, the number of scientists publishing n papers is roughly proportional to $1/n^2$ where the proportionality constant varies with the discipline. Supposing the constant to be 10 000, then the number of scientists producing n contributions would be $10^4/n^2$. That gives us 10000 scientists with just one publication, 2500 with two, 1111 with three, and 100 with ten publications. Only one scientist would contribute as many as 100 papers. According to Price's law, if K represents the total number of contributors to a given field, then $\sqrt{K}$ will be the predicted number of contributors who will generate half of all contributions. The larger the discipline, the more elitist the outcome (Zhao and Jiang, 1985), although the law may cease to hold at extreme values of K.

These laws do not only apply to scientific productions, but have far wider applicability. Dennis (1955) found similar distributions in the publication of secular music and in the books represented in the Library of Congress. Simonton demonstrated its applicability to classical music (Simonton, 1984a,b; 1987). As he points out, about 250 composers account for all the music heard in the modern repertoire, but a mere 16 are responsible for creating half of the pieces heard ($16 \cong \sqrt{250}$). Whether Price's Law can be derived from Lotka's Law is immaterial (Allison *et al.*, 1976); there is a general agreement on the major outlines of the function.

The causes of creative achievement

The very different distribution of creativity as a trait (originality), which is approximately normal (Hovecar and Bachelor, 1989; Michael and Wright, 1989; Woodman and Schoenfeldt, 1989), and of achievement, which is approximately J-shaped, suggest a theory which may indicate their proper relationship. Fig. 1.4 will suggest this relationship. It will be argued that *creative achievement* in any sphere depends on many different factors: (1) *cognitive abilities*, like intelligence, acquired knowledge, technical skills, and special talents (e.g. musical, verbal, numerical); (2) *environmental variables*, like political-religious factors, socio-economic factors, and educational factors; and (3) *personality traits*, like internal motivation, confidence, non-conformity, persistence and originality. All or most of these, in greater or lesser degree, are needed to produce a truly creative achievement, and many of these variables are likely to act in a multiplicative (synergistic) rather than an additive manner. This point will be argued presently; let us consider for a moment the variables listed; they are not claimed to be a complete set, but are merely indicative of the many different variables that have been suggested in the past (Amabile 1983a,b; 1985; Glover *et al.*, 1989).

How then would we judge Galton's argument in the light of the information gained in the 120 years that have elapsed since the publication of his book? The view of 'connectedness', i.e. no absolute, qualitative gap between genius and near-genius, has certainly been supported; continuity is most likely the rule. But Galton was almost certainly wrong in his enthusiasm for the normal curve, and its application to 'eminence' and genius. Ever since Quetelet fell in

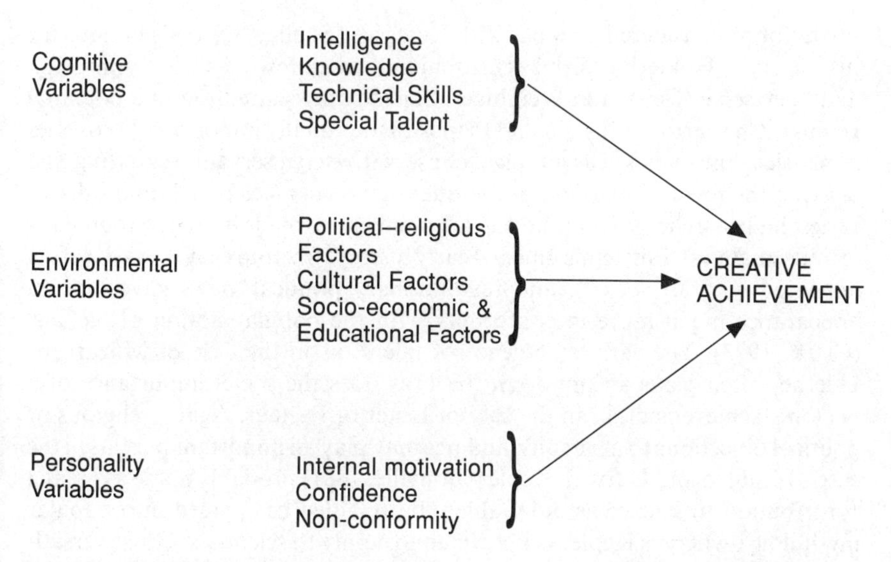

Fig. 1.4 Creative achievement as a multiplicative function of cognitive, environmental and personality variables.

love with it, and used it to define the 'average man' (Stigler, 1986), the Gaussian curve has exerted a fatal influence on social scientists. As Gabriel Lipman told Poincaré, *à propos* the curve: 'Les experimentateurs s'imaginent que c'est un théorème de mathématique, et les mathématiciens d'être un fait experimental'. (Experimentalists think that it is a mathematical theorem, while mathematicians believe it to be an experimental effect.) Galton's arrangement may serve well enough as a picture of the distribution of intelligence, but certainly not of eminence. (Even as far as intelligence is concerned, Burt (1963) has suggested that a Pearson Type IV curve fits the case better than the normal curve!) Micceri (1993) has analysed a large sample of curves of distribution that would normally be expected to be 'normal', finding practically all of them to deviate markedly from statistical normality. As Geary (1947) said long ago: '. . . normality is a myth; there never was, and never will be, a normal distribution' (p. 241).

Other features of genius

But is outstanding achievement all that determines 'genius' status? There is a popular expectation that genius is also characterized by certain unusual behaviours or accidents of life. Keats and Shelley died young, as did Schiller; Wordsworth lived to a comfortable old age. To many people the image of crabbed old Wordsworth, reactionary and shorn of his early powers, makes him less likely to be considered as a 'genius'; had he died in his thirties he would no doubt be considered on a par with Keats and Shelley. Byron is famous for his life-style, Holderlin for his schizophrenic breakdown, as were Van Gogh and Nietzsche, Kleist, and Weininger. Suicide draws attention to a potential genius – Chatterton. Einstein and Planck made equally important discoveries in physics, but Planck was a typical conservative civil servant, regretting and denying the revolutionary nature of quantum mechanics based on his discoveries; high starched collars and old-fashioned dress made him appear unlikely 'genius' material. Einstein's unusual early history, free and easy dress (he often appeared without socks!), and revolutionary political views gave him an appearance much more in conformity with the popular notion of 'genius' (Clark, 1973). The early emergence of talent, as in the case of Mozart, or Goethe, often plays an important part, as does the social importance of a person's achievement, as in the case of Lister or Pasteur. Again, religious or political objections to a person's findings may play an important part, as in the case of Galileo and Darwin. Kepler (Bonville, 1981) probably made as great a contribution to science as did Galileo, but Galileo has passed into popular mythology, whereas Kepler is known almost only to scientists. Great versatility is another popular criterion, Leonardo da Vinci, Leibnitz, Descartes and Newton are examples.

A good example of how important irrelevant considerations may be is the status of Stephen Hawking, whose sad disability has captured the public imagination (White and Gribbin, 1992), and has propelled his poorly written popular *A Brief History of Time* (Hawking, 1988) into the best-seller list. It also identified him in the popular mind as a 'genius', which is patently over the top; he is an outstanding mathematical physicist, equal perhaps to Hoyle; first-rate, but no genius. To give but one instance of the how the label of 'originality' or 'creativity' can accrue to the wrong person, consider his early stance in denying that the surface area of a black hole is actually a measure of entropy; if true, that would mean that a black hole had a temperature, defined in terms of its surface area. Now anything that has a temperature must radiate energy, and therefore cannot be a black hole!

A young Californian research assistant called Jacob Bekenstein suggested, in a series of publications, just that: the surface area of a black hole *was*, indeed, a measure of entropy, and black holes *do* have temperatures related to their surface area. Furious, Hawking attacked Bekenstein's interpretation of the equation Hawking had originally published. Later, Hawking paid a visit to Moscow and learned about the work of Yakov Zel'dovich on the way black holes interact with light. He became convinced that black holes must indeed emit radiation, and did have temperatures! He completely changed course, and advocated what earlier he had condemned, and now 'Hawking radiation' is considered one of the great achievements of the past 50 years of physics, combining as it does general relativity and quantum mechanics in one package. Poor Bekenstein, the original discoverer, is forgotten. As Sir William Osler said: 'In science the credit goes to the man who convinces the world, not to the man to whom the idea first occurs'.

Genius or Zeitgeist?

A few additional points will be dealt with here rather briefly; although important in their own right, they are somewhat tangential to the main theme of this book, and also because they have been dealt with in some detail elsewhere (Simonton, 1988a,b). The first point concerns the suggestion that the notion of genius, particularly in the sciences, is a superfluous concept because the Zeitgeist creates conditions which make discovery inevitable (Kroeber, 1917; Ogburn and Thomason, 1922; Merton, 1961a,b and 1973). This view is usually put forward by sociologists and anthropologists, as one might expect; it is mostly based on the facts of multiple discoveries. This notion forms part of a general finding that scientific creativity is more apt to appear in particular political, cultural and ideological settings. Peacetime is more favourable than wartime (Simonton, 1980a,b); so is a cultural ideology favouring empiricism, materialism, nominalism, individualism and determinism, as opposed to mysticism, idealism, statism, and the doctrine of free will (Simonton, 1976c; Sorokin, 1937–41). Another important condition contributing to the growth of scientific knowledge is economic prosperity (Rainoff, 1929; Schmorkler, 1966), although we may remember Lord Rutherford's recipe for scientific success: 'We have no money, so we will have to think!'.

Actually there are many problems with the idea of multiple discovery, in particular that of establishing the identity of the discovery. Newton and Leibnitz both discovered a method of analysis that we now call 'calculus', but their methods were in fact very different, and the continental adoption of the much more practical Leibnitzian system put their mathematicians at a marked advantage over the English mathematicians who had to contend with the much more cumbersome Newtonian system. Simonton (1988a) gives a long discussion of a number of alleged multiples, showing that in most cases the discoveries were in fact very dissimilar, although covered by the same name. It may be useful to illustrate this point further.

In more modern times, Schwinger's quantum electrodynamics and Feynman's may have been mathematically the same, but one represented a mathematical style doomed to grow fatally overcomplex, while the other used a new style of visualization involving and leading to a search for general principles (Gleick, 1992). Or consider the empirical equivalence of Heisenberg's and Schroedinger's version of quantum mechanics – in terms of matrices or wave function, or the widely different view of Gell-Mann and Feynman about particle physics. Do we accept and follow Zweig's 'aces', Gell-Mann's 'quarks', or Feynman's 'partons' – three paths to the same destination? Reality can be presented in many different ways, not only in art but also in science, and the method of presentation is strongly determined by personality and other factor specific to the scientist or the artist. Scientific achievements may bear the personal mark of the originator just as much as artistic achievements.

Another problem noted by Simonton is that of independence vs. antecedence. Often the simultaneity is stretched over a considerable period, so that one 'discoverer' might very well have heard of the work of another; to prevent this happening scientists in the olden days often communicated their discovery to others in some indecipherable code, so that they might prove priority later on, after proper publication, by revealing the code. Simonton gives many examples where independence is highly questionable, in cases quoted as 'multiple'. Many multiples are not simultaneous discoveries, but may be rediscoveries, plagiaries, or partly co-operative efforts.

Finally, Simonton argues that Merton's notion of certain discoveries being 'historically inevitable' is clearly an overstatement; he suggests that instead of using the word 'inevitably' he should have said: 'eventually'. 'The accumulation of pre-requisite knowledge does not *make* a discovery happen but only allows it to happen, eventually' (p. 145). Again, the point is buttressed by many examples where conditions were ripe, but discovery was delayed dozens, hundreds or even thousands of years. Somewhat related is another problem that will be discussed in more detail later – when is a discovery not a discovery? Copernicus is credited with discovering that the earth went round the sun, but almost 2000 years earlier Aristarchus put forward the same idea. Archimedes anticipated Newton and Leibnitz in putting forward the central idea underlying the calculus. Long before Darwin, ideas about evolution were circulating among the learned. It is sometimes quite difficult to say how original original ideas are, or were!

Coming back to the problem of multiple discoveries. Simonton (1988b) presents a statistical analysis which clearly disproves the Zeitgeist interpretation. First, he distinguishes multiples according to their 'grade', i.e. the actual number of investigators who are reported to have independently made a given discovery or invention. There are thus doublets (grade 2), triplets (grade 3), quadruplets (grade 4), and so on. Second, he goes on to show that these different grades show a quite specific pattern, namely that of the Poissonian distribution (Prince, 1963), i.e. a distribution resembling an inverted J-curve, with singletons having much the greatest number of discoveries, doublets many fewer, triplets, fewer still, with higher grades having very few instances. As Simonton points out, having

> this conception of multiple originations dovetails more closely with a chance-configuration theory than it does with the traditional viewpoint. In the chance permutation of elements, a huge succession of unstable mental aggregates must be sifted through before a stable permutation can emerge, and there is no guarantee that the desired chance configuration will appear at all, given how improbable and unstable permutations are. Thus, the isolation of a chance configuration requires many trials, each with a miniscule probability of success (p. 151).

By chance, multiple discoveries should thus occur roughly in the proportions observed.

Probably the most remarkable instance of the influence of one man on the development of science is the development of science itself. What we in the West call 'science' is a relatively late growth, in large measure the heritage of Galileo Galilei. The ancient Greeks had isolated triumphs, but nothing systematic like modern science; of course Archimedes was a genius, but he produced isolated bits and pieces, not a unified structure. Even more astonishing is the failure of the Chinese to construct such a unified science. They clearly had the ability; they anticipated a large number of 'modern' discoveries; but they did not possess anything resembling what we call 'science'. Mendelssohn (1976) has most eloquently made the case that the Chinese had technology of a high order, but not that systematic search after truth we call science. It is impossible to prove that it was one man who threw the switch and got the train to change direction, but clearly something remarkable happened 400 years ago in Italy that never happened in China in over a thousand years of high-level civilization. (Wolpert (1992) would agree that the Chinese excelled in technology, but never created science as conceived by Galileo; he would argue, however, that the switch was thrown by Thales in ancient Greece, and that the Greeks originated the concept of science in its modern meaning. I do not wish to debate the point here; what is clear is the immense influence of one man in creating something entirely new that might never have existed otherwise.)

This argument may also throw some light on the question of whether artistic achievement is fundamentally different from scientific achievement, in that the artistic product is *unique*, while scientific discoveries would have been made in any case by somebody else. Mendelssohn's and Wolpert's arguments and historical discussions show that even the very discovery of science as a major human activity did not occur as a necessity, and regardless of who might have invented it; the Chinese had a much better opportunity, but they did not! Would impressionism have conquered the art world if it had not been for Monet (or Cezanne, or Renoir, or Manet)? Such questions are of course impossible to answer, although we can approach them scientifically and quantitatively, as Martindale (1992) has shown (his contribution will be dealt with in a later chapter). I do not claim to have given the correct answer; the argument is presented simply to make the reader question the traditional opposition between scientific and artistic creativity. The opposition is certainly nothing like as absolute as is often argued; there may be little left when we know more about historical developments than we do now.

Conclusions

What have we learned about the conception of 'genius'? Is it something *sui generis*, apart from the world of talent, categorically removed from all others, sitting on a solitary throne, or is it rather *primus inter pares*, the (marginally) best in a group of peers, the top of the ladder? Both views have their adherents, and there can be no 'true' answer; where definitions are concerned, we cannot design an experiment to judge between different points of view. Yet empirical investigation is possible, and such studies as those quoted in this chapter may throw some light on the matter. However that may be, I shall here indicate how the term will be used in this book, and I shall give my reasons for using it in this fashion.

Consider the term 'giant', a term used to describe unusual physical size, but an analogy also used to describe 'mental giants'. We may use the term to describe Goliath, and imagine that Goliath categorically differentiated from all others in his army. Yet we can apply a measuring scale, graded in inches, centimeters, or what not, and show that while Goliath scores top marks (7 feet?) on this scale, he is so many inches taller than the next man, who in turn is somewhat taller than the third, and so on. There is a continuum, and if the second tallest was 6 feet and 11½ inches tall, we would have difficulties arguing for a *categorical* separation of Goliath and the rest of the army. We have seen that among musical composers, the differences between the top man and the second, or the second and the third, are all minute; indeed, there is a regular sequence of showing tiny differences all along the way. Clearly eminence is a continuum; there are no abrupt differentiations, except when as in Fig. 1.3 it has been artificially introduced to show the difference between classical composers and popular modern lightweights like MacDowell and Victor Herbert.

This reduces our choices; eminence is indeed graded, as Galton suggested. What, then, of the notion of genius? Are we all geniuses, but in the lower reaches of the ladder? It is here that subjectivity comes in. In the measurement of size, at what point of the continuum do we stop calling a person 'tall'? Is someone just over 6 feet like myself 'tall'? In my generation, in Europe, the answer would have been in the affirmative; at present, in the USA, it would probably be negative. There are so many six-footers that they would simply be regarded as average. In my generation someone 6 feet 4 inches tall would have been called a giant; among modern basketball players they would at best be average. Clearly, where you make your cut in the continuum is a subjective matter, depending on individual preference and surrounding conditions. Everyone would agree that Mozart, or Beethoven, or Bach was a genius; how about Mendelsohn, or Chopin, or Grieg? If you say no, I can give you a list of names intervening in the ladder which, by tiny and statistically insignificant steps, lead from the acknowledged genius to the lower-placed composer. There is no point where the sequence ends in any clear-cut fashion; the point where we say: Enough is enough! is determined subjectively. Clearly a man 5 feet 8 inches tall is not 'tall'; clearly Schieri is not a musical genius. We must recognize the existence of greater and lesser geniuses; probably any composer who after one hundred years or more is still played by the leading orchestras deserves to be called genius, even if a minor one. The same would probably apply in science; the Gilberts, Hookes, Huygens and Brahes are still remembered, form part of our scientific history, and have achieved enough creative results to deserve the title of genius.

Why, it may be asked, are there no modern geniuses in science after Einstein and Planck? One reason may be that science has been breaking up into numerous specialties; no man can any longer be creative in many different fields, like Newton, and universality is one of the characteristics of genius. There is now so much to learn and know in each sub-branch of physics, or astronomy, or chemistry that it is difficult to become an expert even in one, let alone in several.

In the second place, there are now vastly more runners in the race than in Newton's time, or even Einstein's – Niels Bohr, Paul Dirac, Enrico Fermi, Richard Feynman, Murray Gell-Mann, Werner Heisenberg, Edwin Hubble, Hendrick Lorentz, John von Neumann, Robert Oppenheimer, Wolfgang Pauli, Isidor Rabi, Ernest Rutherford, Andrei Sakcharov, Erwin Schroedinger, Eugene Winger, Chen Ning Yang – in less crowded times every one of them would have been called a genius; they suffer from a surfeit of talents. If genius is *primus inter pares*, too many *pares* make it difficult to recognize the *primus*!

In the third place, there is usually only one outstanding jewel in the crown, and if you miss the right moment to pluck it, there will never be another. Laplace mourned that Newton had pre-empted any claims to having solved the riddle of the solar system, and while it seems likely that Laplace could have succeeded equally well, he came over a hundred years too late. Einstein came at the right time to lead the post-Newtonian revolution. The next Great Advance presumably will be a Grand Unified Theory that incorporates gravitation as well as the electromagnetic with the strong and weak nuclear forces. Task and man have to be matched to produce a supreme genius; the man alone is not enough.

The truth of this dictum is apparent when we consider Newton. He was the supreme genius in physics, but although he spent much of his time in the field of alchemy/chemistry, his contribution there was negligible (Westfall, 1980). The discipline was just not ripe for the wide-ranging integration Newton brought to physics; it was pre-paradigmatic! Much the same is probably true of psychology today; I have met several psychologists who thought they had been selected by providence to be the Newton of psychology, but turned out rather to be the Newton of Alchemy!

Possible exceptions are two men to whom the title 'genius' might rightfully be applied: Galton and Pavlov. Both were instrumental in inaugurating new lines of enquiry, fashioning the instruments needed for advance, and furnishing the theoretical underpinning necessary for the development of the disciplines of the correlational investigation of individual differences (Galton), and the experimental study of the laws of learning, conditioning and association (Pavlov). The next generation contained many outstanding scientists who elaborated these beginnings (Spearman, Thorndike, Köhler, Hull, Tolman,

Watson, Skinner, to name but a few), but none can be regarded as a genius. The present generation seems exceptionally lacking even in men of their achievement; good, rigorous experimenters abound, but creative new ideas are conspicuous by their absence. Perhaps my generation is preparing the ground for the appearance of the genius who will pull all these disparate strands together, but the Messiah is surely a long time a-coming!

What may be underlying this kind of development may be a universal trend. The genius who originates a large-scale new development is followed by high-ability people who develop new ideas within the field newly opened up; they are successful because everything is new and controversial. The next generation has to work in a field now partitioned and governed by orthodoxy; work now is along tramlines laid down by venerated predecessors, and creativity is frowned upon. Ordinary science is the watchword, until the Kuhnian revolution starts the process all over again. I shall quote some historical support for this view of present-day psychology in a later chapter; history may of course show that this view of our position is completely mistaken! However, mistaken or not, few would argue that psychology is bursting with geniuses comparable with those this century has thrown up in physics and astronomy.

At the other end, the notion that there is no clear-cut development and evolution in the arts is also false; Martindale (1990) has shown quite plainly that the arts show a clearly marked pattern of change, so that certain advances come in a regular fashion. I will discuss this in some detail later, here let me only note that in principle there is no reason to believe that what is true of science would not be equally true of the arts.

2 Genius and intelligence

Errors using inadequate data are much less
than those using no data at all.

Charles Babbage

Intelligence – a dispositional variable

Few writings on genius have neglected to give intelligence a very high place, and indeed it is difficult to think of leading philosophers, scientists, writers, statesmen and artists other than highly gifted intellectually. But the opposite does not follow; not all people who are highly gifted intellectually turn out to be geniuses – were it otherwise the world would be overrun with geniuses! Intelligence is a *dispositional* variable, i.e. it enables a person with that ability to solve certain problems, produce certain results, achieve certain aims, but it does not guarantee success. Very early in the history of psychometric testing, Alexander (1935) analysed the relation between ability and school grades and discovered an 'X' factor, which was found to run through all the school subjects but through none of the ability measures. Alexander stated: 'We are suggesting that X must be interpreted as a character factor which exercises an important influence on success in all school subjects. If we were to attach a name to this factor, we should be inclined to call it persistence.'

The distinction between a *dispositional variable* and what we might call an *achievement variable* (e.g. school success, production of a work of genius) is absolutely vital in understanding psychological analyses of abilities and traits. The distinction currently made between *trait* and *state*, say of anxiety, embodies the distinction. The dispositional hypothesis states that some people are more likely than others to react with anxiety to situations perceived as dangerous, and to perceive as dangerous situations which to others may not appear to be so. But a *state* of anxiety may be induced even in non-anxious persons by presenting them with a very real danger, while even those high on trait anxiety may be quite relaxed upon occasion when no trace of danger looms. Cicero, in his *Tuscularum Disputationes* (1943), produced an excellent argument marking this vital difference:

> Ut sunt alii ad alios morbos procliviores, itaque dicimus gravedinosos quosdam, quosdam torminosos, non quia iam sint, sed quia saepe, sic alii ad

metum, alii ad aliam perturbationem: ex quo in aliis *anxietas*, unde anxii, in aliis *iracundia* dicitur, quae ab *ira* differt, estque aliud iracundum esse, aliud iratum, ut differt *anxietas* ab *angore*; neque enim omnes anxii qui anguntur aliquando nec qui anxii semper anguntur, ut inter ebrietatem et ebriositatem interest aliudque est amatorem esse, aliud amantem. Atque haec aliorum ad alios morbos proclivitas late patet; nam pertinet ad omnes perturbationes.

(Translated freely: As some people are more prone to one disease, others to another, we may say that certain people are liable to catch colds, others to suffer attacks of colic, not because they are suffering at the moment, but because they frequently do so. In the same way some people are prone to fear, others to other disorders. As a result we may sometimes speak of an anxious temperament, and of anxious people. In other cases we talk about *irascibility*, which is not the same thing as *anger*. It is one thing to be irascible, another thing to be angry, just as an anxious temperament is different from feeling anxiety. Not all people who are at times anxious are of an anxious temperament, nor are those who have an anxious temperament always feeling anxious. The difference is like that between intoxication and habitual drunkenness, or between being gallant and being in love. This proneness of some people to one disease and others to another is of wide application, for it applies to all disorders mental and physical.)

Thus does Cicero distinguish between trait (*anxietas*) and state (*metus*), or trait (*iracundia*) and state (*ira*), just as being liable to frequent colds is different from having a cold at the moment! (Cicero actually anticipated in part a modern finding that introverts are more liable to colds than extroverts!) The distinction here made is absolutely vital to an understanding of the theory underlying modern psychometric testing. It might be though that the distinction would be too well known, and indeed too obvious, to require being spelled out in detail, but as Fig. 0.1 shows, well-known psychologists (Sternberg and Wagner, 1986) still define intelligence in terms other than dispositional; the notion of 'practical intelligence' they advocate is in terms of achievement, i.e. in terms of the product of many variables of which dispositional intelligence is only one. In this chapter we shall be concerned with intelligence as a dispositional variable (IQ), and consider achievement as a component variable correlated with, but separate from, intelligence as such. (See Eysenck (1988) for a discussion of this point.)

The Sternberg–Wagner thesis is in fact a revival of the discredited view originally presented by McClelland (1973) advocating 'testing for competence' rather than intelligence. 'Competency', as conceived by McClelland (but never properly defined by him), may be regarded as 'an underlying characteristic of a person in that it may be a motive, trait, skill, aspect of one's self-image or social role, or a body of knowledge which he or she owns' (Boyatzin (1982) p. 21). This multiplicity of factors is very similar to the multiplicity of factors in Sternberg's (1985) theory, a complexity which immediately rules such a concept out of consideration as a meaningful scientific concept.

The major points originally made by McClelland were: (1) School achievement does not predict occupational success. (2) Intelligence tests do not predict occupational success or important life outcomes. (3) Tests of academic performance only predict job performance because of an underlying relationship with occupational status. (4) Such tests are unfair to minorities. (5) 'Competencies' would be better able to predict important behaviours (life success) than would traditional tests. A great deal of evidence collected both before and after the appearance of the McClelland article decisively disproves each one of these points (Barrett and Depinet, 1991), and renders the whole argument nugatory.

Intelligence and achievement

If we take income as an index of the life achievement which adherents of the 'practical intelligence' notion might accept, we are at once faced with the contrast between the normal distribution of IQ, which more or less follows the Gaussian curve, and the J-shaped distribution of income, which shows a few having huge incomes, and a long tail of those having very small incomes. These differences in distribution can be demonstrated quite well by computing the appropriate functions of the higher moments (beta function). For the normal curve, $\beta_1 = 0$, $\beta_2 = 3$, and these values are approximated by the majority of IQ distributions which give β_1 values between 0.0 and 0.2, while for β_2 they lie between 2 and 4. For curves of income in modern capitalist countries β_1 is around 1.2, and β_2 is around 50 000 or more (Burt, 1943)!

Pareto (1897) attempted to express this universal 'law of inequality' in terms of a simple mathematical equation:

$N = \frac{C}{x^a}$, where N is the number of persons whose income exceeds x units. C is a constant, while the exponent a measures the inequality of the incomes, with a value usually around 0.5, but apparently declining since Pareto's day. Burt differentiated Pareto's equation to obtain a formula more familiar to psychologists:

$Y = \frac{aC}{x^{a+1}}$, where Y denotes the proportionate number of persons having an income of:

$£ = (x \pm \frac{1}{2}dx)$.

The resulting curve is J-shaped belonging to Pearson's Type XI. Burt applies various forms of Pareto's formula to a variety of data, including an early version of citation count, and concludes

> that individual output . . . does not follow the normal curve, although individual ability conceivably may. But I venture to suggest that the apparent inconsistency between the two distributions vanishes directly we recognize that the functional relation between output (as effect) and capacities (as causes) may be of many different kinds, and indeed is more likely to be indirect and complex than immediate or simple (p. 96).

The combination of factors producing output, or income, or achievement generally, is more likely to be multiplicative than additive. If we take a simplistic example, namely Alexander's (1935) picture of scholastic achievement as a product of IQ and persistence, it is clear that if either is zero, or very low, the product will be zero or very low, in spite of the fact that the other variable may be very high; thus simple addition makes no sense. To take multiplication, rather than addition as the more likely way of combining factors, has the additional advantage that this will lead from a normal curve for the components to a J-shaped curve for the products. Assume that both IQ and persistence are distributed into five classes, with marks of 0, 1, 2, 3 and 4 respectively, and that the distribution follows the binomial law, with frequencies proportional to 1, 4, 6, 4, 1. Multiplying marks now gives us a distribution where 50% have a score of 0, 36% of 1, 11% of 2, 3% of 4, and 0·4% of 5. This is the usual J-shaped curve of output, income, or creative achievement so frequently found and documented in the Price–Lotka law discussed in Chapter 1, which indeed derives from Pareto's equation.

Some criticisms of the IQ

Intelligence conceived as a dispositional variable has often (erroneously) been criticized on two grounds. One is that psychologists do not agree on the meaning of the term; the other that IQ tests measure only the subject's ability to do IQ tests. There is no ground for either assertion. As regards the first, Snyderman and Rothman (1987) questioned over 600 experts in the field on this (and many other) points, and found that 99.3% agreed that intelligence was concerned with abstract thinking or reasoning, 97.7% agreed that it was concerned with problem-solving, 96.0% agreed that it was concerned with the capacity to acquire knowledge. These figures do not indicate large areas of disagreement.

The notion that IQ tests do not measure vitally important ingredients of worldly success is equally mistaken (Eysenck, 1979; Vernon, 1979; Barrett and Depinet, 1991). IQ tests predict extremely well scholastic success, both at school and university; they predict occupational success, both for blue-collar and white-collar jobs; they predict success in the armed forces, the police and in government jobs generally. As Ashenfelter and Kreuger (1992) have shown, each additional year of education adds on average 16% to an individual's pay packet, and of course the correlation between educational duration and success is very close. Indeed, it is difficult to think of any area allowing for free competition where IQ tests fail to be predictive, and more so than alternative measure.

Furthermore, and this is an equally strong argument, *IQ has a strong biological basis*. Genetic factors account for some 70% of the total variance (Eysenck, 1979), and important links have been established between IQ and psychophysiological variables, such as EEG and averaged evoked potentials (Eysenck and Barrett, 1985; Barrett and Eysenck, 1992, 1994). It is vacuous to imagine that nature had installed a large physiological apparatus in mankind merely to foster excellence in IQ test solution!

Some other objections to the concept of intelligence as measured by IQ tests are of a philosophical kind. Thus, Keating (1984) argues that those who believe in the usefulness of the concept of intelligence adhere to a belief 'that it is a thing that exists in the head of a person' (p. 2). He and many others would argue that intelligence does not exist and that, hence, all efforts to measure it must be useless. This argument is erroneous on both counts. Spearman and his followers have never posited the *existence* of intelligence; they have regarded it as a scientific *concept*, analogous to such concepts as gravitation, humidity, or mass. These are all scientific concepts and, as such, they carry no implication of existence just as little as does intelligence. It is possible to assert the existence of pigs, or psychoanalysts or the yeti, although philosophers might have a great deal of fun arguing about even that. But, as a scientific concept, no reputable psychologist would have attempted to reify it or to assert its existence in this sense. Intelligence is a scientific concept that may be useful or useless; the main point of this argument is that it is *useful* and, at the moment, accounts for the known facts better than any other concept or collection of concepts (Carroll, 1993).

The second criticism often voiced refers to the lack of agreed theory concerning intelligence. It is said that, in the absence of such a theory, intelligence cannot be regarded as a useful scientific concept. Such a view would certainly run counter to anything that the history of science can teach us. Concepts develop for centuries before agreed theories arise, and often the theories in which they are based are known from the beginning to have faults. Gravitation is a good example. Newton's Action at a Distance theory was already known by him to be absurd, but it served a very useful purpose. Even now, 300 years later, there is no agreed theory of gravitation. What we have are three quite dissimilar theories between which it is impossible to make a rational choice. On the one hand, we have Einstein's view according to which gravitation is a distortion of the space-time continuum, and on the other, we have the quantum mechanics interpretation in terms of particle interaction (gravitons) . And as a third alternative, we have Sakharov's theory of gravity as a long-range Casimir force; i.e. the postulation that gravity might be an effect brought about by changes in the zero-point energy of the vacuum, due to the presence of matter.

Much the same may be said about the theory of heat, where we have the thermodynamic and the kinetic theories side by side. Thermodynamics deals with unimaginable concepts of a purely quantitative kind: *temperature*, measured on a thermometer; *pressure* measured as a force exerted per unit area; and *volume*, measured by the size of the container. Nothing is said in the laws of thermodynamics about the nature of heat. This on the other hand, is the foundation stone of the kinetic theory of heat, using Bernoulli's view that all elastic fluids, such as air, consist of small particles which are in constant irregular motion and which constantly collide with each other and with the walls of the container, their speed of motion creating the sensation of heat. Many formulae are quite intractable to kinetic interpretations even today but yield easily to a thermodynamic solution. The unified theory here, as elsewhere, eludes physics, after centuries of endeavour. Should we expect psychology to do better? The unified theory appears at the end, not at the beginning, of scientific search, and to demand such a theory before a concept is taken seriously is to make impossible all scientific research.

A third objection relates to the absence of an agreed definition of intelligence. To read such a set of views as those contained in *What is Intelligence?* (Sternberg and Detterman, 1986) or its forerunner, the classic symposium published in the *Journal of Educational Psychology* some 65 years ago under the title 'Intelligence and its Measurement' (1927) is to see quite clearly that indeed there is no agreement on definition. This could hardly be expected in the absence of an agreement on theory; definition follows theory. However, what is interesting and important is the failure of the majority of participants in the symposia to consider for a moment the *nature* of definition in a scientific context.

When we look at the usual definitions offered by psychologists, these turn out for the most part to be *examples* of what intelligence might be expected to do, rather than *definitions* of any underlying concept. Consequently, definitions in terms of learning, remembering, problem-solving, following instructions, educational success, worldly achievement, or even in terms of Spearman's education of relations and correlates, are not even attempts at definitions of intelligence. What would we think of a physicist who tried to *define* gravitation in terms of the apple falling on Newton's head, the shapes of the planets, the occurrence of tides or black holes, the equatorial bulge of the earth, planetary movements, the laws of gunnery, or the formation of galaxies? These are all examples of the *operation* of gravitational forces; they are not *definitions*, and any physicist who tried to define gravitation in those terms would be laughed out of court. So would any critic of the concept of gravitation who argued that because different physicists used different examples of the effect of gravitational forces, the concept therefore was useless!

This is a brief and unsatisfactory discussion of the concept of intelligence, but it must suffice in the context of a book not devoted to intelligence, but to creativity and genius. One further clarification, however, is necessary. General intelligence ('*g*') is not the only factor in the field of cognitive functioning; there is now general agreement that in addition to Spearman's '*g*' there are a number of special abilities or factors, such as verbal ability, numerical ability, visuospatial ability, and so forth (Spearman and Jones, 1950; Thurstone and Thurstone, 1941). Carroll (1988, 1993) gives a list of some 20 such factors, although he admits that some are better defined than others, that the list is probably not exhaustive, and that some factors may ultimately coalesce with others. The important point here is that genius (or even high achievement) in any field is likely to be linked not only with '*g*' but also with special ability in a given field. Indeed, in the case of music, geometry and the visual arts, these special abilities may play a particularly important role, perhaps even overshadowing '*g*'. Some psychologists, e.g. Gardner (1983), posit a variety of different 'intelligences', but this is simply a return to the early theories of Thurstone (1938) when he postulated a set of 'primary abilities' but omitted '*g*' (Eysenck, 1939); he later agreed that the concept of '*g*', *in addition* to the notion of primary abilities, was essential. Ideally we should always measure the major primaries as well as '*g*', but time constraints make that difficult and often impossible. In considering the results of psychometric tests, however, we should always bear in mind that the measurement of '*g*' alone does not exhaust the field of cognitive abilities. Indeed, creativity, as we shall see, may be considered one of the 'primaries', more important perhaps than most; curiously enough it has not usually been viewed under that label (but see Carroll, 1993). It has been customary, rather, to contrast the measurement of '*g*' by *convergent* and *divergent* tests, as we shall see in a later chapter; it may be more useful to conceive of creativity as a separate primary ability.

Genius and intelligence: measurement

What, then, about the intelligence of geniuses? The only proper study of the field has been contributed by Cox (1926), whose classic work has been cited more frequently perhaps than any other book on genius. She proceeded in a series of steps. First, find some geniuses. She started with a list of 1000 men of genius published by Cattell (1930a); he identified these by measuring the space allotted in biographical dictionaries. (Havelock Ellis (1926) carried out a similar study of British genius, also following, as had Cattell, Galton's usage.) Those included might indeed be considered eminent, but the largely European bias is undesirable, and ensures that many eminent Japanese, Indian, Chinese and Arabic scholars and artists are not represented. This is a handicap, but not really a vital objection; as long as all the men chosen from this list by Cox are eminent, and recognized as geniuses, the study is still valuable, leaving it open whether the conclusions would apply to other cultures and other times. (Hart (1993) has been rather less provincial in his choice of the 100 most influential people in history.)

Cox's selected 301 subjects from Cattell's list for her study using three criteria. (1) They should reach a standard of unquestionable eminence. (2) They should be as far as possible persons whose eminence was the result of unusual achievement and not as a consequence of fortuitous circumstances, such as accidents of birth (e.g. royalty). (3) They must be persons for whom adequate records were available, i.e. records upon which reliable ratings of early mental ability might be based. No subject born before 1450 was selected, because of paucity of records, and the majority were British, French and German. Subjects were divided into 10 different fields, according to their major accomplishment: (1) Writers (poets, novelists, dramatists); (2) Statesmen and politicians; (3) Writers (essayists, critics, scholars, historians); (4) Scientists; (5) Soldiers; (6) Religious leaders; (7) Philosophers; (8) Artists; (9) Musicians; (10) Revolutionary Statesmen. (Numbers in each category decline from 1 to 10.)

The second step taken by Cox was to calculate an IQ for each of her subjects, based on known behaviours or achievements properly authenticated. It is known what the average child can do at various ages; comparing the feats achieved by each genius at a given age it is possible (depending on the availability of such information) to calculate an IQ for that particular subject. The method may be illustrated by reference to Terman's (1917) attempt to calculate the IQ of Sir Francis Galton.

Cox actually made two IQ assessments – one for development to the age of 17, the other for development from 17 to 26. In the Galton records information is included by Terman with respect to (1) the earliest period of instruction, (2) the nature of the earliest learning, (3) the earliest productions, (4) the first reading, (5) the first mathematical performance, (6) typical precocious activities, (7) unusually intelligent applications of knowledge, (8) the recognition of similarities or differences, (9) the amount and character of the reading, (10) the character of various interests, (11) school standing and progress, (12) early maturity of attitude or judgment, (13) the tendency to discriminate, to generalize, or to theorize. To these may be added (14) family standing, which is, in the case of Galton, besides being well known, implicit in the report of the kind of school attended. The nature of the items under some of these heads are briefly reviewed and illustrated in the following paragraphs dealing first with childhood.

(1) *The earliest period of instruction.*

From earliest childhood Galton was under the instruction of his sister, Adele, herself a mere child. She taught him his letters in play, and he could point to them all before he could speak. She never had him learn by heart, but made him read his lesson, bit by bit, eight times over, when he could repeat it. He could repeat much of Scott's *Marmion*, and understood it all by the time he was five.

(2) *The nature of the earliest learning.*

Francis knew his capital letters by twelve months and his alphabet by eighteen months; he could read a little book, *Cobwebs to Catch Flies*, when two and a half years old, and could sign his name before three years.

(3) *The earliest production.*

Galton's mother assures us that the following letter preserved from her son's fourth year was written and spelled by Francis himself without assistance:

'My
dear
Uncle/
we have
got Ducks. I know
a Nest. I mean
to make a
Feast.'

A second letter gives further evidence as to Galton's early ability in composition and also contributes information under some other heads:

(4) *The first reading*
(5) *The first mathematical performance*
(6) *Typical precocious activities.*

The letter, written to his sister the day before his fifth Birthday runs as follows:

My dear Adele,

I am 4 years old and I can read any English book. I can say all the Latin Substantives and Adjectives and active verbs besides 52 lines of Latin poetry. I can cast up any sum in addition and can multiply by 2, 3, 4, 5, 6, 7, 8, [9], 10, [11].

I can also say the pence table. I read French a little and I know the clock.

FRANCIS GALTON,

Febuary 15, 1827.

Terman comments upon this letter as follows: 'The only misspelling is in the date. The numbers 9 and 11 are bracketed above, because little Francis, evidently feeling that he had claimed too much, had scratched out one of these numbers with a knife and pasted some paper over the other!'

(7) *Unusually intelligent applications of knowledge*, and
(8) *The recognition of similarities or differences.*

The fact that Francis' reading at the age of five years was intelligent and not of the mechanical kind, is demonstrated by his ability at that age to offer quotations which would fit a given situation. For example, when he was five years old, a boy friend asked his advice as to what he ought to say in a letter to his father, who, it seems, was in danger of being shot for some political affair. Little Francis replied immediately from Walter Scott:

'And if I live to be a man,
My father's death revenged shall be'

Again at the age of five, he was found holding a group of tormenting boys at arm's length, shouting meanwhile,

'Come one, come all. This rock shall fly
From its firm base, as soon as I.'

(9) *The amount and character of his reading.*
By six, under the tutelage of Adele, Galton had become thoroughly conversant with the *Iliad* and the *Odyssey*.

'... He is 7 next February and reads Marmion, The Lady of the Lake, Cowper's, Pope's, and Shakespeare's works for pleasure and, by reading a page twice over, repeats it by heart.'

(10) *The character of his interests other than reading.*
It seems that Adele also taught Francis a good deal about entomology, and at six and seven years he was active and persistent in collecting insects and minerals, which he is said to have classified and studied in more than a childish way. (It has been shown that collections of an analytical and classificatory type are not common before 12 or 13 years.) Here, again, we find evidence of an intelligence quotient not far from 200.

Francis' interests were not wholly literary, for at the age of 13 he gave us 'Francis Galton's Aerostatic Project.' It seems this was a series of drawings representing a flying machine. It was to work by large, flapping wings with a sort of revolving steam engine, and was supposed to carry five passengers, a pilot, and an engineer. All this, and much more information, can be gained from Pearson's (1914) four-volume life of Galton. Terman (1917) adduced further evidence from Galton's development. Terman estimated the IQ of Galton as around 200. I shall presently discuss the evidential value of the Cox and Terman methodology; first let us look at the IQs so estimated of our 301 geniuses. Note that ratings of IQ were made independently by several raters.

Genius and intelligence: 301 geniuses

Intercorrelations between raters were usually between .70 and .75, and retest correlations somewhat higher; these are acceptable reliabilities. There was an obvious correlation between the size of the estimated IQs and the size of the reliability coefficient for the data on which the rating was based (0.77). In other words, the more data are available, the higher is the IQ estimate. Cox concludes

Table 2.1. *Original and corrected IQs of geniuses for 10 professional groups in childhood (A1) and youth (A2)*

	A1 IQ		A2 IQ	
	Obtained	Corrected	Obtained	Corrected
Artists	122	140	135	160
Writers (P.N.D)	141	160	149	165
Musicians	130	145	140	160
Writers (E,H,C,S)	139	160	148	170
Soldiers	115	125	125	140
Scientists	135	155	152	175
Philosophers	147	170	156	180
Revolutionary Statesmen	140	160	144	170
Religious Leaders	132	150	145	170
Statesmen	135	155	142	165

Source: From Cox (1926), with permission.

> that more reliable data would in general form a basis for higher IQ ratings throughout the study. As none of the reliability coefficients are above .82, it appears that all of the IQ ratings are probably too low . . . Hence it appears that the highest IQs in the group are somewhat too low, while many of the lowest may be much too low, and that the *true* IQ for the group . . . is distinctly above the estimated ratings of this study, since the estimated ratings are dependent upon data whose unreliabilty introduces a consistent reduction of the estimated IQ from its true value (p. 81).

Observed correlations were therefore corrected for attenuation, and both the original and the corrected values are given by Cox. I will delay a critical assessment of the correlations undertaken by Cox.

Table 2.1 shows both the original and the corrected IQs for the 10 professional groups, both for the young period (A1 – up to 17) and the older period (A2, 17–26). The average IQs so obtained are 135 for A1 and 145 for A2. The corrected values are 155 and 165. As Cox says:

> The result of the correction indicates that the true IQs of the subjects of this study average above 160. It further indicates that many of the true IQs are above 180, while but few of them are below 140. Analysis of the data from which our first estimated IQs were obtained and careful re-examination of the original items in the case studies offer no contradiction to these findings (p. 85).

There is no question that the Cox study was carefully and conscientiously done, and is of great importance for every student of the subject. Yet it is imperative to resist the temptation to take the actual figures too seriously.

There are many reasons for this warning. The first is that IQ ratings of children can be improved dramatically by what is known as 'teaching the test'. In other words, if the test is a vocabulary one, then obviously a child will get a much higher score if we actually teach him beforehand the meaning of the words in the test. The value of the test is that it contains a random sample of words of different difficulty level, so that we may get an idea of the difficulty level at which the child can still answer correctly. But the words used by young Francis are not randomly selected; they had been taught him by his sister, just as John Stuart Mill received similar backing from his father, at a very young age. No wonder that Francis and John knew words the average child, not so taught, would not know; they had in effect been 'taught the test', because the test was in fact (inevitably!) based in what they had learned! This criticism should not be taken too far; many of the things they did, no ordinary child could have done, regardless of teaching, and I am not doubting for a moment that these geniuses were in fact very bright; I am seriously doubting that a precise IQ estimate can be given on that basis.

Another problem in this connection is the assumption underlying the IQ concept that intelligence is normally distributed. Burt (1963) has argued, on the basis of large-scale data, that the true distribution of intelligence follows a Pearson Type IV curve, and 'that the usual assumption of normality leads to a gross underestimate of the number of highly gifted individuals' (p. 175). His argument is too technical to be presented here, but it may be interesting to present the estimated proportions of the population reaching or exceeding the borderline specified. If we set the borderline at IQ 200, then only one in a thousand million would reach that level on the assumption of a normal distribution, but 6.2 in a million would do so with a Type IV distribution, i.e. a disproportion of 1 to 6000! For IQ 160 the disproportion would be 1 in 10, i.e. 32 vs. 342 per million for the two distributions. Burt cites much additional evidence to strengthen his case. I do not here wish to assert that he was right; his assertion and his data have not really been properly criticized and tested – most psychologists simply assume the truth of the normal curve, even though its extension above the 140 or so mark is based more on assumption than on factual demonstration, there being too few children at the highest level to arrive at any reliable frequency assessment. Leta Hollingsworth (1942) in her work with children with IQs over 180 found far more than compatible with a normal distribution, and so did Terman (1925) in his search for children above 140 IQ. I merely want to indicate that the issue of the proper distribution of IQs is still *sub judice*, and that the very meaning of an unusually high IQ depends on this outcome.

A third reservation derives from the finding that the mean IQ has been growing by something like 20 points in Europe and the United States in the past 50 years (Flynn, 1987; Storfer, 1990). It is not necessary to agree with Flynn or Storfer about the *reasons* for this rise, or its meaning for the future of the concept of IQ; the very fact that in our lifetime such changes can take place makes it rather problematical to extend our standardization of IQ test items to mediaeval times.

In some cases one must doubt the relevance of the facts used to derive a given genius's IQ. Little is known about Mozart's childhood achievements over and above his startling musical performance. Yet to use this to argue for a high IQ would seem to beg the question. Cox might have been well advised to reject a few of her geniuses for lack of evidence.

One final word concerning Cox's correction for attenuation. This is based on the correlation between size of IQ estimated, and amount of data for making the estimate for a given person – the more data, the higher the estimated IQ. Cox bases her argument on the incontrovertible truth that if there is *no* information, the estimate would have to be the mean of the population, i.e. 100; the more you know, the higher the estimate is likely to be. Einstein (Clark, 1973) and Faraday (Williams, 1965) were undoubtedly geniuses, but there is very little childhood evidence of extremely high IQ. There may be good reasons for lack of information; perhaps the child never did or said anything notable! In other words Cox *assumes* that lacking data are responsible for low IQ estimates; it is also possible that low IQs may be responsible for absence of data! Again I do not want to exaggerate the importance of this objection, but it makes it difficult to assign any great accuracy to the disattenuated IQ values.

What can we conclude from this brief review of the Cox study? I think that she has demonstrated beyond any doubt that geniuses in many different lines of endeavour have uniformly IQs well above the average; indeed, as all the different occupations which led to their achievements obviously used considerable mental powers any other result would have been unbelievable. But there must be considerable doubt about the actual IQ values assigned to individuals, or to groups; these might be higher or lower than given by Cox, depending on how one evaluates the criticisms and considerations I have outlined. On the whole they might be thought to balance out, and to give us a final estimate of the typical genius as having an IQ some three to four standard deviations above the mean. In the long run such precision does not matter very much; the main point is that high IQ is one of the features of genius, and apparently a universal, and hence probably a *necessary*, one. That of course does not mean that it is a *sufficient* one; there clearly are far more individuals of very high IQs in the population then geniuses! Fortunately there is more strictly scientific evidence on this point than such simple observations, namely the observation and follow-up of children with very high IQs, and we must next turn to a discussion of this evidence.

High IQ does not equal genius

Here, too, the major evidence comes from the Terman *Genetic Studies of Genius*, published in five volumes. The first of these (Terman, 1925) describes the early stages of the study of *Mental and Physical Traits of a Thousand Gifted Children*, and Volumes 3, 4 and 5 describe follow-up studies at various points of time (Burks, Jensen & Terman, 1930; Terman and Oden, 1947, 1959). (We have already discussed the Cox volume, which was number 2 in the series.) In the study that concerns us now, 1528 children (857 male and 671 female) were chosen from a large IQ tested group in California on the basis of having a high IQ – the mean IQ was 151 for both sexes, with 135 the lowest. Different tests were employed, making comparisons difficult, but more than two-thirds were selected on the basis of the Binet, only recently translated and standardized by Terman, and of this group 77 subjects scored 170 or higher. A large amount of additional information was acquired on family background, physique and health, educational history, childhood interests and preoccupations, personality and character traits. (An even later follow-up by Janos (1987) does not add anything of importance to the data here considered.)

A brief summary of the findings may be useful. The following description is excerpted from *The Gifted Child at Midlife*, by Lewis M. Terman and Melita H. Oden (*Volume V*, *Genetic Studies of Genius*, edited by Lewis M Terman) with the permission of the publishers, Stanford University Press. © 1959 by the Board of Trustees of the Leland Stanford Junior University.

> Although there are many exceptions to the rule, the typical gifted child is the product of superior parentage – superior not only in cultural and educational background, but apparently also in heredity. As the result of the combined influence of heredity and environment, such children are superior physically, to the average child of the general population.
>
> Educationally, the typical gifted child is accelerated in grade placement about 14 per cent of his age; but in mastery of the subject matter taught, he is accelerated about 44 per cent of his age. The net result is that during the elementary-school period a majority of gifted children are kept at school tasks two or three full grades below the level of achievement they have already reached.
>
> The interests of gifted children are many-sided and spontaneous. The members of our group learned to read easily and read many more, also better books than the average child. At the same time, they engaged in a wide range of childhood activities and acquired far more knowledge of plays and games than the average child of their years. Their preferences among plays and games closely follow the normal sex trends with regard to masculinity and femininity of interests, although gifted girls tend to be somewhat more masculine in their play life than the average girls. Both sexes show a degree of interest maturity two or three years beyond the age norm.
>
> A battery of seven character tests showed gifted children above average on every one. On the total score on the character tests the typical gifted child at age nine tests as high as the average child at age 12.
>
> Ratings on 25 traits by parents and teachers confirm the evidence from tests and case histories. The proportion of gifted subjects rated superior to unselected children of corresponding age averaged 89 per cent for four intellectual traits, 82 per cent for four volitional traits, 67 per cent for three emotional traits, 65 per cent for two aesthetic traits, 64 per cent for four moral traits, 61 per cent for two physical traits, and 57 per cent for five social traits.

> Only on mechanical ingenuity were they rated as low as unselected children, and this verdict is contradicted by tests of mechanical aptitude.
>
> Three facts stand out clearly in this composite portrait: (1) The deviation of gifted children from the generality is in the upward direction for nearly all traits; there is no law of compensation whereby the intellectual superiority of the gifted is offset by inferiorities along non-intellectual lines. (2) The amount of upward deviation of the gifted is not the same for all traits. (3) This unevenness of abilities is no greater for gifted than for average children, but it is different in direction; whereas the gifted are at their best in the "thought" subjects, average children are at their best in subjects that make the least demands upon the formation and manipulation of concepts.
>
> Finally, the reader should bear in mind that there is a wide range of variability within our gifted group on every trait we have investigated. Descriptions of the gifted in terms of what is typical are useful as a basis for generalizations, but emphasis on central tendencies should not blind us to the fact that gifted children, far from falling into a single pattern, represent as almost infinite variety of patterns (pp. 15–16).

Thirty-five years after Terman's original study, Terman and Oden (1959) found out what was happening to *The Gifted Child at Midlife*. They concluded that 'the superior child, with few exceptions, becomes the able adult, superior in nearly every aspect to the generality' (p. 143). The superiority of the group is greatest in intellectual ability, in scholastic accomplishment, and in vocational achievement. Physically and with regard to emotional adjustment the group are if anything slightly superior to the majority, but comparisons are more difficult. But it is with respect to intellectual ability that the group shines most brilliantly; on tests like the Concept Mastery Test specially devised for high-fliers, the majority scored close to the top percentile; if anything, they showed a gain in IQ over the period covered.

As regards achievement, this is remarkable. Some 70% graduated from college, with two-thirds of the men and nearly three-fifths of the women continuing for graduate study. Occupational status and different kinds of achievements are difficult to compare across sexes, so they have to be discussed separately. Taking women first, seven of them were listed in *American Men of Science* (!), two in the *Directory of American Scholars*, and two in *Who's Who* in America, all before reaching the age of 43! Publications included five novels, five volumes of poetry, and some 70 poems appearing in literary and popular journals. They contributed 32 technical, professional, or scholarly books; around 50 school stories; four plays; more than 150 essays, critiques, and articles; and more than 200 scientific papers. At least five patents have been taken out by gifted women. These figures do not include the writings of reporters and editors, nor a variety of miscellaneous contributions. For considering these achievements, it should be remembered that all this took place prior to World War Two, i.e. long before the age of feminism, and that the majority of women became housewives rather than blossoming out into career women.

For the men, 86% were found in the two highest occupational categories, i.e. the professions and semi-professions and higher business. No gifted men are found in the lowest level of the occupational hierarchy, whereas 13% of the total urban population are in these categories.

A number of men have made substantial contributions to the physical, biological, and social sciences. These include members of university faculties as well as scientists in various fields who are engaged in research either in industry or in privately endowed or government-sponsored research laboratories. Listings in *American Men of Science* include 70 gifted men, of whom 39 are in the physical sciences, 22 in the biological sciences, and nine in the social sciences. These listings are several times as numerous as would be found for unselected college graduates. An even greater distinction has been won by the three men whom have been elected to the National Academy of Sciences, one of the highest honours accorded American scientists. Not all the notable achievements have been in the sciences; many examples of distinguished accomplishments are found in nearly all fields of endeavour.

Some idea of the distinction and versatility of the group may be found in biographical listings. In addition to the 70 men listed in *American Men of Science*, 10 others appear in the *Directory of American Scholars*, a companion volume of biographies of persons with notable accomplishment in the humanities. In both of these volumes, listings depend on the amount of attention the individual's work has attracted from others in his field. Listings in *Who's Who in America*, on the other hand, are of persons who, by reasons of outstanding achievement, are subjects of extensive and general interest. The 31 men (about 4%) who appear in *Who's Who* provide striking evidence of the range of talent to be found in this group. Of these, 13 are members of college faculties representing the sciences, arts and humanities; eight are top-ranking executives in business or industry; and three are diplomats. The others in *Who's Who* include a physicist who heads one of the foremost laboratories for research in nuclear energy; an engineer who is a director of an aeronautical laboratory; a landscape architect; and a writer and editor. Still others are a farmer who is also a government official serving in the Department of Agriculture; a brigadier general in the United States Army; and a vice-president and director of one of the largest philanthropic foundations.

Several of the college faculty members listed in *Who's Who* hold important administrative positions. These include an internationally known scientist who is provost of a leading university, and a distinguished scholar in the field of literature who is a vice-chancellor at one of the country's largest universities. Another, holding a doctorate in theology, is president of a small denominational college. Others among the college faculty included one of the world's foremost oceanographers and head of a well known institute of oceanography; a dean of a leading medical school; and a physiologist who is director of an internationally known laboratory and is himself famous both in this country and abroad for his studies in nutrition and related fields.

Additional evidence of the productivity and versatility of the men is found in their publications and patents. Nearly 2000 scientific and technical papers and articles and some 60 books and monographs in the sciences, literature, arts, and humanities have been published. Patents granted amount to at least 230. Other writings include 33 novels, about 375 short stories, novelettes, and plays; 60 or more essays, critiques, and sketches; and 265 miscellaneous articles on a variety of subjects. The figures on publications do not include the hundreds of publications by journalists that classify as news stories, editorials, or newspaper columns, nor do they include the hundreds, if not thousands, of radio, television, or motion picture scripts. Neither does the list include the contributions of editors or members of editorial boards of scientific, professional, or literary magazines. There have also been a sizable number of scientific documents reporting studies in connection with government research which are restricted publications.

The foregoing are only a few illustrations of conspicuous achievement and could be multiplied many times. They by no means represent all of the areas or types of success for there is scarcely a line of creditable endeavour in which some member of the group has not achieved outstanding success. There are men in nearly every field who have won national prominence and eight or 10 who have achieved an international reputation. The latter include several physical scientists, at least one biological scientist, one or two social scientists, two or three members of the United States State Department, and a motion picture director. The majority, though not all so outstanding as those mentioned, have been highly successful vocationally from the standpoint of professional and business accomplishment as measured by responsibility and importance of position, prestige, and income.

It should always be remembered that these men were just over 40 years old when the data were collected. Clearly the next 20 or 30 years are likely to increase the products, honours and achievements by a factor of three or more, but already we may say that these children have been outstandingly successful on the whole. What about those who were less successful? Terman and Oden (1947) formed an achievement group of 150, and compared these men with another 150 low achievers (Group A and Group C). Personality traits were self-rated, spouse-rated and parent-rated. Scores from each rater ranged from good (1) to poor (5), so that a high score means relative failure. Table 2.2 gives an overview of the major results, without going into the detail which Terman and Oden present. I have simply given a + or + + to the group showing a given trait more clearly than the other, depending on the degree of differentiation.

The results clearly show that personality has a powerful influence on achievement with IQ held constant – the two groups showed little difference in mean IQ. The successful men are less moody, impulsive, or conformist; they show more self-confidence, sociability, perseverance, integration towards goals, absence of inferiority feelings, and commonsense than do the relatively unsuccessful; in addition they have more friends, are more talkative, are

Table 2.2. *Table showing personality traits of successful (Group A) and less successful (Group C) children in Terman's group of 'geniuses'*

	Group A	Group C
Happiness	+	
Moodiness		+
Impulsiveness		+
Self-confidence	++	
Emotionality	+	
Conformity		+
Good Nature	+	
Sociability	++	
Perseverance	++	
Integration towards goals	++	
Absence of Inferiority Feelings	++	
Number of friendships	+	
Talkativeness	+	
Common Sense	++	
Popularity	+	

Source: From Terman and Oden (1959), with permission.

happier, show more good nature and emotionality, and are more popular. These ratings fit in well with the addition of a variety of personality traits.

These ratings were furnished by the subjects themselves, spouses and parents. A possibly more objective assessment was made by the field assistants who became personally acquainted with the subjects. Table 2.3 shows their mean ratings; those where the difference is not significant (CR less than 2) may be disregarded. It seems that the more successful are superior in appearance, attractiveness, poise, speech, alertness, friendliness, attentiveness, curiosity, and originality. Note that the successful group was rated higher on originality, and lower on conformity (Table 2.2); this suggests they were more *creative*, as compared with equally bright controls. This point will be discussed later on in connection with the personality traits of creative people and geniuses.

Two further points deserve discussion. The first relates to a criticism often made of this study, namely that the crucial variable in the success story of our high-IQ subjects is not IQ, but social standing of their parents. This is the so-called 'sociological fallacy'; it fails to recognize that as IQ makes for success, and as the IQ is largely inherited, the parents of high-IQ children are very likely to also have high IQs, and to be successful. Hence the success of the children is largely due to their IQ, with social and economic family status being just a correlate, not a causal factor. The truth of this interpretation can be documented by the fact that within the same family, the brighter children go

Table 2.3. *Table showing ratings by field assistants of successful (Group A) and less successful (Group C) children in Terman's group of 'geniuses'*

	Group A (N=81) Percent	Group C (N=115) Percent	CR
1. Appearance	71.6	35.6	5.4
2. Attractiveness	76.2	45.2	4.7
3. Poise	75.3	43.5	4.8
4. Speech	79.0	53.0	3.0
5. Freedom from vanity	12.3	20.0	1.5
6. Alertness	91.4	74.8	4.9
7. Friendliness	75.3	53.0	3.0
8. Talkativeness	48.1	40.9	1.0
9. Frankness	60.5	56.5	0.6
10. Attentiveness	91.4	70.4	4.0
11. Curiosity	63.7	40.3	4.5
12. Originality	75.6	45.4	4.5

Source: From Terman and Oden (1959), with permission.

up in social standing, the duller ones go down (Eysenck, 1979). Another indication that the sociological argument is mistaken comes from comparisons made by Terman and Oden (1959) between the participants of their study and students generally, i.e. a group of higher social and economic standing than the average, and also much higher in IQ. This comparison shows the study participants very much higher in achievement than students in general.

The other point to be emphasized is that this group contained many children who grew up to be eminent enough to enter Galton's category E, and possibly F (2423 and 233 in each million of the same age); perhaps one or two will get into category G (14 in a million). Certainly none appears to belong to class X ('which includes all cases beyond G, and which are unclassed' – Galton, p. 33; this of course is the genius class proper). (The only testee who might go into class X (and certainly into class G) was Shockley, the Nobel Laureate; unfortunately he scored just below the IQ needed for inclusion!) In other words, in this high IQ group, most if not all of whose members had sufficiently high general intellectual ability to become geniuses, none in fact achieved this apotheosis, or even approached it. The facts leave little doubt that high IQ is not a *sufficient* factor for genius status.

Nor is high IQ necessarily a predictor of creativity. The large-scale studies of Weiss (1993) of mathematical ability in the East German (communist) state looked at the children who emerged as the winners from 32 rounds of talent tournaments in which at the beginning 2.8 million children took part. The children were followed up and special note taken of their creative achievements (inventions, patents, discoveries, etc.). Just about 50% showed creativity of this kind; the other 50% of the high achievement group failed completely to do so. High IQ does not guarantee creativity! It had been planned to look at the differences in personality between the creative and the non-creative, but for political reasons this was not allowed!

Particularly impressive are individual case histories. Best known, perhaps, is the sad story of W. J. Sidis, a remarkable prodigy in his early youth, but a genius *manqué* in his later life (Wallace, 1986). He was reading the *New York Times* at eighteen months; at three, Homer in the original Greek. His ultimate failure to achieve anything makes an enthralling story, demonstrating the importance of other factors than IQ. Many such prodigies are picked up by the media, only to sink into oblivion later. In 1967, Marcello Carlin made the headlines as a four-year-old who read fluently at three, enjoyed G. B. Shaw plays, and was obsessed with the classical age, devouring Livy's *Early History of Rome* at five. Now, at 31, he is working as an office manager, with a low salary; he was a drop-out at university, spent two years on the dole. Nick Hawksworth had an IQ of 170 at the age of 11; now he is a van driver with a basic pay of £9500. Like Carlin, he had little ambition and didn't know what he wanted to do. Jocelyn Lavin also had an IQ of 169, she passed six A-levels, all Grade A. Now, at 29, she is teaching maths and runs a band – she had won a place at Chetham's School of Music at the age of nine, and is a virtuoso on piano and oboe. She failed her degree at university. She feels she could have done more with her life, but lacked ambition. Lois Cody at 19 months could count up to 20, say the alphabet, recite poems and sing songs. She could spell her own name, tell you where she lived, and talk on the telephone. She failed her A-levels and is now a full-time mother. As an adult student, she got a degree with second-class honours. She is very content with her life. These and many similar stories emphasize the importance of ambition, hard work, scholarly values, and significant aims as at least equally important with IQ for even moderate achievement.

Some criticisms of the Terman studies

I have not dealt in any detail with possible criticisms of Terman's work; Shurkin (1992) may be consulted for a rather more objective account than that given by Terman himself. In the 1920s when his work began, research planning was much less sophisticated than it has become in the past 70 years, and one would not be justified in criticizing Terman for neglecting precautions that would now be considered essential – such as not actively interfering in the careers of his 'geniuses' by writing testimonials for them. These criticisms do not touch the major one I have made of his work, namely to show that *IQ is not enough*, and consequently are not reviewed in detail. Another important study, although confined to 12 children and their follow-up history, is Leta Hollingsworth's (1942) account of exceptionally high-IQ children, tested between the ages of seven and nine; the borderline here was an IQ of 180. These children when retested in early adulthood showed similarly high ability, some 'going through the ceiling' of available tests. Unfortunately she died before being able to follow these children to maturity, but none is known to have developed to a 'genius' level, or anything approaching it.

The hypothetical IQ values assigned to Terman and Cox's geniuses may be compared with actual IQ values of 23 scientists, all active research workers, highly regarded but not in the Nobel prize élite (Roe, 1953). There were actually three tests, comprising estimates of verbal, spatial and numerical ability. On the verbal test, the median IQ equivalent was 166, ranging from 121 to 177. (The test had a ceiling which was probably too low for some of the participants, hence 177 may be an artificial 'highest' score; with a better test the ceiling might have been higher.) Theoretical physicists scored highest, followed by anthropologists, psychologists and biologists, with experimental physicists lowest. (For the sake of comparison, it may be noted that graduate students averaged IQ 123, honours graduates 133, on that test and at that time.)

On the spatial test the median IQ was only 137, but this test was highly correlated with age; had the scientists been tested 20 years earlier their scores would undoubtedly have been much higher. Experimental physicists here do better than biologists and anthropologists, with theoretical physicists on top. On the mathematics test the median IQ equivalent was 154, ranging from 128 to 194. For such a high-scoring group, the correlations between the tests are quite low; this may explain why some of these scientists had fairly low scores on one or other of the tests, not very much above the graduate mean. Presumably different subjects of study call for different combinations of abilities, and if you need high mathematical ability, the fact that your verbal ability is not particularly high is not so serious. But note that none of these scientists score below 125 or so on any test; even the worst score, on the lowest-scoring test for a given person, was still almost two standard deviations above the popular mean! Roe (1953) is certainly right in saying that 'the average ability of the scientists is very great' (p. 30). These results are in good agreement with those of Cox. (Roe also administered so-called projective tests, i.e. the T.A.T. or thematic apperception test and the Rorschach Inkblot test to her subjects; and 'unexpectedly' these 'tests' revealed very little; not really surprisingly so as they are known to have little reliability and less validity – Zubin, Eron and Schumer (1965).)

Johnson (1955), too, has provided much evidence to show that virtually all the persons who made major creative advances in science and technology in historic times have possessed very great general problem-solving powers, but that of course does not answer the question; high intelligence may be a necessary but not a sufficient trait in the production of creative results, as argued above.

There is one criticism of these studies which may be labelled the 'Educational Fallacy', to take its place beside the 'Sociological Fallacy' already noted. According to this fallacy, the existence of *idiots savants* (Howe, 1989), and the appearance of very high performance on complex tasks by relatively low IQ subjects, or the absence of a correlation between success of such tasks and IQ, limits the usefulness of the IQ concept, and makes doubtful its application to genius (Ceci, 1990).

This fallacy is related to the error already noted of identifying *intelligence* with real life success ('practical intelligence'); Ceci clearly believes that IQ specialists consider, as he puts it in the title of his Chapter 8, that: Biology = IQ = Intelligence = Singularity of Mind = Real-World Success. None of these equalities would be endorsed by any of the experts I know. Biology (i.e. heredity, and the physiological mechanisms mediating its effect on behaviour) contributes powerfully to measured IQ, but is not *identical* with it; equally, IQ is a good measure of psychometric intelligence, but only a fair measure of biological intelligence. Finally real-world success is only very partially determined by IQ (Eysenck, 1979); there are many other factors, as we have seen, which are instrumental in such success, i.e. motivation, personality, education, etc. (see Fig. 0.1).

Let us look in more detail at just what causes Ceci and many others to doubt the importance of IQ in determining individual performance on complex cognitive tasks. Among the studies he quotes is one by Murtaugh (1985), who noted that female shoppers managed to perform complex mental operations relating to price and quality of goods on sale, but found no relationship between performance on a mental arithmetic test and subjects' shopping accuracy. Similarly, Scribner (1984) noted a remarkable accuracy of blue-collar workers in filling large boxes according to a complex plan minimizing bending over; white-collar workers who occasionally performed the same task were less successful, and there was no correlation with IQ.

Ceci follows this up by demonstrating that in experimental tasks involving accurate estimates of time (Ceci and Bronfenbrenner, 1985) and making accurate estimates of a moving object (Ceci, 1990), success depends on the context in which the task was done. He also refers to two other studies which are meant to illustrate the irrelevance of IQ to success in complex tasks. Thus Dorner and his colleagues (Dorner and Krenzig, 1983; Dorner, Reither and Standel, 1985) computerized the problems facing a city manager of a mythical town called Lothausen; over a thousand variables are identifiable in this task, interacting with each other, and the subject's success is measured in terms of the amount of revenue at the end of the simulation. They found no correlation between success in this task and IQ.

However, the Dorner studies are hardly surprising; the reliability of the task is probably quite low, and the range of ability quite limited. One could not have expected a high correlation with IQ, particularly as such a very lengthy task would require high motivation, and probably cognitive abilities other than IQ, but not tested.

Finally, Ceci and Liker (1986) studied the prediction of post-time odds at the racetrack, apparently a very complex cognitive task in which they found that 'the experts with low IQs always used more complex, interactive models than did non-experts with high IQs, and their success was due in large part to the use of these complex interactive models' (p. 43, Ceci, 1990). Ceci concludes that 'a view of performance has begun to emerge suggesting that cognition is tied to specific knowledge contexts and to motivation' (p. 44). Actually reliance on this interactive mode of reasoning was uncorrelated with IQ, and Ceci and Liker's conclusion that individuals of low IQ are able to reason in a complex way to solve real-life problems does not contradict *g* theory, provided there has been lengthy practice. Other criticisms have been made by Detterman and Spry (1988) and Brody (1992), to indicate that Ceci and Liker's conclusions do not follow from their results.

The acquisition of outstanding specific abilities

Underlying the Ceci argument is a failure to recognize the assumption governing the IQ construct, and the course of long-term learning (overlearning) of cognitive processes. To take the latter first, such learning embodies chunking, automation, and strategy development. 'Chunking' means that elements which at first are used in isolation become joined into complex chunks which now constitute the units. Few people can repeat more than 7–9 letters presented at one second intervals, e.g. S.B.X.I.N.O.E.T.D. But how about I.N.T.E.R.D.E.N.O.M.I.N.A.T.I.O.N.A.L? We now have only one chunk to remember, instead of 19 units. Again, consider automation. The youngster learning to read has to put together laboriously the letters of a word: C-A-T; the adult does so automatically and reads CAT.

All this of course is well known, but the use of strategies is perhaps not so well known, and the improvement in performance that can be mediated in this way is not always realized. Consider the measurement of digit or word span, often used as a test of IQ; few people can repeat more than nine digits, letters, or words. Yet quite average IQ people can be taught to succeed at levels 20 S.Ds above this level, equalling IQ equivalents of 400 or thereabouts! Let us look at such studies as those of Baltes and Willis (1982); Kliegl, Smith and Baltes (1986, 1989); Baltes (1990); Baltes and Kliegl (1986); and Staszewski (1989).

Consider two strategies which subjects are taught. One consists of associating numbers with non-numerical facts or objects; the other consists of helping the subject remember the actual sequence of the 'chunks' reproduced. This is

Table 2.4. *Examples of the History-Dates and Digit-Noun models (Kliegl et al., 1987)*

Knowledge System (History–Dates Model)	Encoding/Retrieval Sequence	Knowledge System (Digit–Noun Model)
Digits to be coded: 492789945 . . . 618		*Digits to be coded*: 407800 . . . 86
Historical Dates	*Method of Loci* 30 Berlin landmarks	*Digit-Noun Pairs*
(1) 492 = 1492 = Columbus	(1) Botanical Garden	(1) 40 = R S = RoSe
(2) 789 = 1789 = French Rev	(2) Museum	(2) 78 = C F = CoFfee
(3) 945 = 1945 = End WW II	(3) Church	(3) 00 = S S = SuSy
.	.	.
.	.	.
.	.	.
(30) 618 = 1618 = Beginning 30-Year War	(30) Fountain	(30) 86 = F SH = FiSH

Note:
In each model the Method of Loci is combined with one of two knowledge systems. In the History-Dates model, using digit triplets, 1000 dates (000–999) would be necessary to encode all digit triplets. The Digit-Noun model, using digit doublets, provides a match to all random sequences based on 100 nouns.

achieved by the so-called method of loci, already taught by Cicero to budding orators. Table 2.4, taken from Kliegl *et al.* (1986), illustrates an actual implementation, namely the use of the *History-Dates* model, and the *Digit-Noun* model.

The two models are identical in the use of the Method of Loci (see below). They differ mainly in the type of permanent knowledge they require. In the History-Dates model, knowledge about historical dates is acquired in terms of the last triplet of each historical unit (1945 = 945, 1492 = 492). One thousand historical dates are necessary to have a full account of all possible variations of random digit sequences (000–999). In the *Digit-Noun* model, digit doublets are recoded into concrete nouns (e.g. 40 = RoSe) using a digit-consonant scheme known as Figure-Alphabet. One hundred words are necessary to have a full account of all digit doublets possible (00, 01 . . . 99). Thus the two models differ in the kind and size of the knowledge necessary to generate all possible combinations of random digit sequences. Table 2.4 illustrates how a random string of digits can be recoded into a random sequence of historical events or concrete nouns, respectively.

In both models, sequences of recoded digit triplets (i.e. historical events) or of recoded digit doublets (i.e. nouns) are committed to long-term memory by means of the mnemonic technique known as the *Method of Loci* (Bower, 1970; Spence, 1984; Volkmann, 1928; Yates, 1966). The subjects work with a list of 30 or 40 locations representing landmarks in the town in which the experiment is carried out, as indicated in the middle column of Table 2.4. During encoding subjects form funny, bizarre or dynamic images or thoughts linking the to-be-remembered items (i.e. historical dates or concrete nouns) with the chosen locality. Because landmarks are mentally visited in an invariant sequence during encoding, they can all serve as retrieval cues when the subject 're-visits' them at time of recall. The local landmarks, their sequence and the Method of Loci procedure thus constitute a highly overlearned part of permanent knowledge. Which particular historical date or concrete noun must be remembered at which location is, of course, trial-dependent.

The sequential chunking of digits into historical events (or concrete nouns) and the formation of images or thought associations between these to-be-remembered items and their corresponding landmarks does not put a heavy burden on working-memory subsystems such as rehearsal loop or visuospatial scratch-pad (Baddeley, 1983). Only central executive functions which serve to integrate relevant knowledge elements are required. Therefore, in the present models, expertise in memory span is not constituted by increasing short-term or working memory, but by invoking long-term memory encoding processes and permanent knowledge during encoding (as proposed by Chase and Ericson, 1982). Once the memory expertise has been acquired, capacity limitations in working memory functioning (e.g. in the capacity of the central executive to integrate short-term memory (STM) and long-term memory (LTM) processes) can be probed, for example, by manipulating the rate at which items to-be-remembered are presented (e.g. reducing presentation rates). Such a probe technique illustrates one theory-guided implementation of the testing-the-limits strategy.

The results of such training, continued over lengthy periods, can be outstanding, depending on the duration of training. The Baltes group used 30 lessons, which was insufficient for many subjects, but achieved remarkable success with some. Later studies used over 250 practice sessions with two subjects who achieved a digit span of 82 and 68 respectively. One of the subjects in the Kliegl *et al.* (1986) study using the History-Dates model with 30 locations achieved the maximum possible span of 90 digits at presentation rates of two seconds per digit, another used the Digit-Noun model with 40 locations, giving him a ceiling of 80 digits, and achieved this ceiling with a presentation rate of five seconds. These results were achieved after some 40–60 sessions.

Thus, given suitable training, perfectly ordinary people with ordinary IQs can achieve extraordinary results with minimal training. It is not difficult to see that the alleged extraordinary feats of race-track tipsters and others, achieved after many years of practice (rather than hours!) appear remarkable, and are not correlated with IQ. The application of everyday training and reward may give rise to strategies and heuristics which produce good results which should never be compared with unpractised achievements, like the digit span test. To use a relevant comparison, we can measure a person's temperature fairly accurately with an ordinary mercury-in-a-glass thermometer. Now suppose the patient holds the bulb over a lighted candle for 30 seconds, a reading will now disclose a raging fever! Of course we know that *under these circumstances* the reading is worthless as an indicator of the patient's temperature; in the same way the every-day achievements of subjects with year-long practice and high motivation do not necessarily bear any relation to measured IQ. One would have thought this to be too obvious to mention, but the advocates of 'practical intelligence' often fail to bear this caution in mind.

The same may be said of idiots savants, i.e. individuals of obviously very low IQ who nevertheless have notable achievements to record. One such case is reported by Anastasi and Lenee (1960). The subject of this study, Z, was a mentally retarded 40-year-old male, whose behaviour was odd, eccentric and childlike. His clothes would have to laid out by his mother, as otherwise he would put on the first piece of clothing to catch his eye. He showed no interest in sex, or other people; only his dog seemed to give him pleasure. He might kiss anything that took his fancy, make odd dance movements, speak into mirrors, belch or fart in public, or lick his plate after meals. He was excessively rigid, the slightest departure from routine causing him great distress. His mental performance was equal of that of a ten-year-old; he did learn to read eventually, but never with any real understanding of the context. He was handicapped to the extent that he had to rely completely on others to provide food and shelter for him. His disorder was diagnosed as due to epidemic encephalitis, contracted soon after birth, and causing permanent damage.

Yet Z was extremely musical, as judged by a number of prominent musicians, and a leading chamber orchestra often asked him to play the piano at rehearsals. Apparently he could read on sight, and also play by ear. His interest in music began very early in life, and he practised up to nine hours a day. His memory was excellent and he knew a great deal about composers and their lives; he could also repeat long passages after a single reading. How can we explain these apparently totally divergent sets of characteristics? First, note that much of his abnormal behaviour is in the field of personality, not intelligence; his social behaviour is abnormal, but sharply contrasts with the average ten-year-old his mental age resembles. Again, musical ability of the executive kind is largely separate from IQ, i.e. the ability to think abstractly, learn easily and solve cognitive problems. Finally, note his tremendous motivation to do well in his chosen field; to practice up to nine hours a day is more than a very small proportion of the population could keep up with. Note also that meningitis destroys certain areas of brain, not others, hence we would expect performances to be extremely varied.

There are many calendar calculators whose IQs are depressingly low (Howe, 1989). Here we have one avenue for low-IQ people to achieve success and excel without needing high ability, High motivation and huge amounts of practice, together with the use of some well-known strategies described by Howe (1989), are sufficient to produce results which in such dull individuals appear miraculous.

Outstanding memory, often for mathematical input, is also found in idiots savants, as well as in normal and bright people. Luria's (1975) study of S.V. Shereshewski is one example: Hunter's (1977) account of the British mathematician A.C. Aitken is another. These, and many others are notable for being able to multiply and divide large numbers in a very short period of time; to discover roots to quite high powers, multiples of primes, etc., with astonishing speed and accuracy; sometimes outperforming computers! How is this possible? Consider a famous anecdote told by the famous British mathematician G. H. Hardy of his even more famous discovery, the untaught Indian genius Srinivasa Ramanujan:

> I remember once going to see him when he was lying ill at Putney. I had ridden in taxi-cab No. 1729 and remarked that the number seemed to me rather a dull one, and that I hoped it was not an unfavourable omen. 'No', he replied, 'it is a very interesting number; it is the smallest number expressible as the sum of two cubes in two different ways.' (Hardy, 1940).

Numbers, to devoted mathematicians, are not just sequences of digits, as they might be to most people; they acquire individuality, and are remembered, and used, as such. Mathematicians do not have to be taught the tricks used by Baltes and his colleagues; the knowledge of chunks, automatization and the development of strategies develop automatically, based upon high motivation and long periods of intensive learning and concentration.

In the same way chess masters do not deal with the individual places of the pieces, but with *positions*, i.e. chunks involving many pieces simultaneously. Thus when pieces are distributed randomly on the chess board, and subjects are allowed to view them for a limited period, chess masters are no better than others at remembering the positions occupied by the pieces. But if the pieces actually illustrate meaningful positions, chess masters produce very much higher scores. To them this is not a meaningless collection of so many different pieces, but the position arrived at after 25 moves in the Capablanca–Tartakover match of 1925. Thus they have to remember one item, not hundreds, as must the unfortunate novice.

Some 50 000 chunks, about the same magnitude as the recognition vocabulary of college-educated readers, may be required for expert mastery of a given field. The highest achievement in scientific disciplines, however, may require a memory store of a million chunks – probably the equivalent of 70 hours of concentrated effort each week for a decade even for a talented student! Without chunking the whole process would be utterly impossible.

Child prodigies and exceptionally early achievers, to quote the title of an interesting book by Radford (1990), seem able to curtail this prodigious expenditure of mental energy; a Mozart, Newton, or Einstein, by combining outstandingly high IQ, special abilities, motivation and creativity may get by with less, and achieve outstanding success at an abnormally early age. But even for them a long period of information acquisition is needed before creativity can emerge to restructure the chunks now available. Because not only do we have to transmute the material in question into chunks, these chunks are themselves tied together with pretty pink ribbons, and the most difficult task of the genius is to undo these ties, and fit the chunks together in a different pattern.

To summarize. Intelligence, which may be defined as innate, general cognitive ability, is a necessary but not a sufficient factor in the genesis of genius. Special abilities (verbal, visuo-spatial, numerical, musical, etc.), persistence, personality qualities and other factors are also required, and probably interact synergistically (multiplicatively) with intelligence, thus producing the typically J-shaped curved distribution of eminence defined in terms of achievement – very few geniuses, a small number of eminent people, and a very large number of ordinary people with no claims to eminence. Precise IQ values of geniuses studied in the past should not be taken too seriously; there are many reasons to doubt their accuracy or meaningfulness. However, the fact of unusually high intelligence in these people cannot be doubted, even if no precise estimate can be given. The existence of large numbers of very high IQ people who are very far from being geniuses demonstrate the fact that factors other than IQ play a large part in producing the geniuses.

The associationist theory of intelligence

We must now go on to take one further step, namely to formulate an associationist theory of intelligence, to broaden in a later chapter into an associationist theory of creativity. This section will be relatively short because I shall refer for the most part to a theoretical conception quite well-known to many readers, namely the theory called by Campbell (1960) the 'chance-configuration theory'. According to this theory, which has recently been adapted and extended by Simonton (1988a,b), the acquisition of new knowledge, and the solution of novel problems, requires some means of producing *variation* by cognition, and it is argued that this variation, to be truly effective, must be fully *blind*. Blindness is defined in terms of variations being correlated to the environmental conditions, including the specific problem, under which the variations are generated (Campbell, 1974). As for the generality of these variations Simonton prefers the adjective *chance*. (This theory was originally offered by Poincaré, whose account is quoted verbatim in a later chapter.)

These heterogeneous variations are subjected to a constant *selection* process to retain only those that exhibit *selective fit*, i.e. those offering viable solutions to the problem at hand. Finally, the variations that have been selected must be preserved, i.e. there must be a proper *retention* mechanism so that a successful variation can represent a permanent contribution to adaptive fitness. Thus this manipulation of mental elements by a process of blind chance, issuing in a selection for problem-solving fitness, leads to *configuration formation*, i.e. stable permutations hanging together in a stable arrangement or patterned whole of interrelated parts. This very brief description fails to do justice to the extended discussion given by the authors of this theory, but it must suffice here.

A similar theory was developed independently and around the same time by Furneaux (1960). He postulates that the brain structure of any individual, P, includes a set of ${}_pN$ neural elements which participate in problem-solving activities. The solution of a particular problem, h, of difficulty D, involves bringing into association a particular set, ${}_DN_h$, of these elements; interconnected in some precise order. When problem h is first presented, single elements are first selected, *at random*, from the total pool ${}_pN$ and examined to see whether any one of them, alone, constitutes the required solution. A device is postulated which carries out this examination – it must bring together the neural representation of the perceptual material embodying the problem, the rules according to which the problem has to be solved, and the particular organization of elements whose validity as a solution has to be examined. It must give rise to some sort of signal, which in the case of an acceptable organization will terminate the search process and will initiate the translation of the accepted neural organization into the activity which specifies the solution in behavioural terms. Alternatively, if the organization under examination proves to be unacceptable a signal must result which will lead to the continuation of the search process. This hypothetical device Furneaux calls the 'Comparator'.

Furneaux goes on to say that if $D \neq 1$, the comparator will reject each of the ${}_pN$ trial solutions involving only a single element, and the search will then start for a pair of elements, which, when correctly interconnected, might constitute a valid solution. If $D \neq r$, then the comparator will reject in turn all the organizations involving from 1 to $(r-1)$ elements. Speed of mental processing (i.e. time to solution) is an essential element of the theory, and Furneaux postulates that there will be a time $\tau \sum_{r-1}^{1} E$ sec within which a solution cannot occur, where τ = the time required for completing a single elementary operation within the search process, and $\sum_{r-1}^{1} E$ is the number of elementary operations involved in the search process up to the level of complexity $(r-1)$.

Similarly, after a time $\tau\sum_{r}^{1} E$ sec all possible organizations embodying r elements will have been examined so that correct solutions to problems of difficulty r will always arise, within the period defined by the two limiting times $\tau\sum_{r-1}^{1} E$ and $\tau\sum_{r}^{1} E$. Extensions of the theory, possible objections and empirical proof are dealt with in the original chapter. Critical discussions are found in Eysenck (1982, 1985a, 1986b, 1987b, 1988), and an experimental test of the theory in Frearson, Eysenck and Barrett (1990).

The similarities between the two theories are obvious. Both postulate a random/chance search process, leading to an organization/configuration which satisfies a comparator/selector. The major difference is that Furneaux deals with *intelligence*, Campbell and Simonton with *creativity*. The difference between the two concepts will be discussed in more detail later, but note that intelligence is defined by Spearman (1923, 1927) in terms of his noegenetic laws, i.e. as creating something new, which is of course also a definition of creativity. Spearman seemingly equates intelligence and creativity, but he also stimulated the work of Hargreaves (1927), who was the first to use tests of divergent ability, and found them to be highly correlated with measures of *g*, although these 'fluency' tests did define a separate factor. These early measures of 'fluency', and the history of their development, are discussed in detail by Eysenck (1970a.) These early studies of what was later to be called 'divergent thinking' (Guilford, 1967, 1981) are sadly neglected in recent writings on 'creativity and intelligence' (e.g. Heansly and Reynolds, 1989). Spearman also suggested that creativity might be a personality characteristic rather than a cognitive one, and Eysenck (1983a) has given some evidence to support this notion which will be discussed more fully later on.

Newell, Shaw and Simon (1962) have also argued against the existence of a division along cognitive lines between intelligence and creativity, and clearly the matter deserves a fuller discussion. It is interesting to note that these authors also postulate a search process involving trial-and-error processes as bulking very large in highly creative problem-solving: as they say, 'at the upper end of the range of problem difficulty there is likely to be a positive correlation between creativity and the use of trial-and-error generators' (p. 73).

The problems of random search

Intuitively, the notion of chance/random/trial-and-error search does not dovetail well with our experience of reasoning and problem-solving; what is characteristic of the process is the adoption of *heuristics* and strategies (Newell *et al.*, 1962). As these authors state: 'We have seen that the success of a problem-solver who is confronted with a complex task rests primarily on his ability to select – correctly – a very small part of the total problem-solving maze for exploration. The processes that carry out this selection we call 'heuristics'.' (p. 96). The employment of heuristics and the strategies built thereupon is found in both human and computer problem-solvers, and, while retaining elements of random search, uses heuristic search techniques such as 'recording' (Miller, 1956) to simplify the choice and availability of elements from which to build a solution.

A good guide to the many different search algorithms that exist can be found in the literature on artificial intelligence (Smith, 1990). The following may be mentioned: (1) A* algorithm, a form of heuristic search that attempts to determine the cheapest path from the initial state to the goal. (2) Alpha-beta pruning, in which nodes not needed to evaluate the possible moves of the top cards are 'pruned'. (3) Band-width search, a search strategy in an ordered state-space search. (4) Beam search, or scheme used in speech understanding systems. (5) Best-first search, in which the move considered next is the most promising in the entire search tree generated so far. (6) Bi-directional search, i.e. state-space search that proceeds both backwards and forwards. (7) Breadth-first search, a strategy applicable to a hierarchy of rules or objects, contrasted with (8) depth-first search, i.e. a search strategy within a hierarchy of rules or objects in which one rule or object at the highest level is first examined. (9) Full-width search, i.e. in which all legal moves from a position are examined; this may lead to alpha-pruning. (10) Generate-and-test, a problem-solving technique that uses a *generator* to produce possible solutions and an evaluator to test whether solutions are acceptable. (11) Heuristics paths algorithm, a generalization of the graph traverser algorithm, giving an ordered state-space search with an evaluation function. (12) Hierarchical search, i.e. an attempt to reduce the problem of combinatorial explosion, which threatens all problem-solvers attempting to use heuristic search in a sufficiently complex problem domain. (13) Length-first search, in which a complete plan for reaching a goal is formed at each node before moving on to any lower level node. (14) Negmax, a technique for searching game trees. (15) Ordered search, a heuristic search that always selects the most promising node as the next node to expand. (16) Uniform-cost search, a type of breadth-first algorithm in which a non-negative cost is associated with every operator. These algorithms are not always clearly differentiated, thus uniform-cost search is reduced to breadth-first search if all operators have equal cost, and (17) state-space search is a generic term for several formalisms already enumerated.

Closest to the random/chance/blind combination model comes (18) blind search, an algorithm that treats the search space syntactically, as contrasted with heuristic models, which use information about the nature and structure of the problem domain in order to limit the search. The search for a solution in state-space search is carried out by making explicit just enough of the state-space graph to contain a solution path. A search is called *blind* if the order in which potential solution paths are considered is arbitrary, and uses no domain-specific information to judge where the solution is likely to be. This type of blind search seems to be what Campbell and Furneaux are suggesting, but the evidence does not lend much support to what would seem a completely unstructured mind working on a random basis.

Newell *et al.* (1962) give an example of the combinatorial explosion involved in blind search. It concerns the Moone and Anderson (1954) study of the problem-solving behaviour of subjects who were given a small set (from one to four) of logic expressions as premises and asked to derive another expression from these, using twelve specified rules of transformation. Assuming (and this is an over-simplification) that each rule of transformation operates on one premise, and that each such rule is applicable to any premise, this particular tree branches in twelve directions at each choice point. A blind trial-and-error search for the derivation would require on the average the construction of 18 000 000 sequences!

Another example is choosing a move in chess. On the average, a chess player has a choice among 20 to 30 alternatives; these can be evaluated by considering the opponent's possible replies, one's own replies to his, and so on. The tree of move sequences is tremendously large; considering just five move sequences for each player, with an average of 25 legal continuations at each stage, the set of such moves has above 10^{14} (one hundred million million) members. No wonder simple power-crunching computer chess players never did well, and were outclassed by machines using heuristic processes, strategies and memory recall – just like humans!

Granted some form of heuristic search, we are faced with a combinatorial explosion less severe than that involved in blind search, but serious enough. In most cases, the domain of possible combinations is so large that the time required to find the (optimum) solution increases exponentially and exceeds the capacity of the computer system, or the human mind. 'Exhaustive search is rarely feasible for non-trivial problems' (Smith, 1990, p. 56). Examining all sequences of n moves, for example, would require operating in search space in which the number of nodes would grow exponentially with n. This is what is meant by the term 'combinational explosion', and it eliminates many of the algorithms considered above, in particular *blind search*.

Considerations of time and the 'combinatorial explosion' make any chance search process extremely unlikely. But in addition we do have a certain amount of firm experimental support for the view that human searching mechanisms adopt a very different mode. This evidence comes from research into linguistics, and into memory. Consider linguistics first. There is considerable evidence from word-association type experiments that verbal mechanisms are strongly constrained; see Cramer (1968); Underwood and Schultz (1960); Bonfield (1953); Bell (1948); Garskof and Houston (1963); Laffel and Feldman (1962); and in particular Osgood, Suci and Tannenbaum (1957). The fact that language and associative mechanisms are heavily structured, and mediate search processes of a highly predictable kind, does not agree with any theory of blind search. For any given verbal problem, this structure immediately suggests sequences of nodes and arcs (to use the language of AI) on the search tree, or in the search space. Each tree is constructed before a search takes place, and includes all states that are theoretically possible. Different problem domains have different search spaces, which may be large or even infinite. It is possible to measure the search space by estimating the number of nodes it encompasses. Given a typical noegenetic verbal problem:

Optimal : Mumpsimus = best : ?

we do not indulge in a blind search, leading to a combinatorial explosion, but deduce the correct solution using the proper heuristic.

Best researched of all search processes is probably that involved in memory retrieval (recall or recognition), which also plays an important part in intelligence testing and problem solving. The generate-recognize model (Watkins and Gardiner, 1979; Norman, 1970; Brown and McNeill, 1966; Jones, 1978) is one example; the spreading activation theory is another (Collins and Loftus, 1975; Quillian, 1967; Collins and Quillian, 1970). Ratcliff's 'resonance' theory is a third (Ratcliff, 1978), and the 'CHARM' model (Nilsson and Gardiner, 1991) is a fourth. It is not the point of this chapter to judge the explanatory value of these models, or to make a choice between them; they are cited to illustrate the point that empirical studies of search mechanisms arrive at a picture that disagrees profoundly with any notion of 'random' search or 'blind chance'.

As Theios (1973) has pointed out,

> in serial human information processing tasks, the short-term stores become completely filled up with representations of the occurring stimuli and responses, and . . . to the extent that there is any structure in the sequence of physical stimuli and required responses, that structure will be mirrored in (or at least affect) the structure and organization of the serially searched, short-term stores (p. 44).

Clearly this organization must determine the nature of the search process, which *cannot* be blind given the degree of organization in the short-term stores.

In Eysenck's (1985a, 1986a, 1987b) theory of intelligence, speed of information-processing is a crucial ingredient, itself a consequence of (comparatively) error-free cortical processing. What, in such a scheme, would be the role of creativity? Any meaningful mental search process which has some empirical support requires qualification of the search domain in terms of *relevance*. Given a particular problem, we only search our memory store in terms of the requirement of that problem. Given the problem:

1 3 6 10 15 21 ?

we do not draw upon our knowledge of the causes of the Peleponesian War, or the quantum mechanics 'graviton' theory of gravitation; we confine ourselves to a heuristic search for numerical solutions fitting the progression indicated by the problem. *Any problem defines its solution horizon, limiting its search to a given, circumscribed area.* While Campbell, Simonton and Furneaux do not formally state such a limitation on their concepts of random or blind search, it does not seem likely that they would disagree and would insist on a truly blind/random search involving the whole of our knowledge background. However, they fail to introduce the important concept of *relevance*, which clearly needs explicit treatment – particularly as it is a vital component of creativity.

Association and creativity

As a preliminary statement, we may consider the usual associationistic approach to creativity (Spearman, 1931), according to which a creative idea results from the novel combination of two or more ideas that have been isolated from their usual association. Mednick (1962; Mednick and Mednick, 1964) has defined the creative process as 'the forming of associative elements into new combinations which either meet specified requirements or are in some way useful. The more mutually remote the elements of the new combinations, the more creative the process or solution' (Mednick, 1962, p. 221). Creativity thus becomes a function of people's 'associative hierarchy', which can be defined as generalization gradients of differing degrees of steepness, with associations to words, percepts or problems ranging from common to unique. Individuals with steep gradients are likely to give common associations at high strength, but few or no uncommon associations; persons with less steep or even with flat gradients are more likely to make uncommon or unique responses.

There is a good deal of agreement that creative contributions in science redefine and reorganize concepts used in past attempts to incorporate certain anomalous findings that were not readily understood within the existing paradigm (Kuhn, 1976). In doing do, bisociation (Koestler, 1964) is involved, i.e. the bringing together of two apparently unconnected associations. Rothenberg (1976, 1979), coining another neologism (homospatial thinking), has attempted to cite a good deal of historic evidence to show that major contributions are often related to such superimposition, or fusion, of ideas, associations, or images. There is good empirical evidence to support some such associationist theory, stressing a reasonably flat associationist gradient (broad associationist horizon) to ensure inclusion of normally remote but relevant associations that would be left unconsidered by a person with a relatively steep associationist gradient (narrow associationist horizon). Some of the studies supporting such a view are: Alissa (1972), Arlin (1977), Barsalon (1982, 1983), Csikszentmihalyi and Beattie (1979), Getzels and Csikszentmihalyi (1976), Glover (1979), Kasperson (1978), Owen (1969), Rothenberg (1986), Rothenberg and Sobel (1980), and Sobel and Rothenberg (1980).

This concept enables us to answer the question sometimes raised: Can computers act creatively? Langley, Simon and Zytkow (1987) answer in the affirmative, Wolpert (1992) in the negative. Langley *et al.* claim to have developed computer programs which, by using their problem-solving approach, can make 'creative' discoveries over a wide range of topics. Of course the use of the term 'discoveries' is not really justified; what they have done is to show that the computer could have discovered universal gravitation or Planck's constant, given the information available to Newton or Planck. But of course by building all the *relevant* information into their program they loaded the dice; knowing already the true solution to the problem, it was easy to know what information was relevant. However, that very recognition may be the most difficult and 'creative' part of the whole exercise.

Consider the theories to be put forward in later chapters. I will argue that the literature concerning concepts like latent inhibition, negative priming, dissipation of inhibition in reminiscence experiments and many others can be used for a proper theory of creativity. Yet these concepts do not appear in the literature covering the concept of creativity; hence they would not be included in the information fed to the computer. The computer, lacking the information, could therefore never arrive at the theory here developed. The major creative act is precisely the *extension* of what is regarded as 'relevant' to concepts, theories and experiments never previously considered in connection with the problems in question. I agree with Wolpert on this point: there is no real evidence that computers can be truly 'creative' in the sense intended, although they may appear to be so when the answer to the problem in question is already known, and can be used to select the information needed to solve the problem. Indeed, often the most creative act is the *selection of the problem*! Such a selection takes into account the importance of the problem, how much is known about it, previous attempts, possible remote sources of information not previously considered, probability that the problem is soluble at the present time, and many more; how could a computer select a 'good' problem when so many of the conditions determining a choice are fuzzy, unquantifiable, and depending on very individual, subjective factors? Historical accidents also play an important part. It would never do to underestimate the power of computers, or to understate predictions of what they might do in the future. But at the moment the likelihood of computers being truly creative in science or art seems remote.

Several of the points dealt with in this chapter will be argued more in detail later on, but provisionally I would state the theory to be developed as follows. (1) All cognitive endeavours require new associations to be made, or old ones to be reviewed. (2) There are marked differences between individuals in the *speed* with which associations are formed. (3) Speed in the formation of associations is the foundation of individual differences in intelligence. (4) Only a sub-sample of associations is relevant in a given problem. (5) Individuals differ in the *range* of associations considered in problem-solving. (6) Wideness of range is the foundation of individual differences in creativity. (7) Wideness of range is in principle independent of speed of forming associations, suggesting that *intelligence* and *creativity* are essentially independent. (8) However, speed of forming associations leads to *faster learning*, and hence to a greater number of elements with which to form associations. (9) The range of associations considered for problem-solving is so wide that a critical evaluation is needed (comparator) to eliminate unsuitable associations. (10) Genuine creativity requires (a) a large pool of elements to form associations, (b) speed in producing associations, and (c) a well-functioning comparator to eliminate false solutions. The following chapters will amplify these suggestions.

3 Creativity: measurement and personality correlates

Everything that exists exists in some quantity
and can therefore be measured

E. L. Thorndike

Creativity: the beginnings of measurement

If intelligence is not sufficient to account for genius, we must look for other factors, and 'creativity' is perhaps more frequently suggested than most. Much work has been done to clarify the nature of this concept (Glover, Ronning and Reynolds, 1989), but much diversity of opinion remains. Taylor (1988) lists six *types* of definition, most of which are somewhat esoteric and unusual. For the purpose of orientation, we may perhaps say that creativity is a dispositional trait or ability which enables a person to put forward ideas, or execute and produce works of imagination, having an appearance of *novelty*, which are immediately or in due course accepted by experts and peers as genuine contributions having social value. In due course we shall flesh out this definition by reference to a large body of experimental and observational studies.

Any such experimental study must of course rely on tests or other measuring devices which can be shown to be valid measures of the concept. In order to do this, we must demonstrate (1) that such tests are not merely measures of *g* or general intelligence, and (2) that they possess some degree of unity or coherence (reliability). One would have thought that the first man to see the precise nature of the problem, find ways of solving it, and be responsible for suggesting how such tests could be constructed would find an honoured place in the history of creativity research. Alas, this is not so; Spearman and the London School were early pioneers of what have become known as tests of *divergent*, as opposed to *convergent*, ability, but the names of those taking part in this adventure are seldom mentioned in the *Handbook of Creativity* (Glover *et al.*, 1989). What Spearman did was first of all to devise a statistical *method* for showing that a matrix or correlations between various tests contained, in addition to the general factor of intelligence, another factor, in our case *creativity*. His method of 'tetrad differences' (Spearman, 1927) was of course primitive compared with modern applications of matrix algebra, but it was sufficient to make possible estimates of statistical significance. One of the first applications of this method was to the delineation of the concept of 'fluency' or 'imagination', an early forerunner of more recent tests of creativity. As Sir William Osler said: 'In science the credit goes to the man who convinces the world, not to the man to whom the idea first occurs.'

This factor was isolated first by Hargreaves (1927) in his studies of 'the faculty of imagination'. He found that a number of tests calling for a large number of imaginative responses tended to correlate together with an average intercorrelation of .3. These correlations fulfilled the demands of the tetrad criterion and were shown not to be identical with '*g*'. These tests called for things seen in an ink-blot, number of words written, number of different completions to an incomplete picture, and so forth. This '*f*' for fluency factor was considered at first as being the reverse of perseveration (the tendency of ideas to perseverate in the mind), but Hargreaves disproved this hypothesis fairly conclusively.

Cattell (1934) took up fluency tests, and found them to have a low but positive correlation with his surgency (extraversion) factor as rated ($r = .30$). The tendency of extraversion (surgency) to correlate with '*f*' has found some slight report in Notcutt's study (1943), who gave five fluency tests which showed an average intercorrelation of .45. Score on fluency was found to correlate −.24 with introversion. Much stronger support for the hypothesis comes from the studies of Gewirtz (1948), who used '*f*' tests on 38 children between the ages of five and seven. Correlations between these tests ranged from .08 to .70 and averaged much the same as the tests reported previously. Gewirtz suggested that the '*f*' factor might split up into two. Different patterns of relationships were found between the tests or word fluency and two types of vocabulary test, one a recall vocabulary test and the other a recognition–definition vocabulary test. The intercorrelation of the word fluency tests and their correlations with mental age and two types of vocabulary tests seem to indicate that there were two abilities involved in word fluency: one involving the rate of word association where there is some restriction imposed, and the other involving the rate of word association where there is little restriction.

In correlating the '*f*' test with ratings made by the use of the Fels Child Behaviour Scales, she found a distinct tendency for the signs of the correlations of the various '*f*' tests to be identical when correlating these tests with each particular item of the Fels Scale. There thus emerges a distinct tendency for the child with high fluency scores to receive high ratings on curiosity, gregariousness, originality, aggressiveness, competitiveness, and cheerfulness, and negative ratings on social apprehensiveness and patience. These results would seem to support the hypothesis of a positive relation between '*f*' and extraversion, and originality, although failure to partial out intelligence and age renders the results less conclusive than one might have wished.

About the existence of such a correlation, there can be little doubt. Benassy and Chauffard (1947) found correlations between intelligence and '*f*' in the neighbourhood of .32 when Cattell's '*f*' test was administered to 282 children and 231 adults. They confirmed, however, that the correlations on which the '*f*' factor is based are partly independent of '*g*', and in the main their results of comparing fluency scores with ratings of temperament bear out the general hypothesis that fluency is a valid measure of extraversion.

In addition to the few studies mentioned, all of which have regarded the '*f*' test as a measure of personality, there are a number of studies in which fluency is related to purely cognitive tests, and as these have been dealt with in some detail by Vernon (1950), the main results only will be mentioned. The work of Thurstone (1938), Johnson and Reynolds (1941), Carroll (1941), Taylor (1947) and Fruchter (1948) tends to bear out the early findings of Holzinger (1934, 1935) that fluency tests have comparatively high saturations on '*g*' and '*v*' (intelligence and verbal ability), but that in addition they have something in common, which gives rise to a separate factor. Thurstone's distinction between his '*V*' (understanding of verbal material) and '*W*' (word fluency) emphasizes this duality; in more recent years he added another factor, '*F*' (ideational fluency with words) to the other two (Carroll, 1993).

Various sub-factors within this general context have been identified by some of the writers mentioned above. Thus, Taylor (1947) distinguishes: (1) verbal versatility or ability to express an idea by several different combinations of words; (2) word fluency, involving no reference to the meaning of words; and (3) ideational fluency, or production of words from meaningful associations. Carroll split Thurstone's '*W*' factor into: (1) speed of word association in restricted context; (2) rate of production of synthetically coherent discourse; (3) naming or ability to attach appropriate names to stimuli. (See Carroll (1993) for the most recent account of the factorial studies.)

One fundamental criticism of all the studies of fluency mentioned so far is that the scores which are being analysed can be shown to be compound rather than simple. It has been shown by Bousfield and Sedgwick (1944) that raw output scores could be rationalized in terms of two semi-independent concepts, C, or total supply of relevant words and m or rate of depletion of this total score. The exponential equation $\mathrm{N} = \mathrm{C}(1 - \mathrm{e}^{mt})$, where N is the number of words produced, t is the time over which the test is carried out and e is the base of the natural logarithms, was shown to fit adequately the rate of output during the test. This finding, if it were to be supported, could revolutionize the factorial study of fluency; we should drop the practice of intercorrelating raw scores and use C and m scores instead.

Later work by Rogers (1956) has not fulfilled this promise of producing a better measure of fluency, and little use has been made of the intuitively attractive innovation of Bousfield and Sedgwick. However that may be, clearly Spearman and his pupils established the importance of *divergent* tests of mental ability as correlated with, but distinct from, convergent tests. The usual convergent test has only one predetermined answer, as in the case of a typical series problem: 1 3 6 10 ?; clearly only 15 will do as an answer. But the problems used by Hargreaves have an infinite number of possible answers, and we can use the actual number of such answers given by an individual as his score, or impose some quality restriction on the answers. Thus we may require them to be unusual, sensible, or non-repetitive, and indeed we may have separate scores for quality as opposed to quantity, although the two are usually found to be correlated.

Guilford and tests of divergent ability

The use of divergent tests received a great impetus when Guilford (1950) gave his Presidential Address to the American Psychological Association on the topic of Creativity, and innumerable studies using such tests have since been done. A careful discussion of their value will be found in Glover *et al.* (1989). Here let us merely note that these studies agree on a number of important points. (1) Divergent tests define an ability correlated but not identical with that defined by convergent tests. (2) This correlation obtains only when IQ, as measured by convergent tests, is (roughly) below 120; above that limit there is little correlation. (3) Different types of divergent tests thus define an ability which we may provisionally label 'creativity', 'originality', or 'imagination', although it is clearly realized that such identification demands proof; we cannot take validity for granted.

Point (2) above, is sometimes called 'Guilford's (1967) triangularity hypothesis', and there is certainly some evidence in its favour (e.g. Guilford and Christensen, 1975; Schubert, 1973); these studies showed that creativity and intelligence were most highly correlated in the lower two-thirds of the population. However, *all* cognitive activities tend to show higher correlation in groups with lower IQs (Detterman and Daniel, 1989), and Runco and Albert (1986) have reported contrary results, throwing Guilford's hypothesis in doubt. It may be true, but only under certain circumstances, and with certain types of test. Sen and Hagtvet (1993) found no correlation between intelligence and creativity in 300 school-children.

An interesting follow-up study by Magnusson and Backteman (1977) has shown that over a time period of three years, i.e. from 13 to 16, creativity tests scores correlated more highly with each other, although different tests were used at the two ages, than they did with intelligence. Thus creativity measures are relatively *stable*, and are clearly differentiated from measures of intelligence. This is an important demonstration, and is relevant to point (1) above.

In the rest of this chapter I will concentrate on describing a few large-scale experiments that have suggested that these tests do have a measurable degree of validity. I have purposely refrained from carrying out a meta-analysis of all previous studies; I have criticized the whole notion of meta-analysis elsewhere (Eysenck, 1984, 1992c), on the grounds that it is pointless to throw together in one overall analysis studies good, and indifferent; having different criteria, different populations and different durations of follow-up; using different tests, different controls, and different methods. One good, careful study, using a well-chosen, large population, appropriate tests and criteria, and extended over a lengthy period, cannot be disconfirmed by any number of short-term, poorly arranged experiments, using inappropriate tests and criteria. As Newton said in his letter to Oldenburg of 18th August, 1676: 'For it is not the number of Exp$^{ts.}$, but weight to be regarded; where one will do, what need of many?'. Hence it seemed more useful to single out the best experiments available, and concentrate attention on their findings.

The view of 'divergent' or 'fluency' tests as measures of creativity has certainly been much criticized. Extraneous factors may determine DT (divergent thinking) test *performance* without measuring DT; any process (instruction, boredom) increasing or decreasing productivity in general would increase or decrease DT scores. Another criticism is that most studies merely count quantity of answers produced, but it may be quality that is important. Worst of all, there is little evidence that DT tests actually predict creative production (Wallach, 1971; Cattell, 1971; Brown, 1989). As Barron and Harrington (1981) say in their review, we still do not have a satisfactory answer to 'the vitally important question of whether divergent thinking tests measure abilities actually involved in creative thinking' (p. 447). Indeed, Barron and Harrington doubted the value of the divergent–convergent dichotomy altogether (p. 443). Hovecar (1979, 1980, 1981), on the other hand, has produced some evidence to show that ideational fluency, in terms of *quantity* underlies originality scores.

What actually is the degree to which the usual tests of DT correlate with each other, and do they correlate at all with what we would ordinarily call 'creativity' or 'originality'? A study by Barron (1963a) may be used to show the kind of answer we may obtain to such a question. Subjects of the study were 100 captains in the United States Air Force, studied in a residential setting by psychologists at IPAR, the Institute of Personality Assessment and Research at the University of California, Berkeley, whose extensive work in the field of creativity will be mentioned repeatedly in the survey. Subjects (Ss) were given eight traditional tests of creativity-originality, and were given independently ratings on 'originality' by staff members who got to know them well during the living-in process of socializing and testing. The list of tests used was as follows (reproduced with permission); all, it will be noted, are tests of divergent thinking or fluency.

Lists of tests

(1) *Unusual Uses.* This test calls upon the subject to list six uses to which each of several common objects can be put. It is scored for infrequency, in the sample under study, of the uses proposed. Odd – even reliability in the sample is .77.

(2) *Consequences B.* In this test S is asked to write down what would happen if certain changes were suddenly to take place. The task for him is to list as many consequences or results of these changes as he can. The responses are scored according to how obvious the imagined consequences are, the less obvious responses receiving the higher scores. Interrater agreement is .71.

(3) *Plot Titles B.* Two story plots are presented, and S is asked to write as many titles as he can think of for each plot. The titles are rated on a scale of cleverness from 0 to 5. The number of titles rated 2, 3, 4, or 5 constitutes the cleverness score. Interrater agreement in this study in .43.

(4) *Rorschach 0+*. This is a count of the number of original responses given by S to the ten Rorschach blots and adjudged by two scorers, working separately, to be good rather than poor forms. Standard Rorschach administrative procedure is followed. Interrater agreement is .72, and only those responses scored by both scorers as 0+ were credited.

(5) *Thematic Apperception Test: Originality Rating.* Two raters, working independently of one another, rate the TAT protocols of the 100 Ss on a 9-point scale, using approximate normal curve frequencies for each point along the scale. Interrater agreement is .70. The S's score is the average of the two ratings.

(6) *Anagrams.* The test word 'generation' is used, and the anagram solutions are scored for infrequency of occurrence in the sample under study. If S offers a solution that is correct and that is offered by no more than two other Ss, he receives one point for originality. Total score is therefore the number of such uncommon but correct solutions.

(7) *Word Rearrangement Test: Originality Rating.* In this test, S is given 50 words which were selected at random from a list of common nouns, adjectives, and adverbs. He is told to make up a story which will enable him to use as many as possible of the listed words. His composition is rated for originality on a 9-point scale,just as the TAT was. Interrater agreement in this instance is .67.

(8) *Achromatic Inkblots.* This is a set of ten achromatic inkblots constructed locally. The S is asked to give only one response to each blot. Responses are weighted according to their frequency of occurrence in the sample under study, the more infrequent responses receiving the higher weights. Score is the sum of the weights assigned to S's responses on all ten blots. Odd – even reliability is .43.

Do the tests intercorrelate with each other? Table 3.1 shows their intercorrelations, as well as their correlations with staff ratings of originality (9) and correlations with composite test scores, i.e. the sum of all eight tests (10).

Table 3.1. *Intercorrelations between 8 tests of divergent thinking*

Test Measures	1	2	3	4	5	6	7	8	9	10
1. Unusual Uses	. . .	.42	.37	.08	.17	.29	.06	.17	.30	.60
2. Consequences B	.42	. . .	.46	−.02	.21	.21	.16	.09	.36	.59
3. Plot Titles B	.37	.46	. . .	.17	.26	.17	.16	.07	.32	.62
4. Rorschach O+	.08	−.02	.17	. . .	.21	.03	−.05	.11	.18	.38
5. TAT originality	.17	.21	.26	.21	. . .	.36	.41	.02	.45	.59
6. Anagrams	.29	.21	.17	.03	.36	. . .	.33	.38	.22	.62
7. Word Re-arrangement originality	.06	.16	.16	−.05	.41	.33	. . .	.09	.45	.51
8. Inkblot originality	.17	.09	.07	.17	.02	.38	.09	. . .	.07	.46
9. Staff rating on originality										.55
10. Composite test originality									.55	

Source: From Barron (1968), with permission.

Clearly, the two Rorschach tests are unreliable and invalid; they do not even correlate with each other, neither do they correlate with the other tests, or with the staff ratings. This would have been predicted; the Rorschach test has not been found useful in any of its many applications (Eysenck, 1959). However, the other six tests correlate together quite well, averaging around .26. All correlate positively with the staff rating of 'originality', and the test composite score correlates a respectable .55 with the staff rating. Thus the battery seems to have shown both reliability (it measures well whatever it does measure), and validity (it does measure what it is supposed to measure), remarkably so in view of the early date when the study was done. The results are not untypical of later work, suggesting that this type of test does have some validity and reliability (Michael and Wright, 1989).

Other tests of creativity

DT tests are not the only ones that have been used; there are many others, differing from them in diverse ways. What, then, are the major ways of measuring creativity as a trait? Hovecar and Bachelor (1989) have given a taxonomy of such measures as have been used in this context. First and foremost we have tests of divergent thinking, of which those of Torrance (1974) and Wallach and Kogan (1965) are perhaps the best known (see Wallach (1985) and Glover *et al.* (1989) for review). Second they list attitude and interest inventories, based on the hypothesis that a creative person will express attitudes and interests favouring creative activities. Third in their list are personality inventories, on the hypothesis that creativity is a set of personality factors rather than a cognitive trait. Fourth come biographical inventories, hypothesizing that past experience may adumbrate future achievement. Fifth come ratings by teachers, peers, and supervisors. Sixth is the judgment of products, seventh the study of eminent people, and eighth self-reported creative activities and achievements.

A recent paper by Amelang, Herboth and Oefner (1991) illustrates this last method of measuring creativity. Called a 'prototype analysis', the method involves first collecting large samples of creative behaviour in ordinary people, putting these together in a questionnaire, and asking subjects if and how often they have shown each behaviour included. Examples of such items are: 'When my car broke down, I managed to get it to the next garage by using a basic commodity', 'I described the problems of a difficult situation in a play and put it on stage', and 'I built my own furniture in order to make it more suitable to the apartment and my personal needs'. The scale was reliable, and correlated well with peer ratings.

Another approach to the problem of measuring creativity is through the finding and solving of real-world problems (Okuda, Runco and Berger, 1991). Children were asked to solve certain problems directly relating to their home and school environment; they were also asked to list as many different problems at home or school as they could. They found that the real-world problem-finding task added significantly to the prediction of creative activities, beyond the prediction made by other tasks, such as standard measures of divergent thinking.

Two important approaches to the measurement of creativity not mentioned by Hovecar and Bachelor (1989) are the use of the word association test, and the use of preference judgments for complex as opposed to simple drawings. There is good evidence, to be reviewed later, that creative people give unusual reactions to the stimuli in a word association test, and that they also prefer complex to simple line drawings. Indeed, it might be said that these two types of test are at least as useful as DT tests in predicting creativity; in addition they are easier to give and to score. It is curious that comparatively speaking much less effort has gone into the development of these tests, as compared with DT tests. In addition to all these types of tests, there are others which are difficult to classify; some of them will be noted below.

Allied to creativity is the process of 'intuition', which theoretically is supposed to be instrumental in mediating creativity. Westcott and Ranzoni (1963) have described a method of measuring intuitive thinking which follows their definition of such thinking as deriving conclusions on the basis of fewer clues than would be used by the average person of similar intelligence – one way of 'jumping' at a conclusion! They would offer their subjects a problem which could only be solved by using a number of clues; these would be offered seriatim to the subject who however could guess at the answer at any stage. The 'intuitive' subject was the one who attempted an answer when only a small number of clues had been received; there was of course no penalty for using all the clues.

The study resulted in four groups, the intuitive/non-intuitive dichotomy being again divided into those who gave the right answer, and those who did not. (We might think of intuitive thinkers who gave the right answers, like Einstein, and intuitive thinkers who gave the wrong answers, like Freud.) Intuition was found to be a persistent trait of subjects in many different experiments (high reliability), and intelligence could be ruled out as being responsible for producing these correlations. It is unfortunate that Westcott did not correlate 'intuition-proneness' with creativity, either as a trait or as achievement; this verification of his theory still lies ahead. The topic of 'intuition' will be taken up again in a later chapter.

Hovecar and Bachelor (1989) report that the different measures of creativity may not always correlate well together, and indeed sometimes hardly show any degree of correlation at all; furthermore most, with the exception of tests of divergent thinking, show little agreement with a variety of criteria. They conclude that 'it is apparent that reliability, discriminant validity, and nomological validity cannot be taken for granted in creativity research' (p. 62). They suggest focussing 'on only those studies that include a measure of real-life creativity' (p. 64). This is the only way of establishing validity, and I will later on review some of the most important studies designed to answer this question.

Is there a factor of divergent thinking?

Before taking up this question of external validity, it may be useful to undertake a more detailed survey of the problem indicated by Hovecar and Bachelor (1989), namely the correlation between different measures of creativity, and the question of whether they give rise to one general factor, and what might be the relation of that factor with intelligence. Such a survey has been undertaken by Carroll (1993), whose summary I shall follow; his is by far the most profound treatment of this area. The types of divergent tests surveyed by him are typified by the following list.

(1) *Clever plot titles.* The task is to write titles for story plots.
(2) *Symbol production.* The task is to produce (by drawing) figural symbols to represent given activities and objects.
(3) *Remote consequences.* The task is to list the consequences of certain hypothetical situations, e.g. 'What would be the consequences if people no longer needed or wanted to sleep?'
(4) *Combining objects.* The task is to name two objects which, when used together, would fulfill a particular need.

(5) *Substitute uses*. The task is to think of a common object used for an unusual purpose (e.g. a shirt used as a sail).
(6) *Making groups*. Given a list of seven words for objects or things, the subject has to specify up to seven ways of grouping or classifying the items.
(7) *Different uses*. The task is to think of up to six different uses for an object, e.g. a brick.

Analysing and re-factoring 42 data sets containing originality/creativity tests of this kind, Carroll found that the bulk of the evidence suggested that this factor 'represents a distinct dimension of individual differences that is linearly independent of other such dimensions' (p. 427).

What is the common element in such creativity tests? They require subjects fairly quickly to think of, and write down, a series of responses fitting the requirements of the task or situation that is presented. Furthermore, the task is such that it is difficult and challenging for subjects to think of responses beyond the more obvious, commonplace ones. This suggests a link with fluency (but not identity!); when a person gives a large number of responses, at least a few of these are likely to be the more 'creative' ones. Simonton (1988b) has brought forward a similar argument to account for the demonstrated correlation in scientists between quantity of production and quality. If a given scientist produces many more ideas than another, some at least are likely to be of a high quality!

Scoring categories for tests of fluency–flexibility–creativity fall into four categories.

(1) *Fluency*. Usually measured in total number of responses.
(2) *Flexibility*. Measured in terms of the number of times a person changes spontaneously from one category of response to another.
(3) *Originality*. Scored according to whether responses are 'unusual', 'clever' or 'original'.
(4) *Elaboration*. Scored according to how 'elaborate' the response is, in terms of multiple detail given.

This enables us to differentiate *fluency* from *creativity*. The *fluency* factor 'is a measure of the tendency of individuals to think of a large number of different responses – whether obvious or non-obvious – to *any* task lending itself to giving of numerous responses' (p. 429), whereas the *creativity* factor 'is a measure of the tendency to give the more unusual or creative responses, when the task permits or requires such responses' (p. 429). Tasks measuring fluency generally do not permit or require unusual responses, whereas the tasks measuring creativity do.

The measurement of the 'unusual' quality of responses may present problems of subjective evaluation. It is easy to give a numerical score in terms of the number of times a given response occurs. Thus when asked about uses for a brick, many subjects talk about hitting a person with a brick or breaking a window by throwing the brick; responses like: 'A small film actor could stand on one to kiss a tall female' would occur very rarely. But there has to be some quality control, too. A response like: 'You could use the brick as a football' would be unusual, but hardly sensible! Ratings of *quality* of responses usually show good agreement, but make scoring tests difficult, time-consuming and expensive.

How valid are tests of divergent thinking?

Carroll (1993) raises the question of *validity* (do tests actually measure what they are supposed to measure?) and points out that such a question cannot be decided by correlational and factorial techniques; all that such techniques can do is to show that a creativity factor 'represents a distinct dimension of individual differences that can be measured with considerable reliability' (p. 431); beyond that we need a different type of evidence.

Some of the most instructive studies of validity were carried out by Torrance (1972a, b, 1981, 1988). One of these longitudinal studies involved high-school students tested in 1959 and followed up seven and 12 years later, while the other involved elementary school pupils tested in 1958 and for five subsequent years and followed up 22 years later. The criteria of 'creativity' used by Torrance were as follows.

(1) Quantity of publicly recognized and acknowledged creative achievements (patents and inventions; novels, plays that were publicly produced; musical compositions that were publicly performed; awards for art works in a juried exhibition, founding a business; founding a journal or professional organization; developing an innovative technique in medicine, surgery, science, business, teaching, etc.).
(2) Quality of creative achievements. Subjects were asked to identify what they considered their three most creative achievements: these data, plus responses to the checklist of achievements, were judged by three judges.
(3) Quality of creative achievement implied by future career image. Three judges assessed this primarily by responses to the following two questions in the follow-up: (i) What are your career ambitions? For example, what position, responsibility, or reward do you wish to attain? What do you hope to accomplish? (ii) If you could do or be whatever you choose in the next 10 years, what would it be?
(4) Quantity of high school creative achievements (used only in elementary school study, similar to item (1), but limited to achievements during the high school years).
(5) Quantity of creative style of life achievements, not publicly

Table 3.2. *Correlations between creativity predictors and creativty measures*

Tests	Quality 1966 $N=46$	Quality 1971 $N=52$	Quantity 1966 $N=46$	Quantity 1971 $N=52$	Aspiration 1966 $N=46$	Aspiration 1971 $N=52$
	Criterion variables					
Fluency	.39	.53	.44	.54	.34	.49
Flexibility	.48	.59	.44	.58	.46	.54
Originality	.43	.49	.40	.54	.42	.51
Elaboration	.32	.40	.37	.43	.25	.41

Source: From Torrance (1974), with permission.

recognized (such as organizing an action-oriented group, designing a house, designing a garden, initiating a new educational venture, painting that was not exhibited, musical composition not publicly performed, etc.; used only in elementary school study).

We thus have three major criteria, namely quality, quantity and aspirations, to correlate with the various scores obtained many years earlier on the Torrance Test of Creative Thinking (Torrance, 1974). Table 3.2 shows the obtained correlation coefficients, all but three of which are at the .01 level of significance, the number of children involved, and the years. Thus clearly for these criteria the tests are valid predictors, and it is interesting to note that the 1971 criteria give somewhat higher correlations than the 1966 ones; clearly creativity needs time to find measurable expression.

Table 3.3 shows results for the second study. Again the correlations are highly significant (all above the .01 level), indicating that predictive validity of the tests was achieved. When predictors were combined, they correlated with combined criteria at .51 (.59 for males and .46 for females). When similar correlations were calculated for the 22-year follow-up, the combined predictors correlated with the combined criteria .63, with males again slightly higher than females. It is noteworthy that in this study, validity coefficients for measures of intelligence ranged from $-.02$ to .34, with a mean of .17; clearly the main burden of the prediction is not borne by general intelligence!

Torrance (1988) drew two conclusions from his results:

(1) Young people identified as creative on the basis of creativity tests during the high school years tend to become productive, creative adults.

(2) At least 12 years after high school graduation appears to be a more advantageous time than seven years as the time for a follow-up of creative adults.

Table 3.3. *Correlations between predictors and criterion measures of creativity*

Predictors	Number	Quantity	Quality	Aspiration
Fluency	254	.30	.30	.27
Flexibility	254	.28	.29	.24
Inventiveness	254	.36	.41	.35
Total (1959)	254	.32	.36	.32
Originality	254	.40	.43	.39
Intelligence	254	.21	.38	.39
Achievement	254	.27	.47	.43

Source: From Torrance (1988), with permission.

The latter of these findings is important because nearly all replications of the original Torrance studies have been for relatively short periods, usually about five years, thus inviting a negative outcome because not enough time is available to demonstrate real-life creativity. It is impossible to accept short-term follow-ups which fail to produce adequate test-criterion correlations as evidence that creativity tests do not work. It is of course realized that long-term follow-up studies are difficult to arrange and finance, and place an extraordinary burden on the investigator. Nevertheless if such studies are undertaken at all, they must be carried out for a long enough period to enable firm deductions to be made from the results; anything shorter than 10 years can only give suggestive answers.

One exception to this criticism is the work of Howieson (1981) who has reported a 10-year follow-up study covering the years from 1965 to 1975. She used a sample of 400 seventh-graders, and employed as a criterion the Wallach and Wing (1969) check-list of creative achievements outside the school. Total score on the Torrance scales correlated .30 with total criterion score, with prediction of male achievements again better than of female achievements.

In another study, Howieson (1984) made use of predictor data collected by Torrance in 1960, in Western Australia. Verbal measures did not predict adult creative achievement at a satisfactory level for the 306 subjects involved, but the figural measures fared better. For her measure of quality of publicly recognized creative achievements, she obtained a multiple correlation of .51; and for quality of personal creativity achievements, one of .44. For quantity of personal (and publicly recognized) creative achievement, the multiple correlation was .33, still highly significant.

In yet another Torrance (1984) study, the Sounds and Images test was used (Khatena and Torrance, 1973) as a predictor in the follow-up study of 92 elementary school children tested in 1961 and scored for criteria in 1980. Almost all the hypothesized relationships were statistically significant, originality in images, strangeness, unusual sensory images, coherent syntheses, colourful images, and movement and action images were positively related to the criterion of young adult creative achievement. It is interesting to note that the number of common images, i.e. an indication of ordinary, non-creative thinking was significantly negatively related to creative achievement (−.30 with quality, −.27 with quantity, −.40 with future aspirations, and −.32 with style!) Similar findings with word association tests will be presented later.

Torrance (1988) ends his survey by saying:

> Although several hundred studies have dealt with the validity of creativity tests, those longitudinal studies with real-life criteria seem to offer the strongest links to test behaviour of creative achievement. We seem to be justified in assuming analogies between test behaviour . . . and 'real-life' creative behaviour.

An alternative to longitudinal testing and follow-up is of course correlational testing using creative and control groups engaged in the same occupation, e.g. architects or writers, or mathematicians. This has been the method used by the Institute of Personality Assessment and Research (IPAR), which has generated a wealth of important information (Barron, 1963a, 1968, 1969, 1972; MacKinnon, 1961, 1962a,b, 1965, 1978). Creativity was assessed by peers, colleagues, and/or experts, and equal status but not creative members of the same profession constituted the control group. The people involved spent several days at the assessment centre in Berkeley, and were interviewed, mixed socially with the assessment staff, and were administered a variety of tests of creativity, intelligence, personality and special abilities. The procedure was adapted from that used by the O.S.S. (Office of Strategic Services) during the war for the selection of agents to be dropped behind enemy lines (Murray *et al.*, 1963), and later used by the British WOSBs (War Office Selection Boards) for the selection of men for officer training (Parry, 1959; Vernon and Parry, 1949).

One of the tests used in these studies was the Symbol Equivalence Test, in which Barron (1963a, 1988) studied the process of transformation of images. Instructions for the test were as follows. In this test, you will be asked to think of metaphors, or symbolically equivalent images, for certain suggested stimulus images. The task can best be made clear by an example.

EXAMPLE:
Suggested stimulus image:
  Leaves being blown in the wind.
  Possible symbolic equivalents:
    A civilian population fleeing chaotically in the face of armed aggression.
    Handkerchiefs being tossed about inside an electric dryer.
    Chips of wood borne downstream by a swiftly eddying current.

Following this, the subject is asked to make up three possible equivalents for each of the images, with a time limit of 20 minutes. The images used are: haystacks seen from an airplane; a train going through a tunnel; the sound of a foghorn; a candle burning low; a ship lost in the fog; a floating feather; the increasingly loud and steady sound of a drum; sitting alone in a dark room; empty bookcase; tall trees in the middle of a field. Scoring was by (1) number of acceptable or admissible, but not original responses, and (2) number of original responses. The former was scored by degree of aptness as 1, 2 or 3; original responses were scored as 4 or 5 according to degree of originality, the total score simply summing the 10 individual scores. Several raters were used to produce an agreed score for the answers.

Barron (1988) gives some examples to illustrate the scoring:

1. Stimulus image:
   A candle burning low
   Admissible responses:
   Life ebbing away (scored 1)
   A basin of water emptying down a drain (scored 2)
   The last drops of coffee going through a filter (scored 3)
   The last pages of a faded book (scored 4)
   The last hand in a gambler's last card game (scored 5)
2. Stimulus image:
   Empty bookcase
   Admissible responses:
   A hollow log (scored 1)
   An empty sack (scored 2)
   An abandoned beehive (scored 3)
   An arsenal without weapons (scored 4)
   A haunted house (scored 5)
3. Stimulus image:
   Sitting alone in a dark room
   Admissible responses:
   Lying awake at night (scored 1)
   An unborn child (scored 2)
   A stone under water (scored 3)
   A king lying in a coffin (scored 4 or 5)
   Milton (scored 5)

To be admissible the response is required to reproduce the main features of the stimulus image, including formal properties as well as functional relationships between them. To be considered 'original' it has to be, first of all, *unusual*. This quality is easy to recognize once a large number of responses have been assembled from many subjects. There are, of course, problems in scoring, as always with divergent tests, but reliability (agreement between raters) can be adequate after training, and of course, validity (prediction of real-life creativity) is the best test of success in this awkward task.

As Barron (1988) point out, 'the validity of the Symbol Equivalence Test . . . proved quite substantial' (p. 89). The test provided a clear rank-ordering by creativity of the various samples studied, which correlated highly with the ranks of the group when rated independently for overall verbal creativity. (In one way, of course, this is a statistical artifact; group correlations must be higher than individual correlations; however, this is only true when there are valid individual correlations.) Famous writers came out on top, followed by famous architects; mathematicians at the Institute for Advanced Studies in Princeton; women mathematicians who had been rated as the most creative in the field in the USA; successful entrepreneurs in Ireland; research scientists; student artists at the San Francisco Art Institute; and finally a group of 'opportunity samples' who had been assessed by test only as control subjects over a period of time.

Next, the test scores were correlated with external criterion ratings of creativity for each subject in relation to all others in the relevant professional group.

> Substantial positive, statistically significant correlations were found for writers, architects, artists and entrepreneurs . . . In brief, a relatively short and simple test, taking half an hour for a trained scorer to evaluate, has demonstrated substantial validity for measurement of a key component of creativity: the ability to make original and apt transformation of a given image, received in words and expressed in words (Barron, 1988, p. 89).

It might be objected that the test is measuring verbal fluency rather than abstract creativity, but this is not so. In another study Barron (1972) used 82 art students as his sample, and administered a large set of tests to them including the Symbol Equivalent Test in its original form, as well as in multiple choice format, scored for 'recognition of originality', in this form subjects had to choose between an original and a commonplace response. Instructions were to choose the more original response. Both versions of the test showed similar correlations with various other tests, and they predicted studio grades with the astonishingly high correlations of .60. Here the criterion is not verbal at all, but is dependent on artistic productivity in the visual modality, yet the predictive accuracy of the test remains unimpaired.

Quite another type of test of creativity is the word association test, originally developed by Galton. Originality or 'commonness' of response to the stimulus word are easy to score by reference to frequency of different responses established on large standardization groups. Thus in Fig. 3.1, the stimulus word 'foot' is responded to by 'shoe'(s)' by 232 subjects but only by one respondent with 'rat', or 'snow', or 'dog'. Similarly, the stimulus word 'Command' is followed by 'order' by 196 subjects, but only one respondent with 'hat', 'book', or 'stern'.

Is it possible to use the word association test as a measure of creativity or

FOOT

Shoe(s)
(232)

Hand (198) Toe (191) Leg (118)

Soldier, Ball (26), Walks (23), Amble (14), Arm (13), Sore (10), Inch(es) (9)

(8)
Rat, Snow, Person, Physics, Dog, Mule, Wall, Shin, Wash, Hat, End

Singles

COMMAND

Order
(196)

Army (102) Obey (78) Officer (65)

Performance (33) Do (27) Tell (27) Shout (26) Halt (23) Voice (20) Soldier (18)

Hat, Polite, Plea, Book, Salute, Fulfill, Obedience, War, Stern

Singles

Fig. 3.1 Frequency of responses to word association test stimuli 'Foot' and 'Command'.

originality, assuming that more creative persons will give unusual answers? The real importance of finding an answer to this question will become apparent later on, when we begin to discuss theory; here let us merely note the answer given in some research carried out by IPAR. The group tested was that of architects already mentioned in connection with Frank Barron's work. There were three groups in all, one of which was selected as having studied and worked with members of this creative group, while a third (Group 3) was highly regarded but not considered creative.

MacKinnon (1962a) starts his account of this experiment with a reference to a study by Bingham (1953), who tested Amy Lowell, the poetess, with (among other tests) the word association test and found that 'she gave a higher proportion of unique responses than those of anyone outside a mental institution' (p. 11). With his architects, MacKinnon found the same; the unusualness of responses correlated .50 with the rated creativity of the architects. Thus group I (the most creative) scored 204; Architects II scored 128; and Architects III, the least creative, scored 114. The overall correlation between creativity and unusualness of responses at .50 is remarkable as the reliability of the test is only moderate; correction for attenuation would lift the correlation to above .65.

Gough (1976) has reported on a similar study done with 60 engineering students and 45 industrial research scientists. The subjects were rated for creativity, and given two word association tests, one a general one, and one using a scientific word list. Both correlated with creativity, but the scientific word list gave rather higher correlations. This is an intriguing finding that ought to be followed up in future research. He also found rare but not unique responses more predictive than unique ones.

There are many other examples of studies which show good predictive accuracy for real-life creativity, but the main purpose of this book is not to paint the lily and reproduce endless evidence on this point; my concern is with *explanation* of the causal chains involved, rather than a detailed review of the evidence by actual creative products. I will, instead, go on to look at another aspect of creativity, namely the traits which characterize the creative person; some of the studies to be considered incidentally contain material also relevant to the point made in the first part of this chapter, namely the connection between tests of creativity and real-life creativity.

Creativity and the Figure Preference Test

Most measures of (trait) creativity, as explained earlier, use some form or other of fluency/divergent thinking measures, which have two major defects. In the first place, such forms of measurement nearly always use *verbal* material, so that non-verbal types are handicapped; numerical or visuo-spatial ability would possibly make such people appear more creative if the tests incorporated material geared to their cognitive strengths. In the second place use of verbal material almost guarantees some correlation with *g*, especially at the lower levels, because to be creative verbally you have to have a large vocabulary which is highly correlated with IQ – at least verbal IQ. Hence particular interest has attached to a very different method of testing creativity, associated with the names of Welsh (1975) and Barron (1953, 1969); see also Barron and Welsh (1952).

Essentially Barron and Welsh are concerned with a perceptual preference factor which seems to have considerable generality in human behaviour. It 'opposes a preference for perceiving and dealing with complexity to a

Fig. 3.2 Sample of *simple* polygons.

preference for simplicity, when both of these alternatives are phenomenally present and when a choice must be made between them' (Barron, 1953, p. 163). Such a perceptual factor was originally isolated by Eysenck (1941), who also provided measures of it (for summaries of this work see Eysenck (1981)). From Eysenck's work with polygonal figures, Figs. 3.2 and 3.3 show typical sets of *simple* and *complex* polygons.

Welsh started out by developing a 400-item Figure Preference Test, in which simple line drawings were presented to the subject, who had to respond with either 'like' (L) or 'Don't like' (DL) (Welsh, 1949). Observation of this WFPT, supported by factor analyses, suggested a dimension opposing liking for complex to liking for simple figures, and because of interesting correlations with personality relevant items were put together to form the Barron–Welsh Art Scale (BW), consisting of 62 items only of the original 400. Scores on this scale are referred to as measures of 'origence', because of the correlation of preference for complex designs with originality and creativity; origence was found uncorrelated with 'intellectence', as measured by IQ tests (Welsh, 1975). In his book Welsh summarizes an enormous amount of research, indicating

Fig. 3.3 Sample of *complex* polygons.

that high origence scorers (a) can be shown to be creative in real life, and (b) have a personality similar to that which has been found to be characteristic of creative people, as defined by their creative achievements. Of all the material available, only some of the most important studies can be cited. Fig. 3.4 shows a typical page from the test, with obvious simple and complex drawings.

Much of the early work was done in the Governors' School in North Carolina, which runs a special summer programme for talented and gifted high school students. (Note that the independence of origence and intellectence in this high-IQ group does not necessarily mean that no correlation would be found in less gifted children. There is a paucity of evidence for below-average IQ groups.) Welsh administered a great variety of personality scales to these students, and arrived at results which are remarkably similar to those reported by Dellas and Gaier (1970), who summarized the literature on that topic. In their summary, Dellas and Gaier (1970) conclude their evaluation of more than two dozen studies of personality characteristics of creative persons by saying that their 'evidence points up a common pattern of personality traits among creative persons, and also that these personality factors may have some bearing on creativity in the abstract, regardless of field' (p. 65). The major 13

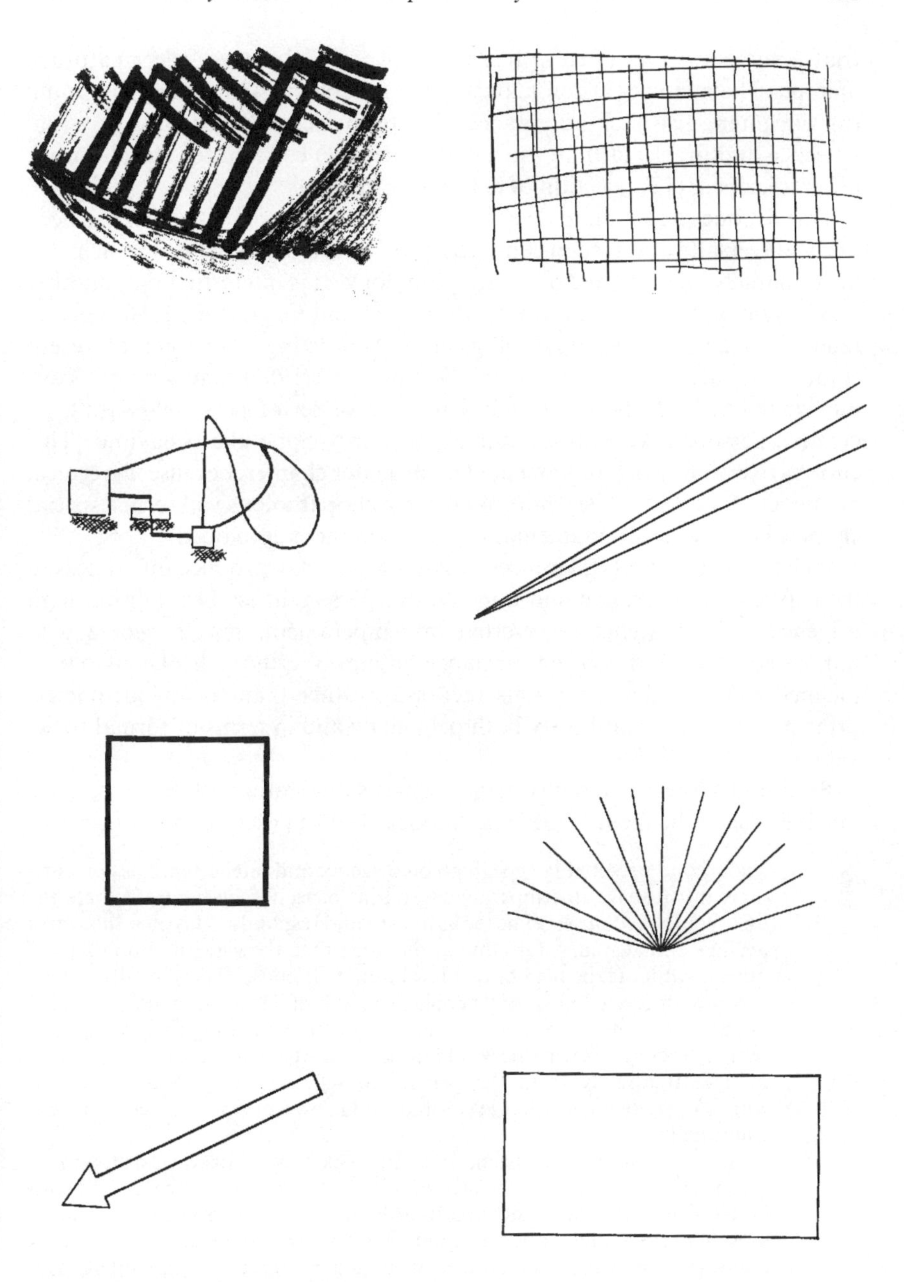

Fig. 3.4 Sample of simple and complex drawings from the Barron–Welsh Art Scale (Welsh, 1949), reproduced by special permission of the publisher, Consulting Psychologists Press, Inc., from *California Psychology Inventory* by Harrison Gough. Copyright 1986 by Consulting Psychologists Press, Inc.

traits they found associated with creativity were: (1) Independence in attitude and social behaviour; (2) dominance; (3) introversion; (4) openness to stimuli; (5) wide interests; (6) self-acceptance; (7) intuitiveness; (8) flexibility; (9) social presence and praise; (10) an asocial attitude; (11) concern for social norms; (12) radicalism; (13) rejection of external constraints.

This list does not contain a very important set of variables which has been fairly universally observed in connection with creativity as defined by achievement, namely a fair degree of psychopathology. It is interesting that another recent summary of the literature (Munchard and Gustafson, 1988) equally failed to mention psychopathological traits. Listing a number of recent studies, they also quoted Barron and Harrington's (1981) summary, but leave out what has been the most widely discovered set of personality traits of creative persons and geniuses, namely their psychotic-like behaviour. This curious oversight will be commented on in a later chapter, because it is typical of one set of theorists. The whole topic of psychopathology will form a special chapter because it is fundamental to the theory here developed.

Welsh indicates that differences in intelligence may produce differences in the expression of creative and non-creative personalities. The dull but high origence student seems extraverted in temperament, rejects people who impose some standards of performance on him or criticize his behaviour, is inclined toward rebellion for its own sake rather than for an ideological principle and rejects authority both personally and in terms of formal social values.

By contrast, the bright high origence scorer shows rather a different pattern, much closer to the Dellas and Gaier model. Barron describes him thus:

> The second personality type, high on origence and intellectence, is contrastingly introversive in temperament and introspective by nature. Where the first type looked outward he looks inward and responds to his own subjective feelings and attitudes, rejecting at the same time the views and opinions of others. Although he may have a few intimate friends, he is generally asocial and is uninterested in having people around him. He is more inclined to act impersonally and to express his views indirectly by writing than by face-to-face interaction; likewise he would rather read what someone has to say than to hear it directly from the person himself. There is an isolative and withdrawing tendency that leaves him to his own devices intellectually and emotionally.
>
> He is playful and persistent and can work toward his own distant goal independently. He rejects the help of others just as he is so preoccupied with his own views that he cannot accept ordinary social values and conventional morality. He would rather do things his own way than to yield to others although he may respond to rational and logical argument of an intellectual kind. He is not affected by emotional appeals unless they coincide with his own values and attitudes. Similarly, he cannot understand why others do not accept his views and ideas as he fails to recognize the need for emotional or social appeals to other persons. He may appear tactless and stubborn because he expects others to recognize as obvious the conclusions that he has arrived at by his own insights which seem so compelling to him. This leads to further

> estrangement from the social world around him and to greater self-involvement. He becomes convinced of the correctness of his own views and is not afraid to take risks because he has confidence in his own ability, If he fails, it is others who do not understand him or appreciate what he is trying to do – they should change, not he (p. 104).

The dull, low origence type also tends to extraversion, like the dull, high origence type. (It should be remembered that the measure used to measure intelligence was the Terman Concept Mastery Test, which is probably more a measure of crystallized than of fluid ability, and hence would favour introverts – Wankowski, 1973.) He is characterized by bland conformance and conventionality. He does not recognize the extent to which his attitudes have been formed by those around him, nor the degree of his dependency on externally imposed rules and regulations. He is a follower, relies on authority figures, prefers the tried and true, and rejects anything new or unproven.

Most interesting is the behaviour of the bright, low origence student, as compared with the bright, high origence student quoted previously. Welsh characterizes him as

> The fourth personality type which is high on intellectence but low on origence is somewhat introversive like the second, but does not seem quite so withdrawn and asocial in orientation, nor is he so introspective by nature. He is much more objective in outlook and responds to people in the world around him and to their attitudes and ideas, although he tends to maintain some social and personal distance from them. Most of his responses are intellectualized or rationalized and he seldom acts impulsively as the first type often does. He seems to believe that the world is an orderly place and that there are rules and regulations to be followed both in daily conduct and in solving problems that may appear. For this reason he finds mathematics and the physical sciences congenial since they are impersonal in nature but challenging intellectually.
>
> He respects his own accomplishments as well as those of other persons but expects them to follow protocol strictly and is resistant to the flashes of insight or the intuitive solutions that the second type may achieve. The nonlogical does not fit into his systematic approach to problems. He follows a well specified code of ethics and expects others to act in the same manner; perhaps he may sometimes seem to be overly strict in his interpretation of moral behaviour. He seems to be more optimistic in outlook than the second type, possibly because of a belief that a desirable and worthwhile outcome may be achieved through hard work and the application of comprehensible principles (pp. 105–6).

Relying primarily on the differences between bright high and low origence students, Welsh states that his findings agree with the Dellas and Gaier summary on 11 out of their 13 points. High origence involves independence of attitude and social behaviour, dominance, introversion, openness to stimuli, wide interests, self-acceptance, intuitiveness, asocial attitude, unconcern for social norms, radicalism, and rejection of external controls. Two traits, flexibility, and social presence and poise, Welsh fails to find in his high origence students; possibly their youth has something to do with this. As he says, 'those who fall into type two, high origence/high intellectence, resemble to a striking degree the composite picture of the creative person inferred from the thirteen-point summary. The prediction seems justified that Governor's School students of this type can be considered to be potentially creative from the standpoint of personality' (p. 107). Of course, several of the 11 traits characteristic of high origence are also found in the low intellectence group, such as independence, asocial attitudes, unconcern for social norms, and rejection of external controls. These may be the most basic correlates of origence. (We should bear in mind that 'low intelligence' in this group is relative, even the low scorers on intellectence would be above the population mean.)

Using samples of university students, Barron (1953) studied his subjects not only by inventories, but also by staff assessment of various behaviour patterns observed in personal contact. Complexity on the Barron–Welsh scale correlated .50 with present tempo, .29 with verbal fluency and −.42 with constriction; thus, the Complex person is more intensely expressive, expansive, and fluent in speech than the Simple person. The Simple person, on the other hand, is seen as being more natural and likeable, and also is more straightforward and lacking in duplicity' (p. 166). Complexity was found to correlate −.44 with naturalness, −.27 with likeability, and .56 with deceitfulness, as rated by the staff. Good judgment and adjustment were also lacking in the Complex person, with r values of −.39 and −.31. (Adjustment was defined as 'getting along with the world as it is, adequate degree of social conformity, capacity to adapt to a wide range of conditions, ability to fit in.')

Originality was also rated by the staff, every subject being rated on the degree of originality he had displayed in his work. The Complex person was seen as more original, both by the assessment staff and by the faculty of his department, with a correlation of .30. A specially designed mosaic construction test was also administered and rated for artistic merit; the rating correlated with Complexity .40.

Other correlations for Complexity were for flexibility (r = −.35 with rigidity defined as 'inflexibility of thought and manner, stubborn, pedantic, unbending, firm'). Impulsiveness showed a correlation of .50 with Complexity, and several MMPI scales of psychopathology showed a correlation of .37 (schizophrenia) and .36 (psychopathic deviate). 'Thus Complexity goes along both with lack of control of impulse (the *Pd*) scale) and with the failure of repression which characterizes the schizophrenic process' (p. 167). Anxiety was also correlated with Complexity (r = .34), Social non-conformity is another correlate of Complexity, staff ratings of conformity giving a correlation of −.47 with Complexity, and self-ratings one of −.53. Submissiveness was not characteristic of Complex persons (r = −.29), but holding socially dissident and deviant opinions was (r with the MMPI F-scale = .36).

Independent judgment was tested by means of the Asch (1956) test. This consists of an experimental social situation in which the subject (S) is put under pressure to conform to a group opinion which is false. There are from eight to 16 ostensible Ss, only one of whom, however, is naïve; the rest are employed by the experimenter. The task is to judge which of three lines of variable length meets a standard line. The Ss, one by one, announce their judgments publicly. The naïve S is so placed as to be one of the last to announce his judgment. On the critical trials, the hired majority gives a pre-arranged false answer. The experimental variable is called Yielding, which is defined as agreeing with group opinion when it is in error. Complexity was significantly and negatively correlated with Yielding in two separate samples.

A rather different kind of complexity ('cognitive complexity') was tested by Charlton and Bakan (1988) and correlated with creativity. Creativity was measured with the 'How do you think?' test (Davis, 1975), a test that assesses attitudes, values, beliefs, motivation and other personality and biographical items shown to be related to creativity, and significantly related to other, more direct creativity measures. For cognitive complexity Crockett's (1965) test was used in which subjects are asked to describe in writing, in any way they wish, four people they know personally. Cognitive complexity was defined as the total number of constructs or descriptions used across the four characterizations. As predicted, high creative subjects were significantly more cognitively complex than low creative subjects. The study did not include a test of intelligence; it seems possible that the discovered correlation might have been due largely to IQ.

Validity of the Figure Preference Test

These findings were replicated on several other samples of students and adults, cited by Barron (1953) and Welsh (1975), but we must now turn to consider evidence that the BW test actually measures creativity, A special study was conducted on a sample of 129 architects at the Institute of Personality and Research (IPAR) who had been rated for professional creativity by their peers; this study has already been mentioned in connection with the Word Association test. Sample I had been nominated as the most outstanding creative practitioners in the United States by other architects and various experts; there were 40 of these who took part in the living-in assessment. A second sample (II) consisted of 43 subjects matched for age and geographic location of practice who had at least two years of working experience with the originally nominated creative architects. A third group, Architects III, was selected as being representative of architects-in-general; they had no connection with the creative architects. Scores on origence were translated into T scores, i.e. scores with a mean of 50 and a S.D. of 10 for the boys at the Governor's School. Architects I, i.e. the ones judged particularly creative, had a mean of 56.90 and a S.D. of 5.73; Architects II, less creative but above the mean, had a score of 50.70, S.D. 6.97; Architects III, the least creative as rated, had a score of 47.71,

S.D. 5.36. Thus degree of professional creativity (creative achievement) was mirrored by origence scores, the most creative architects scoring about twice their own S.D. above the least creative (who of course would be well above the large group of architects who would fall below the mean for achievement – architects III did not contain any failures or poor performers). Overall mean creativity ratings for the individual architects was significantly correlated with origence scores, $r = .47$ ($p < .001$). This correlation of course underestimates the 'true' (i.e. unattenuated) value. Using estimates of .75 for reliability of measurement for origence and creativity, the corrected value is in excess of .62, i.e. very respectable indeed. (It is only fair to mention that in results for a much smaller group of 45 unselected scientists there was little agreement of ratings and origence score, but apparently none were specially creative or non-creative. This sample should be seen as being in the middle range of the Lotka–Price continuum, and hence showing too little variance to expect correlations between ratings and creativity and origence, although even here the group of scientists rated most creative had had a mean score of 49.47, as compared with the rest at 46.76.)

From what has been said so far, we would expect to find positive correlations between the complexity score, word association unusual responses, and certain personality traits such as psychopathology, impulsiveness, and lack of conformity. Such congruence has been found by Eysenck (1993b); the details of this study will be discussed in a later chapter.

The original work of Barron on complexity–creativity was taken up by Eisenman (1968, 1969), using polygonal figures rather than line drawings, as in the Barron–Welsh Art Scale. Using 302 students, Eisenman (1969) found a correlation of .55 between his measure of complexity and the BW Scale. Both scales correlated about equally with several tests of fluency and originality, derived from the 'unusual uses test'. In another study of 232 undergraduate students, Eisenman, Borod and Grossman (1972) found correlations of .51 and .59, in males and females respectively, between the BW Art Scale and the Eisenman complexity score for polygons. Some further correlations were of interest. For males, but not for females, authoritarian attitudes correlated negatively with the two measures of complexity, and with a questionnaire measure of creative attitudes; this questionnaire, in turn, correlated positively with the two complexity measures.

Eisenman and Robinson (1967) later demonstrated that liking for complexity correlated with creativity, using the polygon test. Grove and Eisenman (1970) demonstrated personality correlates of complexity preferences similar to those found by Barron and Welsh. It was also found that first-born males preferred complexity, a result that was reversed for females where first-born females preferred simplicity. This finding was tested in a later study (Eisenman, 1987), where 200 adult males (100 first-born and 100 later-born) constituted the sample. Three creativity measures were used, one of them the preference for complexity test, as well as an IQ test, and a risk-taking questionnaire. First-born subjects were superior on all the creativity tests, but not the IQ test; they also scored significantly higher on the high-risk option. This is in good agreement with studies of eminence and birth-order, where past geniuses and present creative achievers have consistently been predominantly first-born or only children. Schachter (1963) has summarized several earlier studies, and Albert (1980), Altus (1966), Helson and Crutchfield (1970), Sutton-Smith and Rosenberg (1970) and Visher (1948) also support this finding. Interestingly enough, and giving support for my decision not to include politicians in my theory, it has been found that eminent politicians do not share with outstanding scientists and artists this trait of being first-born. Goertzel, Goertzel & Goertzel (1978), and Albert (1980) have shown this to be so, and Walberg, Rasher and Parkerson (1979) found revolutionaries to be less likely than other eminent people to be first-born. Finally, Clark and Rice (1982) found that winners of peace and literature Nobel prizes were less likely to be first-born than Nobel Laureates in science.

How do actual psychotics perform on the Eisenman complexity test? Apparently (Eisenman, 1965, 1966), schizophrenics prefer the less complex polygons. In a later study, Eisenman (1990) showed that schizophrenics, as compared with hospital employees, preferred simple polygons and wrote less creative stories, a clear indication that persons actually suffering from schizophrenia are significantly less creative than normal hospital employees, themselves not particularly creative. It seems that preference for complexity, whether measured by line drawings or by polygonal figures, is a reliable index of creativity (Eisenman and Coffee, 1964).

In yet another experiment, using 196 undergraduates, Taylor and Eisenman (1986) found that the personality of students who preferred complex polygons was similar to that of subjects who had shown independence in the Ash test of conforming behaviour.

I have already quoted some personality traits which correlated with creativity in the account of the work on the Figure Preference Test. The main reason for discussing the evidence for a close relation with complex polygons was similar to that of subjects who had shown independence in the Ash test of conforming behaviour.

Creativity and personality

The extensive work of IPAR, some of which has already been mentioned, has used ratings, interviews, questionnaires and other methods to investigate the personalities of creative people in many different occupations. Barron (1969) has pulled together the many strands to indicate the major traits characteristic of truly creative people. Thus creative women mathematicians were described by the staff as individualistic, original, preoccupied, artistic, complicated, courageous, emotional, imaginative, and self-centred, while their non-creative controls were considered cheerful, active, appreciative, considerate, conventional, co-operative, helpful, obliging, organized, practical, realistic, reliable and sympathetic.

Other characteristics of the creative women, emerging from a clinical Q-sort (ranking statements in order of applicability) were: unconventional thought processes, rebellious and non-conforming, self-dramatizing and histrionic, fluctuating moods. On the other hand, items which did not fit the creative women were: dependable and responsible, sympathetic, conservative, and moralistic.

Some of these terms occur again and again. Thus in the staff evaluation of creative writers, among the dominant items are: independence, autonomy, aesthetic appreciation, productivity, has high level of aspiration, wide-range of interests, unusual associations and ways of thinking, is an interesting, assertive person.

Scores for various IPAR groups on the California Psychological Inventory (Gough, 1957) are given in Table 3.4. Results on the various scales are given for creative and non-creative (representative) samples, to make comparisons easier. Clearly the creative groups are more flexible, but lower on socialization! But on the whole the CPI is a disappointment; it does not disclose much in the way of personality attributes of creativity.

Yet another approach to the personality correlates of the creative person is via the Myers–Briggs Type Indicator (Myers and McCaulley, 1985), widely used in the IPAR studies (Thorne and Gough, 1991). This inventory aims (with moderate success) to incorporate Jung's (1921) theory of personality in the form of a psychometric inventory. The main dimension in the MBTI is called E–I, or extraversion–introversion; this is mostly a sociability scale, correlating quite well with the MMPI social introversion scale (negatively) and the Eysenck Extraversion scale (positively) (Eysenck and Eysenck, 1985). Unfortunately the scale also has a loading on neuroticism, which correlates with the introverted end. Thus introversion correlates roughly (i.e. averaging values for males and females) −.44 with dominance, −.24 with aggression, +.37 with abasement, +.46 with counselling readiness, −.52 with self-confidence, −.36 with personal adjustment, and −.45 with empathy. The failure of the scale to disentangle Introversion and Neuroticism (in fact there is no scale for neurotic and other psychopathological attributes in the MBTI) is probably its worst feature, only equalled by the failure to use factor analysis in order to test the arrangement of items in the scales.

The next variable in the MBTI is the SN scale, with high scores called 'intuitive' as opposed to 'sensing'; correlations with other scales and ratings are −.60 with obsessional personality, −.42 with preference for economic and +.42 with aesthetic values, +.50 for rule rejection and .49 with sensation-seeking. The TF scale (thinking–feeling) gives a high score to the feeling type, i.e. people do tend not to base their decisions on analytic or pragmatic values; they also tend to avoid conflict by giving in to others. Finally, the JP scale

Table 3.4. *Scores on the scales of the California Psychological Inventory (published in 1956) by creative and representative groups*

CPI scale	Creative architects	Representative architects	Creative writers	Representative writers	Creative women mathematicians	Representative women mathematicians
Dominance	59	56	55	54	46	50
Capacity for status	60	57	60	57	52	54
Sociability	48	51	52	49	42	47
Social participation	58	53	60	57	52	52
Self-acceptance	61	56	63	54	44	51
Sense of well-being	48	54	41	48	50	50
Responsibility	51	54	52	50	55	55
Socialization	47	52	42	46	45	48
Self-control	45	53	45	52	51	53
Tolerance	50	54	53	47	56	56
Good impression	43	52	44	51	46	47
Communality	48	53	49	51	41	47
Achievement through conformance	50	56	50	54	46	54
Achievement through independence	59	58	63	60	65	64
Intellectual efficiency	51	54	54	52	54	55
Psychology-mindedness	61	57	60	59	68	65
Flexibility	59	51	60	55	69	56
Femininity	57	52	62	55	53	49

Source: From Gough (1957), with permission, after Barron (1965).

contrasted judging and perceiving; perceiving types seek intensity of experience and augmentation of stimuli, whereas judging types seek to control and minimize affect-arousing sensations.

Thorne and Gough (1991) use three criteria of creativity to correlate with the MBTI. These are the creativity scale from the Adjective Check List, already discussed; an in-house criterion based on ratings made by IPAR staff on the basis of observations made during the one-to-three-day assessments; and an external criterion, made up of ratings obtained from professional peers, faculty members, journal editors, etc. Correlations between these criteria and the MBTI scales are shown in Table 3.5.

Clearly creativity is correlated only with the 'intuitive' type on the SN scale; there are isolated correlations with the 'perceiving' types, but too low and isolated to reward discussion. This agrees well with the Westcott studies of intuition, mentioned earlier, and also with the correlates of this type mentioned earlier – negative correlation with obsessionality, dislike of economic values, rejection of rules, sensation-seeking and preference for aesthetic values. In as far as they go, these correlations then agree well with the general picture of the creative personality derived from other sources.

The personality of the genius

How do the various traits characteristic of creative people compare with personality traits shown by acknowledged geniuses? The only study to give this question an empirical answer (as far as non-pathological traits are concerned) is the investigation by Cox (1926) of the intelligence of 301 geniuses, already discussed at length in an earlier chapter, In one of her chapters (Chapter XII) she gives an analysis of character traits of 100 youthful geniuses – the number is small because not all geniuses could be rated for personality traits because of paucity of evidence. Cox used a number of tests, in particular ratings on interests and of character traits.

For interest, she used a seven-point scale on which 0 was the average for unselected youths, +1, +2 and +3 showing degrees of interest, and −1, −2 and −3 indicating degrees of interest below the average. Results may be summarized as follows.

(1) No individual rating on *intellectual interest* is below +1; i.e., more than average interest in mental activities is evident in each case. (2) In *social* and *activity interests* the group rates considerably above the average. (3) *Breadth of distinct interests* is slightly above the average; *breadth of related interests* is considerably above the average. (4) *Intensity of two or more interests* is slightly above the average; *intensity of a single interest* approaches the upper limit. (5) *Intellectual interest* and *breadth of distinct and of related interests* correlate positively with intelligence. (6) *Social* and *activity interests* show a slight negative correlation with intelligence, but none of the ratings fall significantly

Table 3.5. *Correlations between Myers–Briggs Type Indicator scales and observational measures of creativity and adjustment (Thorne and Gough, 1991)*

Observational measure	Sex	N	MBTI Scale EI	SN	TF	JP
Creativity						
In-house	M	223	−.07	.33**	−.04	.28**
	F	199	−.08	.28**	−.03	.10
ACL cluster	M	374	−.07	.17**	.01	.02
	F	240	−.09	.28**	.07	.22**
External	M	310	−.03	.29**	.02	.25**
	F	68	.08	.28*	−.06	.09
Adjustment						
Efficacy	M	374	−.22**	−.06	−.10	−.08
	F	240	−.31**	.07	−.10	−.09
Soundness	M	361	−.12*	−.20**	−.03	.14**
	F	240	−.31**	−.07	.23**	−.09

Notes:
*$p<.05$, **$p<.01$

below average. Cox concludes that factors other than the intellectual make for eminence, and that a greater social or activity interest may imply, in the achievement of eminence, a lesser interest in intellectual matters.

For personality ratings Cox used an adapted form of a scale originally used by Webb (1915), a student of Spearman who demonstrated that intellectual and personality traits give rise to different factors. He extracted a factor he called 'w' (for *will*); it seems in many ways the opposite of emotionality, anxiety or neuroticism. The matrix of intercorrelations was re-analysed by Garnett (1918) who extracted a third factor which resembles extraversion; Cox did not mention the works of Garnett, or any of the many other psychologists who re-analysed Webb's original matrix, led to do so by the excellence of his methodology. A detailed account of all this work, which is hardly ever mentioned in textbooks of personality, is given by Eysenck (1970a).

Cox again used a 7-point scale for the 67 traits used. The average reliability of the ratings is only .53 – too low for individual judgments, but adequate enough for group comparisons. The average rating of the 100 geniuses for all the 67 traits is 1.2, 'indicating that the members of the group are perceptibly above the common average in the possession of 67 good traits' (p. 177).

Cox combined appropriate traits to form aggregates, for which she calculated mean scores for the whole group. Strength or force of character (2.0) comes top, followed by Activity (1.0), Intellectual (1.6), Self (1.5), Social (1.1), Balance (1.0), and Emotional (0.7). Cox concluded that the boyhood characteristics of young geniuses indicated

> that they rate high in *intellectual*, *social*, and *activity traits*, while the *emotional* side of personality is, on the whole, not distinctly other than that of an unselected group. *Forcefulness* or *strength of character as a whole*, *persistence of motive*, and the *intellectual* traits rate conspicuously higher. The high scores on all traits containing the *persistence of motive* factor, and the *intellective* factor indicate that young geniuses possess these traits to an unusual degree. These and the summation trait of *strength* or *force of character as a whole* are the traits in which our subjects score the highest ratings; they appear to be peculiarly characteristic of dynamic vigour of character and an innate assurance of superior ability in all of the members of the group. Finally, it is perhaps significant that the single trait that rates highest among our representative youthful geniuses is desire to excel.

Cox went on to compare childhood personality of members of the different professional groups, and also the brightest with the dullest, and the most eminent with the least eminent. The professional comparisons are not of great interest, partly because of the small numbers in each category, which make comparison difficult. The results of the other comparisons however are of some interest. Comparing the 10 most eminent with the 10 least eminent in her group, Cox found that the most eminent were above the least eminent 'in all trait groups except the emotional' (p. 184). When IQ was taken into account, Cox came to a startling conclusion: 'High but not the highest intelligence, combined with the greatest degree of persistence, will achieve greater eminence than the highest degree of intelligence with somewhat less persistence' (p. 184).

When allowance is made for differences in nomenclature, these findings of the personality and interests of geniuses compare reasonably well with the personality traits of very ordinary individuals who do well on standard tests of creativity, and those of creative writers, architects, mathematicians, etc., who are well short of genius. These findings thus support the hypothesis that creativity is a widespread trait, shown in different degree by different people, somewhat independent of IQ, and correlated with certain traits of personality. Originality, lack of emotionality, individuality, imagination, rebelliousness, independence, a radical approach – these are found in common with creativity at all levels. What particularly characterizes the high achiever, the genius, in addition to creativity and high intellect is persistence, perseverance, strong activation, strength of character, forcefulness. A more modern term embracing many of these traits is 'ego-strength'; we will encounter this concept again shortly.

These results, as far as genius is concerned, link up with many common sayings, some already noted in my first chapter. Edison's stress on genius being 99 percent perspiration and only one percent inspiration; Buffon's view that 'le genie n'est qu'une grande aptitude a la patience' and many others bear witness to the public recognition of persistent, patient application as an important aspect of genius. Gruber's (Gruber and Barrett, 1974) book on Darwin well illustrates the application and hard work over many years required to produce a product of genius.

Genius, creativity and psychopathology

We must finally turn to an aspect of creative personality and genius which has already been mentioned in my first chapter, namely psychopathology; as Diderot said: 'Oh! how near are genius and madness! Men imprison them and chain them, or raise statues to them'. Early investigations of this 'divine madness' among Greeks and Romans have already been quoted; the modern period may be said to have been opened by Lombroso (1891), although several other, earlier writers like Ficino, Esquirol, Morel, and Moreau de Tours are quoted by Ochse (1991), as are opponents of this view, like Maudsley. Lombroso set the tone, typically right in principle but excessive in his claims, by declaring a close correspondence between insanity and genius. He drew up a long list of mentally disturbed geniuses, as did Lange-Eichbaum (1956), Kretschmer (1931), Ellis (1926), and more recently Prentky (1980). Prentky has described the symptoms experienced by a number of highly creative people, as well as showing how these disorders would probably be diagnosed in terms of contemporary nosology. He criticized the diagnoses made by earlier biographers and researchers, but concluded that in most cases they were quite justified in considering the behaviour of these geniuses abnormal.

Many modern writers, on the other hand, have followed Maudsley in believing that, as Kesel (1989) has put it: 'Genius, the supreme flower of human endeavour, stands so opposed to the lowering conditions of mental illness, that we can scarcely contemplate their association' (p. 198) – fine words to confront a large body of data contradicting that conclusion. Others who held the same opinion, like Adler (1927), Fromm (1955), Maslow (1976), Rank (1945), or Rogers (1976) viewed creativity as a natural tendency towards personal balance, mental health and 'self-actualization' – terms lacking proper definition and measurement even more than the concept of psychopathology. They simply refused to see that historically there was a positive correlation between genius and pathology, and rested content with using denial as a reaction to the stress imposed by the facts; thus Maslow (1976) simply classed many individuals, with pathological characteristics who had made outstanding artistic or scientific contributions as 'uncreative'! Arasteh and Arasteh (1976) found that in many studies psychopathology and creativity, defined as flexible functioning in everyday life, were indeed negatively correlated, but the studies in question dealt with schoolchildren and students who had never created anything of cultural value. There is simply no good evidence of a negative relation between pathology and genius.

To say this is not to deny that self-actualization is indeed related to creativity. Runco, Ebersole and Mraz (1991) have adduced positive evidence for a positive correlation; this review of earlier work indicates that apart from their own work there is little empirical support for the hypothesis. However, as we have seen, many studies have found a positive correlation between creativity and personality traits which form part and parcel of the notion of 'self-actualization'. The paradox is that these traits exist *together* with obvious psychopathology.

More convincing are arguments denying any relationship one way or another, or at least denying a *causal* relationship. Thus Brain (1960) and Storr (1983) maintain that the mental experiences of creative people are such that ordinary people are unable to understand them, and hence consider them pathological. Becker (1978) thought that geniuses seem to be mad because they upset widely held beliefs. Kessel (1989) has argued that Lombroso and his followers committed a 'Type Two' error, accepting the hypothesis when it was false, by failing to take into account the many geniuses who were sane. Or, according to Gedo (1972), perhaps proponents of the psychopathology argument, have deliberately chosen obviously disturbed personalities for their studies. Others have argued that creative people may feign madness in order to stand out from the rest (Becker, 1978; Nicolson, 1947), or take drugs which may simulate pathological states (McKellar, 1957).

All of these may play a small part, but they do not explain the high incidence of abnormality. More acceptable are arguments that the behaviour of eminent people is more closely scrutinized than that of ordinary people whose pathology might go undetected (Frosch, 1987; Nicolson, 1947). This also means that no non-eminent controls are available to make a proper comparison. Finally, there may be special causes for the apparently abnormal behaviour of geniuses. Newton may have suffered from lead poisoning in the paint used in his house; Faraday from the mercury which he used in large quantities in his experiments; Nietzsche and Beethoven from a syphilitic infection. Taking all their arguments into account, the careful studies of Prentky (1980) and Richards (1981) leave little doubt that highly creative people suffer a greater degree of psychopathology than would be found in the ordinary population.

The latest, and in many ways the best, attempt to apply psychiatric diagnosis to famous men on the basis of their biographies has been reported by Felix Post (1994). Basing himself on the biographies of 291 world famous scientists, composers, politicians, artists, thinkers and writers, he attempted to apply the standards of DSM-III, the widely accepted Diagnostic and Statistical Manual of Mental Disorders. His methods are subject to some criticisms which he himself acknowledged, but it is doubtful if these could be said to invalidate his conclusions. Scientists, he found, had the lowest prevalence of psychic abnormalities, but even in their case these were absent or trivial in only one-third. The amounts of psychopathology increased steadily from composers, politicians, artists and thinkers through to writers. Severe psychopathology, in the sense of interrupting work, requiring periodic rest and sometimes treatment, exceeded the incidence of less disabling disorders in the case of artists, composers and writers. The percentages of *severe* psychopathology was 18% for the scientists, 31% for the composers, 17% for the politicians, 38% for the artists, 26% for the thinkers, and 46% for the writers.

However, it is important to realize that although quite severe psychopathology is apparently much more common in these geniuses, actual functional psychoses are not. 'The low lifetime prevalence of functional psychoses (1.7%) is remarkable, as is the absence of schizophrenia' (p. 30). Thus actual functional psychoses were almost entirely absent in this group. Actual psychosis is destructive; high psychopathology short of diagnosed psychosis is apparently beneficial. This would suggest that among the geniuses considered, there would be a great deal of psychopathology in the family history, and among relatives, and Post reports results that are similar to those of Iuda (1949), She had concluded that the families of her 294 German geniuses had a much higher incidence of psychoses than the average population, and Post also found a high incidence of schizophrenia, depression, suicide, alcoholism and various other kinds of instability in the family histories of his subjects. Incidence of psychopathology is high in all the groups – scientists, politicians, composers, thinkers, artists and writers.

An important aspect of Post's findings is that his geniuses do not only show considerable psychopathology, but also marked *ego-strength*. As Post points out, 'they were powerfully driven by the urge to create. All had in common exceptional industry, meticulousness, and perseverance' (p.31). These are characteristics also noted by Cox (1926) in her 301 geniuses, and ego-strength has been found allied with psychopathology in creative individuals well below the genius class in many studies discussed in a later chapter.

Given that there is a correlation between psychopathology and genius, we may still ask the question of what is cause, what effect? Richards (1981) suggests that there are several possible patterns: (1) Galton and Esquirol suggested that creativity may lead to psychopathology (Tsanoff, 1949; Cattell, 1971), perhaps through the high degree of tension involved. (2) Psychopathology may lead to creativity – because strange ideas and distorted cognitive processes may become vital in creative works. Or psychopathology may produce the emotional drive essential to genius. (3) A third view is that both psychopathology and creativity have a common origin, presumably constitutional and genetically transmitted. My own theory lies along these lines (Eysenck, 1992a) and evidence for it will be cited later on. (4) A fourth interpretation is that stress is a factor which historically often underlies both psychopathology and creativity, hence their correlation. We shall discuss the evidence for this hypothesis later. These various theories are not incompatible, and may all play some part. The theory here preferred will be discussed in a later chapter. But before doing so it may be useful to look at more experimental and empirical studies of the problem of the pathology–creativity correlation. Just one point, however, may be raised before doing so.

Historical studies use modern psychiatric nosology in order to classify historical figures in terms of recently developed diagnostic entities. This is not a very useful occupation, in view of the very low reliability of such diagnoses, i.e. the poor agreement between psychiatrists diagnosing the same patient. Even more, the diagnostic categories used are of very doubtful value, and have little scientific backing. Both points will be discussed in a later chapter, so we will here only mention them *en passant*. They are, however, vital in uncovering the *meaning* of all the historical studies cited, as we shall see. It may be argued that the recent advent of DSM-III, an agreed diagnostic system published in the USA, has made diagnosis respectable and highly reliable; alas, the evidence is very much opposed to such opinions (Kirk & Kutchins, 1992).

An exception to this generalization is Jamison's (1993) recent book examining the literature linking artistic creativity with manic-depressive insanity. She acknowledges that there is no sharp dividing line between manic-depressive illness and a cyclothymic temperament, thus endorsing the continuity hypothesis (i.e. that there is a continuum between normal personality types and functional psychoses) which forms the background of my own theory, but she fails to extend this from manic-depressive illness to schizo-affective disorder and schizophrenia, as I have done (Chapter 6). In any case, she, as well as earlier writers (e.g. Ludwig, 1992; Martindale, 1972; Trethowen, 1977) makes a very strong case for a definite link between artistic creativity and manic-depressive illness. (Possibly scientific creativity is more closely related to schizophrenia and schizothymia, if we can consider cyclothymia to be more closely related to extraversion, schizothymia to introversion, as the associated behaviour patterns might suggest?)

Let us now look at more contemporary studies of psychopathology and creativity. It is certainly a frequent finding in studies of genuinely creative *living* people (achievement criterion) that they show evidence of what is often called 'psychopathology'. Thus Barron (1969), comparing creative groups (writers, mathematicians, architects) with representative (average, non-creative) groups, states that 'the creative groups consistently emerge as having *more* psychopathology than do more representative members of the same profession. The *average* creative writer, in fact, is in the upper 15 per cent of the general population on *all* measures of psychopathology furnished by this test (the MMPI)' (p. 72). (The Minnesota Multiple Personality Inventory is a well-known measure of psychopathology.) Thus creative writers have averaged MMPI scores of 63 (for Hypochondriasis), 65 (for Depression), 68 (for Hysteria), 65 (for Psychopathic Deviate), 61 (for Paranoia), 64 (for Psychasthenia), 67 (for Schizophrenia) and 61 (for Hypomania) – as compared with a score of 50 for the general population. Representative (non-creative) writers also had scores for psychopathological scales above the average, but below the creative writers. On ego-strength they were near average, and well below the creative writers, who showed considerable ego-strength. The concept of ego-strength is best interpreted as emotional stability, i.e. the opposite of neuroticism. It is similar to the action of 'strength of force of character' Cox found in her geniuses. (The mean population score on any MMPI scale is 50, with a standard deviation of 10.)

Andreasen (1987) looked at the rate of mental illness in 30 creative writers, 30 matched control subjects, and the first-degree relatives of both groups. The writers had a substantially higher rate of mental illness, predominantly affective disorder, with a tendency towards the bipolar type (manic-depressive illness). There was also a higher prevalence of affective disorder and creativity in the writers' first-degree relatives, suggesting that these traits run together in families and could be genetically mediated. Writers and controls had IQs in the superior range, with the writers only excelling on the WAIS vocabulary subtest, confirming the view that intelligence and creativity are independent variables once a threshold value of about 120 has been achieved.

A series of studies by Richards and her colleagues have also given some support to this thesis. Having constructed a 'peak creativity' index based on raters' assessments (the Lifetime Creativity Scales, Richards *et al.*, 1988a), these authors tested 17 manic depressives, 16 cyclothymes, and 11 normal first-degree relatives, and compared their creativity scores with that of 33 controls with no personal or family history of major affective disorder, cyclothymia or schizophrenia. Oddly enough only 15 controls were normal, while 11 'carried another diagnosis' (p. 281); the nature of this diagnosis is coyly hidden, and it is not made clear why subjects were included in the control group who were not 'normal'. All this makes the results difficult to interpret. However, it appears that creativity was related to normality – manic-depressive psychosis in a curvilinear manner; normals and manic-depressives having low scores, cyclothymes and normal first-degree biological relatives of cyclothymes and manic depressives having the highest scores. Significance levels are not impressive, but the ordering of creativity scores is certainly in line with theory – actual psychosis lowers creativity, psychopathology raises it.

Jamison (1989) contributed a study of neurotic British writers and artists showing that 38% of her subjects had been treated for mental disorder; three-fourths of these had been hospitalized, or given lithium or other antidepressants. Playwrights had the highest total rate of treatment for depression (63%). These figures should be compared with lifetime prevalence rates for manic-depressive and depressive illness in the general population of 1% and 5% respectively. Jamison's (1993) book discusses this and many other similar studies in greatest detail.

MacKinnon (1962b) found correlations with creativity in his sample of architects of MMPI 'psychopathic deviate' (.22) and 'schizophrenia' (.19). As he points out, 'the meaning of these correlations for such an effective reality-relating sample as our 124 architects, are not those which would apply in psychopathological groups. In the present context they are indicative of greater unusualness of thought processes and mental content and less inhibition and freer expression of impulse and imagery' (p. 34). This may be so, but comparing creative with non-creative architects, MacKinnon (1962b) finds the former lower than the latter on sense of well-being, responsibility, socialization, self-control, good impression, communality, achievement via conformance and sociability (as shown by the California Psychological Inventory – Gough, 1957). All these correlate with psychopathology.

Barron (1969) found that his creative writers and architects showed superior ego-strength (score of 58 for writers, 61 for architects on there MMPI), a pattern which as he points out is quite unusual; the ego-strength scale usually correlates *negatively* with the psychopathological scales (between −.50 and −.60). A similar pattern was found for the CPI scores, linking psychopathology in creative subjects with personal effectiveness. Barron contrasts psychosis and the 'divine madness' of the artist by saying that the latter is not, like psychosis, something subtracted from normality; rather, it was something added. 'Genuine psychosis is stifling and imprisoning, the divine madness is a liberation from "the consensus"' (p. 73). All this may be true, but it does not furnish us with a criterion of 'genuine psychosis' as contrasted with 'divine madness', other than the creative achievement – but of course that is what we have to explain!

The ego-strength side of the creative personality is also highlighted in the Gough (1979) Creative Personality Scale, which forms one scale of his Adjective Check List (Gough and Heilbrunn, 1983). It represents the outcome of his research with the IPAR, already mentioned many times, and contains both positive and negative items. (The scale is discussed and reproduced in Chapter 6.) These 30 items give a mean adult score of 48, according to the Manual. This scale was administered by Kaduson and Schaefer (1991) to 10 exceptionally creative women who had first been seen at high school, and retested after a 25 year follow-up (Kaduson and Schaefer, 1991). Their raw scores on the Gough test averaged 60 + / − 11 points, which is more than one standard deviation higher than the population mean, a very significant difference ($p < .01$). These results bear out those of Domino (1974), Welsh (1975) and Gough (1979); all found positive correlations between the Gough scale (or early versions of it) and ratings of creativity. Gough reports data on seven male groups and five female groups, including highly creative groups like architects, mathematicians, and research scientists, as well as relatively undistinguished groups; there are in all 1113 males and 588 females in these groups. All the correlations between the test and creativity ratings are positive and range from .15 to .42, with a mean around .30.

Another important study has been reported by Cattell and Drevdahl (1955), using Cattell's personality inventory on 144 eminent researchers in physics, biology and psychology. They are characterized by high intelligence, tender-minded sensitivity, radicalism, high self-sufficiency, low guilt-proneness, introversion, lack of inhibition, bohemianism, and a high degree of schizothymia, to represent the psychopathological element. In a similar study, 153 writers of creative literature were found to have a profile on the Cattell scales very similar to that of the scientists. The same was true of artists taken from persons listed in *Who's Who in American Art* (Drevdahl & Cattell, 1958). Cattell concludes that the most relevant aspects of proven creativity in science and art were intelligence, ego-strength, schizothymia, radicalism, introversion, dominance and self-sufficiency (Cattell, 1963). This is in good agreement with the other studies reviewed.

A number of genetic studies have supported the view that psychopathology and creativity are associated, probably through a genetic link. Heston (1966) studied offspring of schizophrenic mothers raised by foster-parents, and found that although about half showed psychosocial disability, the remaining half were notably successful adults, possessing artistic talents and demonstrating imaginative adaptations to life to a degree not found in the control group. Karlsson (1968, 1970) in Iceland found that among relatives of schizophrenics there was a high incidence of individuals of great creative achievement. McNeil (1971) studied the occurrence of mental illness in highly creative adopted children and their biological parents, discovering that the mental illness rates in the adoptees and in their biological parents were positively and significantly related to the creativity level of the adoptees.

These findings clearly support the theory developed by Hammer and Zubin (1968) and Jarvik and Chadwick (1973), to the effect that *there is a common genetic basis for great potential in creativity and for psychopathological deviation*. These studies also make it clear that actual *psychosis* works in ways that are inimical to creativity and achievement; it appears to be *psychopathology in the absence of psychosis* that is the vital element creativity.

We end this chapter with a paradox, or even several. We have found that several authors associated genius and creativity with psychopathology, often quite serious in degree (Lombroso, Ellis, Lange-Eichbaum), while other associate genius and creativity with self-actualization, mental health and personal balance (Adler, Fromm, Maslow, Rogers). Empirical evidence exists for both these contradictory conclusions, and, as the IPAR work illustrates, the usual *negative* correlations between ego-strength (which may serve as an indicator of the positive personality traits associated with creativity) and psychopathology of $-.60$ is turned into a *positive* one when creative individuals are concerned. How is this possible?

Consider the scatter diagram in Fig. 3.5. It plots a .60 correlation between ego-strength and pathology. This implies that the majority of cases fall into the high ego-strength, low pathology and the low ego-strength – high pathology quadrants. A much smaller number of cases falls into the high ego-strength – high pathology quadrant ('creative'), and the low ego-strength, low pathology (non-creative). By choosing our population to contain large numbers of

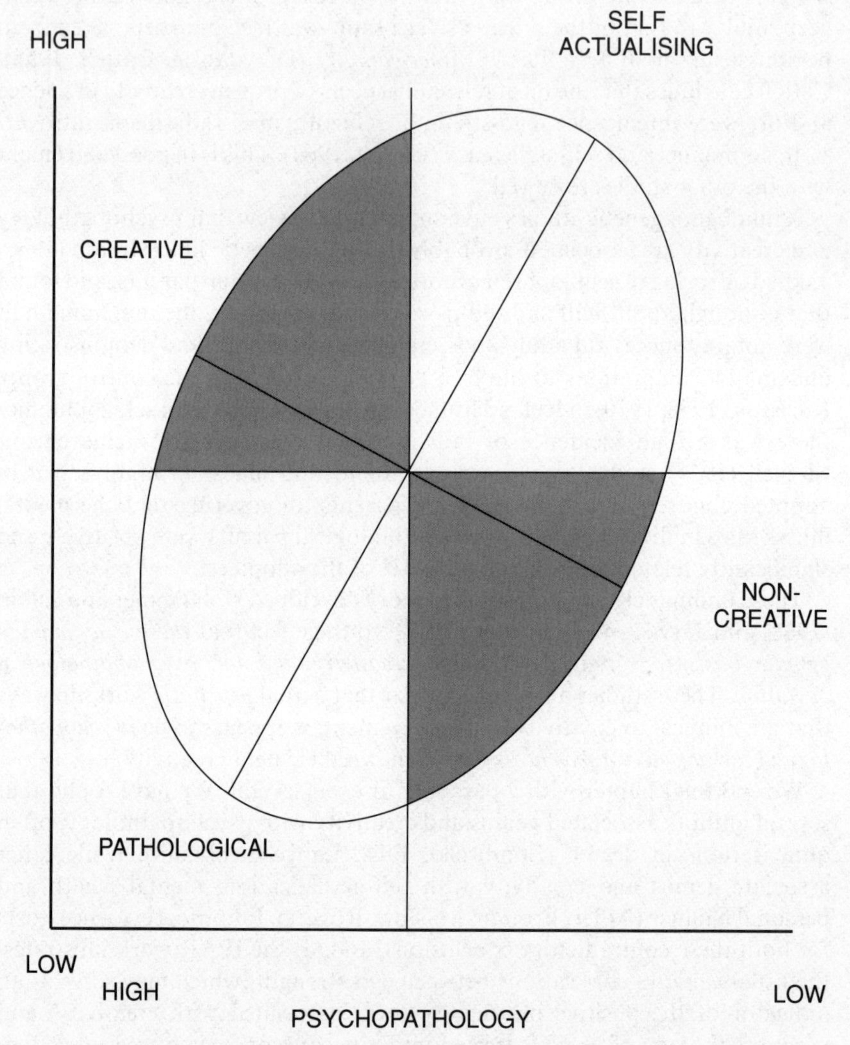

Fig. 3.5 Correlation surface of psychopathology and ego-strength.

creative people, we can change the sign of the correlation quite easily. Thus the paradox is not in fact a paradox; it appears that creative people combine in an unusual way two normally disparate traits of personality.

A similar combination of ordinarily negatively correlated variables has been studied by McKenzie (1989) and McKenzie and Tindell (in press – 1994). They found a *negative* correlation between neuroticism, and super-ego control of −.32, a high *positive* correlation between neuroticism and academic achievement in high super-ego groups, and a negative one for a low super-ego (.53 vs. −.05). In other words, just as high P needs high ego-strength to counter-balance the more pathological aspects of P, so high N needs high super-ego-strength to counter-balance the more pathological aspects of N. This suggests that simple linear correlations between personality factors and achievement are not enough; in the MacKenzie studies, and others quoted by him, these zero-order correlations are all insignificant. What is needed is a *combination* of scores based on a clearly formulated theory; given that quite reasonable and significant results emerge.

This unusual correlation of ego-strength and psychopathology may be one answer to those who fail to find schizophrenia unusually frequent among geniuses (Juda, 1949; Maudsley, 1909); thus clinical schizophrenia would be in the high pathology – low ego-strength quadrant. This is not the whole story, however, and two later chapters (6 and 7) deal with my own explanation of this particular paradox, namely that psychotics are not creative, but geniuses, and creative people generally, show quite marked psychopathology. I will also try to answer the causal question, i.e. does pathology produce creativity? And finally I shall try to answer the crucial question of *why* and *how* this is done. Without such a theory (briefly stated in the Introduction) *explaining* the observed correlations our efforts to explain genius and creativity would fall far short of the demands of science.

4 Conditions for excellence

Genius is sorrow's child.

John Adams

Works of genius depend on the confluence of certain personality variables (intelligence, creativity, persistence, etc.) and certain social conditions; Newton, Mozart or Shakespeare would not have been able to show their true genius in a primitive culture. It is the purpose of this chapter to look at a number of social and other conditions, largely external to the individual, which facilitate or hinder the development of genius. It is not always easy to say whether these conditions have a direct causal influence on the production of works of genius, or the development of the genius himself. In epidemiology we distinguish between risk factors and causes. Risk factors are variables which are significantly correlated with specific diseases, but may not be *causally* related to them. Causes must be shown to be necessary and sufficient for the disease to develop. Koch's famous postulates were enunciated to give precision to this distinction, and make it clear just what was needed for a particular factor (such as the tubercle bacillus) to be regarded as a *cause* of tuberculosis, and not just as a risk factor.

In the scientific study of complex phenomena, this distinction is absolutely vital; we search for *causal* elements, but are only too often satisfied with what are mere risk factors. An example may make clear the distinction. In Denmark there has been observed over many years a correlation between the birth of a baby, and the presence of a stork's nest on the roof of the house where the baby was born. The presence of the nest is a risk factor, but has no causal relevance; prospective parents often put up such a nest in the hope that a stork will choose it for his own. The difficulties of making the distinction may be illustrated by reference to cancer of the cervix. This is correlated with smoking, but is smoking a causal factor (Eysenck, 1980)? Cancer of the cervix is also correlated with sexual promiscuity; it is quite frequent in prostitutes, absent in nuns. Is promiscuity a causal factor? Smoking is correlated with promiscuity, perhaps one or the other is correlated indirectly with cancer of the cervix only because it is also correlated with the true cause. Or perhaps both are innocent and are linked with cervical cancer through the agency of extraversion, which is correlated both with smoking (Eysenck, 1965) and with promiscuity (Eysenck, 1976)? Extraversion, in turn, is correlated with diet, life style and many other variables, any or all of which might exert a causal influence – or might not! The problem is an extremely difficult and complex one, and medical people often treat it the way Alexander the Great dealt with the Gordian Knot – by simply slashing at it. Thus smoking has been elevated to the *cause* of cancer and coronary heart disease, although the evidence only makes it a *risk factor* – there is no direct evidence to demonstrate causality (Eysenck, 1991a).

The same difficulty arises when we discuss the conditions which affect genius. As we shall see, middle-class origin is closely correlated with the emergence of genius, but *why* does such a relationship exist? Is it that middle-class and professional parents have a higher IQ than working-class parents (Eysenck, 1979), and that this IQ is inherited by the prospective genius? Or does the environment – presence of books, intelligent conversation, better schooling – contribute a causal shove in the right direction? Or is it the presence of financial independence which is important in making lengthy study possible? Rost (1993) compared gifted and non-gifted children equated for socio-economic status and found little difference on any of the environmental factors investigated. The statistical association is merely the first step in leading to a proper understanding; unfortunately many investigators are content with leaving things there once a statistically significant correlation is obtained, disregarding the fact that such a correlation presents us with a *problem*, not an *answer* – what we should be looking for is a causal explanation of the observed correlation.

These few words are intended as a warning not to take the correlational studies to be cited as necessarily indicative of causal influences; to establish causation we need to go beyond correlations, and find something like Koch's postulates to guide our way. In the absence of such postulates, we shall have to do the best we can in assessing the value of the material presented. Indeed, it seems very unlikely that any such postulates exist. Tuberculosis is produced by the tubercle bacillus; hence it makes sense to say that it is never produced in its absence, but always in its presence. Even this is not quite correct; in healthy persons, living in a healthy environment, the bacillus may be present without causing tuberculosis – the bacillus needs a proper soil in which to grow. But single-cause effects are rare in psychology and the social sciences generally; I have tried to indicate the complexity we encounter in our discussion of genius in Fig. 1.4. Where there are many factors contributing, factor A's abundance may partly offset factor B's poverty. These difficulties should always be borne in mind. The factors to be discussed are different from such personality attributes as intelligence, creativity and persistence, in being largely risk factors – correlated in varying degrees with genius, but with possible causal links unexplained.

Socio-economic status (S.E.S.)

There is much agreement that for all historical epochs of which we have some knowledge, creative people, in most disciplines and in different societies, have in the main come from the so-called middle and upper-middle classes, more particularly from professional-class homes. Albert (1975), Cattell and Brimhall (1921), Chambers (1964), Cox (1926), Moulin (1955), Raskin (1936), Simonton (1984b) and Zuckerman (1977) provide the evidence for this generalization. Thus the fathers of Cox's (1926) sample of historically eminent men were noblemen or professional men in over half the cases; skilled workmen, lower businessmen, semi-skilled and non-skilled fathers all added up to only less than 19%. Raskin's (1936) sample of eminent writers and scientists in the eighteenth century showed 54% of fathers to come from nobility and professional classes, only 17% from skilled and clerical workers, farmers and artisans, or unknown.

These are historical samples. For present times, Roe (1951a,b, 1953) also found 53% of fathers coming from the professional classes, in her sample of contemporary creative scientists. Zuckerman (1977) looked at fathers of American Nobel Laureates, again finding 54% of professional men. (This compares with a percentage of 29% among science doctorates, and 3.5% among employed males.) There is little doubt about the existence of a *correlation* between success and parental status; the problem is to account for it.

It is often suggested that IQ, inherited from high IQ parents, may be an important factor. There is much evidence of high IQ in the professional classes (Eysenck, 1979), and also of the high heritability of IQ (but of course subject to regression effects – Eysenck, 1989b). We may safely assume that IQ plays an important part in this association. Provision of books and other intellectual support is possibly also related to parental profession and IQ, as is stress on educational success.

Financial status of the parents is another possible factor. There are reasons to doubt this. (1) Among people with doctorates in science, far fewer come from professional homes than would be true of Nobel Laureates, yet they too must have had sufficient financial background to study extensively. (2) The recent trend to award bursaries, grants and fellowships, to able students with limited means has increased the number of scientists coming from working-class families, but has not had such an effect on Nobel Laureates. (3) Seventy-five per cent of Jewish Nobel Laureates came from lower socio-economic backgrounds (Zuckerman, 1977), suggesting that it was the motivational properties of the family, rather than their financial status, which produced the drive to eminence of the Nobel Laureates.

Gender

Creativity, particularly at the highest level, is closely related to gender; almost without exception, genius is found only in males (for whatever reason!). Illustrations abound. In the list of geniuses studied by Cox, (1926) there are no women. There are no women among Roe's (1951a,b, 1953) eminent scientists, and very few in American Men of Science, or among members of the Royal Society; none in a list of the leading mathematicians (Bell, 1965), and none would be found among the 100 best known sculptors, painters, or dramatists. Simonton (1992) found no women in a list of the most famous 120 composers, from the Renaissance to the twentieth century, and hardly any women among his scientists. It is only among poets and novelists that a small proportion of women would be found in the top class. Over (1982) and Rushton (1989, 1992) have published figures for productivity for men and women in Departments of Psychology, and also for number of citations, showing very marked and in fact increasing disparity between the sexes. In Great Britain, the mean number of publications for men was 1.5 per annum, for women 0.9. Citations averaged 12 and 6 respectively. The divergence between males and females was larger than in Over's (1980) previous study. Of the 26 most cited authors, only two were women.

Reasons for this disparity are not all that obvious. It is sometimes said that being married and having children is the responsible factor, but Cole and Zuckerman (1987) found that 'women published less than men, but marriage and family obligations do not generally account for the gender difference. Married women with children publish as much as their single female colleagues' (p. 119).

A more likely reason is psychopathology. Psychoticism (Eysenck, 1992a) is conceived as being a dispositional trait underlying schizophrenia and manic-depressive illness; males usually score twice as high on it as compared with females. As we shall see in a later chapter, psychoticism is also closely related to creativity, so that the lower creativity of women may be due to their lesser psychopathology. (Women do have higher scores on neuroticism, which is conceived as a dispositional trait underlying proclivity to neurotic disorders, but these are quite different from psychotic-type disorders, and probably not related to creativity.)

One possible contributor to the problem of male superiority in genius-like performance may lie in a field seldom mentioned, namely general intelligence. Recent work by Ankney (1992) and Rushton (1992) has shown that males have on average larger brains than females, even when adjustments are made for body size, and brain size is positively correlated with intelligence (Willerman *et al.*, 1991). Yet most textbooks assert that there are no sex differences in IQ. Thus Matarazzo (1972) points out that the conclusion that no differences exist 'was reached by all previous reviewers as well as by the present writer' (p.

352). As he points out, 'from the very beginning developers of the best known individual intelligence scales (Binet, Terman, and Wechsler) took great care to *counterbalance* or *eliminate* from their final scale any items or subtests which *empirically* were found to result in a higher score for one sex over the other' (p 352). Thus on a test like the Wechsler its very mode of creation would seem to rule out any differences in IQ between the sexes. How can we resolve this apparent contradiction?

Lynn (1994) decided to submit to the test the assumptions that on the Wechsler test females and males did indeed have equal overall scores, as well as the often made hypothesis that males achieve higher scores on spatial abilities, but lower scores on verbal abilities than females (Halpern, 1992). If such differences in IQ could be found in a test which was constructed so as to eliminate any such differences, this would surely indicate a *minimum* difference between the sexes. Lynn gives the result of eight standardization groups for the Wechler tests in various samples and countries (USA, Scotland, China), and found that

> in all eight studies males obtained a higher Full Scale IQ than females and in each case the difference is statistically significant. Males also obtained higher means than females on both verbal and performance IQ, contrary to received opinion which holds that females have higher verbal abilities than males. The medians of the male advantage in IQs are 2.35 (Full Scale), 3.15 (Verbal) and 1.00 (Performance).

This overall advantage in IQ can be increased by 40% by weighting the subscales in terms of their *g* loadings, i.e. giving greater weight to the scales providing a better measure of intelligence (Jensen & Reynolds, 1983).

Lynn reports results from four methods of measuring sex differences in intelligence; all methods show that males have higher mean IQs than females, with the range of the male advantage lying between 2.35 and 4.65 IQ points. He also calculates that the advantage accruing from greater brain size should lie between 2.9 and 4.1 IQ points. Allowing for the fact that Wechsler made every effort to *equalize* IQs between the sexes, and rejected items and subtests which violated the assumption of equality, we may perhaps say that an IQ difference of four points would be a conservative estimate of the *true* difference. This may appear too small a difference to matter, or be relevant to the problem of male genius predominance, but there are two considerations which argue against such a view.

The first point is that small differences around the mean become greatly increased at the extremes of the distribution. What would a difference of four points around the 100 IQ mark mean at an IQ of 160, which Cox decided was the mean of new geniuses? In such a calculation we must take into account the well-attested fact that females have a lower variance than males, i.e. they tend to congregate around the *mean* and have fewer very high or very low scores. Assuming a perfectly normal Gaussian curve of distribution, a mean difference of four points of IQ, and a standard deviation of 15 for the males, 14 for the females, the expected probability of scoring 160 and above is 0.0004743 for the females and 0.0055167 for the males. In other words, in 10000 individuals randomly selected, there would be five females, but 55 males with an IQ as high as that or higher. A difference of a whole order of magnitude would certainly be of considerable importance in relation to producing works of genius.

Equally important is another consideration. I have argued that the many factors involved in producing works of genius interact multiplicatively or synergistically; hence even relatively small differences on any *one* factor could assume great importance when using it as a multiplier for a whole series of other factors. These two considerations combined suggest that it would be quite wrong to dismiss sex differences in intelligence from consideration. Neither should we assume that the argument presented here may not be subject to informed criticism. All that is asserted is that the usual assumption of equal IQ from the two sexes may be flawed, and that even apparently quite small differences may have far-reaching consequences in relation to past events, such as the production of a genius.

This, of course, is not all. Benbow and Lubinski (1993) have taken mathematical reasoning, a discipline in which gender differences are particularly large (in favour of males), and have compared large samples of young men and women who scored equally in the top ½% of the population. They then investigated vocational interests and personal value orientation. It is known that investigative interests and theoretical values are among the most salient personal preferences of physical scientists. Also it is known that a high theoretical interest complemented by a higher aesthetic orientation is correlated with scientific creativity (Southern and Plant, 1968). Benbow and Lubinski (1993) found that theoretical values, which are characteristic of physical scientists, are more characteristic of gifted males than gifted females (Lubinski, Benbow and Sanders, 1993). Social values, which are negatively correlated with interests in physical science, are more characteristic of gifted females than males. The interest pattern of males were primarily focussed around investigative areas, those of females were more evenly distributed across investigative, artistic and social areas (Benbow and Lubinski, 1992). Mathematically talented females tend to have interests in people, males in things (Lubinski and Humphreys, 1990). Furthermore, there were huge differences in committment to full-time work: 95% of the gifted males versus 55% of the gifted females planned to work full-time until retirement (Benbow and Lubinski, 1992). As Benbow and Lubinski (1993) state:

> This finding must be taken into account when addressing all forms of gender differences in achievement and career advancement; as long as the genders differ so markedly in their committment to full-time work, marked differences in achievement and promotion are sure to remain, even among males and females whose relevant ability/preference profiles are equivalent (pp. 6–7).

Even if the data quoted by Lynn should turn out to be in error, there clearly are (still) powerful differences between males and females which point the males in the direction of greater achievement. These differences are often said to be due to social and cultural pressures, and that may well be true. The possibility of a genetic differentiation along these lines should not be ruled out; there is a good deal of evidence in that direction (Eysenck and Wilson, 1979). Much research will need to be done before the facts are all in; at the moment no definitive statement is possible.

Religious denomination

While many scientists are agnostics or atheists (Chambers, 1964; Helson and Crutchfield, 1970; Lehman and Witty, 1931), the religion practised by their parents shows quite high correlations with achievement. There is no doubt that Catholics are less likely to achieve eminence in science than Protestants, and Protestants than Jews. The studies of Arieti (1976), McClelland (1963), Roe (1951a,b, 1953) and Zuckerman (1977) amply testify to this conclusion. Four per cent of Americans are Catholics, but only 1% of Nobel Laureates have been. Only 3% of Americans are Jewish, but 27% of American Nobel Laureates have been Jewish (Zuckerman, 1977). This is a disproportion of 1:36; in other words, your chance of winning a Nobel Prize in science is 36 times higher if you come from a Jewish home than if you come from a Catholic home. A similar disproportion exists when we compare Protestant countries (England, Scandinavia, Northern Germany) with Catholic countries (Italy, Spain, France).

Religion *per se* is clearly not the major determinant, seeing that the scientists themselves are seldom religious. We have a choice between differences in intelligence and differences in culture. As Ochse (1990) puts it: 'The question remains as to whether the acquisition of (intellectual) skills depends on native intelligence, or on social transmission of a sense of values that inspires the young to develop their intellect' (p. 63). She argues against the former hypothesis, on the grounds that in the Terman longitudinal studies of gifted children, it was found that gifted children who were eventually most successful in their university and professional achievements were not distinguished from the less successful on the basis of IQ or academic achievement at school (Terman and Oden, 1947); the difference lay in their family backgrounds. The successful children came from families with a much stronger educational tradition; in addition, the successful contained proportionately three times as many Jewish children as did the unsuccessful group. Terman argued that Jewish children are under pressure to succeed, so that following this pressure they do better in life.

It would be unreasonable to deny the importance of such cultural factors; motivation is undoubtedly vital to achievement, and parental pressure may increase motivation. On the other hand, it may also have the opposite effect; as we shall see, externally induced motivation may be much less effective than internal motivation, and may actually be harmful. The available data certainly do not justify us in rejecting the importance of IQ; we must take into account the fact that the *range* of IQ in the Terman group is severely restricted, so that IQ correlations with external criteria would be lowered drastically. Also of course none of the children was a genius, or even a Nobel Laureate! We now know enough to ask the appropriate questions in this field; we certainly do not know enough to give confident answers. It is relevant to note, however, that Jewish children have notably higher IQs than non-Jewish children (Storfer, 1990).

Home environment

A more detailed look at the effects of home environment will make clear the importance of this statement. The evidence to be reviewed immediately reveals a paradox. On the one hand, creative achievers typically receive much intellectual stimulation in their childhood homes (Albert, 1978; McCurdy, 1957). On the other hand, a disproportionate number of creative achievers lost one or both parents in childhood (Albert, 1971; Eisenstadt, 1978). While not completely at variance (it is possible for a child to receive intellectual stimulation *before* the parent dies), early bereavement does not fit well into the idyllic picture of a home devoting much energy to the stimulation of the early intellectual groping of the young child. Yet both sides of the paradox are well documented.

Thus Eisenstadt (1978) studied 699 famous historical figures and found that one in four had lost at least one parent before the age of 10. By the age of 15 the loss had exceeded 34%, and 45% before the age of 20. These losses almost certainly exceed those suffered by the average citizen of those times, although estimates of life expectancy are hard to come by until quite recent times. We can compare these figures with those covering the beginning of the twentieth century, and find that death of mother or both parents by the age of 15 was three times more frequent in the sample of eminent people than in the general population. The conclusions of Albert (1980) were similar, for artistic and scientific achievers as for politicians and eminent soldiers. He found a threefold rate of parental loss, as compared with twentieth-century populations. (Delinquents and suicidal depressives achieve approximately the same proportion of childhood bereavement – Albert, 1980.)

Figures from past centuries, impossible to compare with accurate population estimates, may not be very impressive, but Roe (1953) also found a three times greater loss of parents in her scientists before the age of 16 than was characteristic of the general population in America 10 years later (26% vs. 8%). The disproportion would probably be even greater if socio-economic status were accurately controlled. Seltzer and Jablon (1977) found that commissioned officers had a 40% advantage over privates in mortality rates in a 23-year follow-up, and many other studies of army personnel as well as civilians have shown considerable advantages for professional and other white-collar groups over blue-collar groups as far as mortality is concerned (Everson and Fraumeni, 1975; Keehn, 1974; Moriyama, Kreuger and Stamler, 1971; Kitagava and Hauser, 1973).

Delinquents and suicidal depressives share with geniuses the personality trait of *psychoticism*, i.e. a dispositional trait leading to psychopathology, as I shall argue later on, and this may be associated with early death of parent(s), either through genetic causes (high-P parents are more likely to die young, and also to hand on their high-P genes to their children), or through the psychological reactions of the children. As Ochse (1990) says, 'the implications are that the death of a parent may have some influence on either adaptive or maladaptive development' (p. 75). Eisenman (1991) has suggested that creative achievement, delinquency and suicide all express dissatisfaction with society; he sees creative achievement, which of course implies criticism of the status quo, as an attack on society. This agrees well with the personality of the high-P scorer (Eysenck, 1992a) who is typically non-conformist and does not believe in society's rules and regulations. This attitude is also characteristic of the creative person and the genius (Eysenck, 1993a).

Other suggestions have been that the pursuit of excellence may be an attempt to over-compensate for the anxiety and the feelings of guilt and unworthiness the child may experience as a consequence of the bereavement (Berrington, 1983). Or the loss may produce a need for power (Albert, 1980; Eisenstadt, 1978); indeed, Veroff *et al.* (1960) claimed that, as compared with males who had been bereaved of a parent before the age of 10, those not so bereaved had significantly *lower* levels of power motivation. Unfortunately this finding is based on the Thematic Apperception Test, which shares elements of subjectivity, unreliability and doubtful validity with other projective techniques, such as the Rorschach Inkblot Test.

Intellectual stimulation

The opposite pole of the bereavement end is intellectual stimulation in the home, usually by parents, but also sometimes by siblings. Albert (1978) and McCurdy (1957) have commented on the large number of geniuses educated by their parents, who often devoted much of their time to the task. John Stuart Mill, Pascal, Mozart and many others are often cited in this connection. Even when there is no direct teaching and guidance, role models may appear in the home, or books displayed which may attract the attention of the clever child. Yet in spite of the impressive nature of the many case-histories illustrating the point, we must remember that for every example of the importance of the home environment, it is possible to cite several negating it. A hostile home environment, inimical to music did not deter Berlioz; the home environments of the great mathematicians cited in an earlier chapter did not usually contain any particular stimulation in the direction of their life interest. No proper study has been reported which would assign home environment a proper place and weight in the histories of geniuses. Mozart was coached from an early age by his father, but Wagner did not even develop an interest in music till he was 15. John Stuart Mill was coached by his father from an early age, but Kant was not. Let me tell the story of a creative genius in the biological sciences which well illustrates how little genius may need the intellectual stimulation of a good professional home. I have on purpose kept this book free from reliance on biographical material, because it can prove but little; however, the life history of George Washington Carver may be illuminative and suggestive. It proves nothing, but may urge more careful consideration of the environmental factors that are claimed to produce geniuses. (I have told this story before in *Psychology is About People* (Eysenck, 1972a), and quote it here, with permission.)

Our hero was born during the American Civil War, son of Mary, a Negro slave on a Missouri farm owned by Moses Carver and his wife Susan. Mary, who was a widow, had two other children – Melissa, a young girl, and a boy, Jim; George was the baby. In 1862 masked bandits who terrorized the countryside and stole livestock and slaves attacked Carver's farm, tortured him and tried to make him tell where his slaves were hidden; he refused to tell. After a few weeks they came back, and this time Mary did not have time to hide in a cave, as she had done the first time; the raiders dragged her, Melissa and George away into the bitter cold winter's night. Moses Carver had them followed, but only George was brought back; the raiders had given him away to some womenfolk saying 'he ain't worth nutting', Carver's wife Susan nursed him through every conceivable childhood disease that his small frame seemed to be particularly prone to; but his traumatic experiences had brought on a severe stammer which she couldn't cure. He was called Carver's George; his true name (if such a concept had any meaning for a slave) is not known.

When the war ended the slaves were freed, but George and Jim stayed with the Carvers. Jim was sturdy enough to become a shepherd and to do other farm chores; George was a weakling and helped around the house. His favourite recreation was to steal off to the woods and watch insects, study flowers, and become acquainted with nature. He had no schooling of any kind, but he learned to tend flowers and became an expert gardener. He was quite old when he saw his first picture, in a neighbour's house; he went home enchanted, made some paint by squeezing out the dark juices of some berries, and started drawing on a rock. He kept on experimenting with drawings, using sharp stones to scratch lines on the smooth pieces of earth. He became known as the 'plant doctor' in the neighbourhood, although still only young, and helped everyone with their gardens.

At some distance from the farm there was a one-room cabin that was used as a school house during the week; it doubled as a church on Sundays. When George discovered its existence, he asked Moses Carver for permission to go there, but was told that no Negroes were allowed to go to that school. George overcame his shock at this news after a while; Susan Carver discovered an old spelling-book, and with her help he soon learned to read and write. Then he discovered that at Neosho, eight miles away, there was a school that would admit Negro children. Small, thin and still with his dreadful stammer, he set out for Neosho, determined to earn some money to support himself there. Just 14 years old, he made his home with a coloured midwife and washerwoman. 'That boy told me he came to Neosho to find out what made hail and snow, and whether a person could change the colour of a flower by changing the seed. I told him he'd never find that out in Neosho. Maybe not even in Kansas City. But all the time I knew he'd find it out – somewhere.' Thus Maria, the washerwoman; she also told him to call himself George Carver – he just couldn't go on calling himself Carver's George! By that name, he entered the tumbledown shack that was the Lincoln School for Coloured Children, with a young Negro teacher as its only staff member.

The story of his fight for education against a hostile environment is too long to be told here; largely self-educated he finally obtained his Bachelor of Science degree at the age of 32, specialized in mycology (the study of fungus growths) became an authority in his subject, and finally accepted an invitation from Booker T. Washington, the foremost Negro leader of his day, to help him fund a Negro university.

He accepted, and his heroic struggles to create an institute out of literally nothing are part of Negro history. He changed the agricultural and the eating habits of the South; he created single-handed a pattern of growing food, harvesting and cooking it which was to lift Negroes (and whites too!) out of the abject state of poverty and hunger to which they had been condemned by their own ignorance. And in addition to all his practical and teaching work, administration and speech-making, he had time to do creative and indeed fundamental research; he was one of the first scientists to work in the field of synthetics, and is credited with creating the science of chemurgy – 'agricultural chemistry'. The American peanut industry is based on his work; today this is America's sixth most important agricultural product, with many hundreds of by-products. He became more and more obsessed with the vision that out of agriculture and industrial waste useful material could be created, and this entirely original idea is widely believed to have been Carver's most important contribution.

The number of his discoveries and inventions is legion; in his field, he was as productive as Edison. He could have become a millionaire many times over but he never accepted money for his discoveries. Nor would he accept an increase in his salary, which remained at the 125 dollars a month (£100 per year) which Washington had originally offered him. (He once declined an offer by Edison to work with him at a minimum annual salary of 100 000 dollars.) He finally died, over 80, in 1943. His death was mourned all over the United States. The *New York Herald Tribune* wrote: 'Dr, Carver was, as everyone knows, a Negro. But he triumphed over every obstacle. Perhaps there is no one in this century whose example has done more to promote a better understanding between the races. Such greatness partakes of the eternal.' He himself was never bitter, in spite of all the persecutions he and his fellow-Negroes had to endure. 'No man can drag me down so low as to make me hate him.' This was the epitaph on his grave. He could have added fortune to fame, but caring for neither, he found happiness and honour in being helpful to the world.

This brief story of a great scientist and a fine human being raises some very fundamental problems. Every year colleges and universities in the USA produce tens of thousands of agriculturalists, biologists, biochemists and other experts in the fields in which George Carver worked. Every one of these has a family background, an education, and a degree of support compared with which Carver's would simply have been non-existent. His father dead before he was born; his mother abducted while he was a baby; born a Negro slave in the deep South, weak and ailing; growing up in a poverty-stricken house with hardly any books, with the white people who brought him up not far from illiterate; denied schooling because of his colour, having to piece together the rudiments of an education while constantly hungry, and having to earn every penny he spent by performing the most menial jobs imaginable; exposed all the time to recurring traumas because of his colour; troubled by a severe stammer assumed to have been brought on by his early abduction under extremely unfavourable weather conditions (to say nothing of his emotional reactions); rejected because of his colour by institutes of higher learning; always having to work his way through secondary school and college; this kind of handicap is practically unknown now – however poor the education given to Negro children in the USA today. And compared with the education of Negro children, that of the favoured white boys and girls who present themselves with shining faces at the commencement ceremonies at American colleges and universities has been exemplary – incorporating all the advances that modern educational science has been able to think up. And all these educational advances are linked with, in most cases, happy, peaceful childhood experiences under the wise guidance and care of loving parents.

On the basis of an environmentalistic hypothesis, what wonders would we not expect these prodigies to perform! Surely soon the world will be completely changed by their discoveries – each one of them many times as productive, as inventive, as sagacious as the poor, ignorant Negro boy with his botched education and his non-existent family life! But reality teaches us that out of all these tens of thousands of molliecoddled youngsters, with all their highly favoured upbringing, their high standard of education, their impeccable family background, not one is likely to achieve even a tithe of what the untutored, self-taught George Washington Carver managed to do. Something, one cannot but feel, has gone seriously wrong; if environment is so all-powerful, then how can the worst imaginable environment produce such a wonderful human being, so outstanding a scientist, and how can the best type of environment that oceans of money can buy and the top brains in education conceive, produce so vast a number of nonentities, with perhaps a few reasonable scientists sprinkled among them?

The argument is sure to be raised that it is precisely the adversity that George Carver encountered which was responsible for his outstanding success; it is this which put steel into his soul. 'All the modern youngsters have had it too easy; what they need is a period of fending for themselves.' Does this amount to a suggestion that our educational system should return to single-teacher shacks; that all children should be made to earn their living while attending school; that children are better separated from their parents at an early age and made to fend for themselves? Unless it means that, as well as making sure that every child is persecuted because of its colour, the hypothesis in question means very little. And in any case it does not account for George's brother, Jim; he too encountered the same degree of deprivation and adversity, but he never learned to read and write, and became a witless shepherd, never showing any sign of even average intelligence. 'The flame that melts the wax tempers the steel.' Does not this saying carry with it the admission that wax and steel are constitutionally different?

George Carver Washington's history may be extreme, but it has often been noted that, contrary to the theories of Adler, Rank, Rogers, Fromm and Maslow, it is not necessary, and may be a negative factor in the development of a creative person, to have loving, supportive parents, a happy home environment and parental stimulation in the cognitive field. The IPAR researchers found all this in bright, well-adjusted individuals of low creativity; their most creative subjects had suffered many traumatic experiences, brutality, deprivation and frustration in childhood. These were contemporary highly creative persons; Goertzel and Goertzel (1962) studied 400 eminent historical figures and found that 75% of them had suffered broken homes, rejection by their parents, many of them over-possessive, estranged or dominating. More than one in four had a physical handicap. In a later study, Goertzel *et al.* (1978) found that 85% of 400 eminent people in the present century had come from highly troubled homes – 89% of the novelists and playwrights, 83% of the poets, 70% of the artists and 56% of the scientists. In a similar vein, Berry (1981) found that literary Nobel Laureates came more often from poor backgrounds, and suffered physical disabilities than did scientific Laureates; emotion recollected in tranquillity needs strong emotions to have been experienced in the first place!

Affection, attachment, warmth, closeness between creative achievers and their parents were certainly more likely to be absent in the scientists studied by Cambers (1964), Roe (1951a) and Stein (1962), the psychologists studied by Drevdahl (1964) and Roe (1953), and the architects studied by MacKinnon

(1962b). Scientists in general were usually feeling distant from their parents, psychologists on the other hand experienced actual open hostility and rejection. Perhaps it is this experience that led them into the study of psychology!

The conclusion is often drawn that it is the treatment received by the child that determines his or her future conduct. This may be true, at least in part, but the facts do not enable us to decide between several alternative hypotheses. The environmentalist hypothesis has just been mentioned; it is almost the only one that is commonly considered. But two others are equally likely, or even more so, and should not be overlooked. The genetic hypothesis states that high-P parents, i.e. parents high on psychoticism, are likely to have high-P children; genetic factors, as we shall see, play a dominant part in a child's personality (Eaves, Eysenck and Martin, 1989). This hypothesis would explain both the distant, cold, rejecting behaviour of the parents, and the rebellious, non-conformist, egocentric behaviour of the child destined to grow up into the highly creative genius. There is much evidence for the inheritance of personality traits, including P; there is no acceptable evidence for the postulated environmental influence of paternal behaviour. Such an influence would be an example of between-family environmental determination, but the evidence clearly indicates that practically all the environmental determination of personality and behaviour is of the within-family type (Eaves *et al.*, 1989)! None of the writers on creativity and genius I have come across have noted these vitally important caveats, yet the literature about them is by now quite large, and almost unanimous. It may be politically correct to disregard it, but hardly in the best tradition of science!

There is a third possibility to account for the hostile or distant behaviour of the parents of creative children. Such children, as we shall see, are likely to be high-P, and hence very difficult to get on with; perhaps the parents' behaviour is their reaction to the behaviour of an exceptionally difficult, undersocialized, non-conformist child resistant to all loving approaches. Of course all three possibilities may have an element of truth in them; I am merely concerned to point out that we should not make a premature and prejudiced choice.

The evidence now available may also be used to throw doubt on another favourite environmentalist belief. It is often assumed and asserted that rigid discipline, authoritarian control and frequent criticism are likely to inhibit originality and the development of creativity (Anderson and Cropley, 1966; Getzels and Jackson, 1963; Nicols, 1964; Weisberg and Springer, 1961). But the children studied were not geniuses, or even Nobel Laureates; in fact they have shown no evidence of creative achievement. When we turn to really creative persons, we find that very many of them perceive and remember their parents as overbearing and tyrannical (Getzels and Jackson, 1962; Illingworth and Illingworth, 1969; MacKinnon, 1978; Roe, 1951a, 1953). Perhaps, as Roe (1953) suggests, such treatment may lead to rebellion and through it to creative behaviour. Neither do the data support the view that parents should foster a child's sense of autonomy in order to make him or her more creative; there is no evidence to support such a view when we consider true creativity. All these assumptions follow from a humanistic, environmentalistic framework which in essence contradicts many known facts. It is time we disabused ourselves of beliefs which may be comforting but which have no basis in reality.

Most of the data quoted have been culled from studies of *male* geniuses; there being a dearth of females in this group. Even when we turn to Nobel Laureates in science and medicine, there are so few females that no generalizations can be derived from them. Even in the more accommodating ranks of university professors, women have been found to publish much less than men. We may note, however, that in many ways creative women resemble creative men. They usually come from professional-class houses, respect cultural values, are stimulated intellectually by their fathers, but seldom form emotionally warm relationships with their parents (Helson, 1971). They tend to be more introverted, aloof and self-sufficient than less creative women (Bachtold and Werner, 1970, 1973). 'It seems, in fact, that the pattern of influences acting upon the development of creative ability in women is similar to that which determines the development of creativity in men' (Ochse, 1990, p. 173).

The age factor

Age as a factor in creativity and the production of works of genius is socially important, and has been studied from two quite different points of view. The first is related to the question of supreme achievement; at what age does genius produce its best work? The second is a more mundane but nevertheless important question; does age at death influence a person's probability of being considered a genius? If the concept of genius is based on social recognition, factors unrelated to the pure value of the work produced may (and do!) play an important part.

The study of age and achievement owes much to the early work of Beard (1874), whose study of more than a thousand biographies of eminent persons laid the foundations for this line of work. He noted carefully at what age they produced their major contributions. He found that achievement tended to increase fairly rapidly from 20 to 40, with a peak just short of 40, and then declined very slowly over the years. This work was taken up by Lehman (1953), whose calculations are more systematic, objective and quantitative, although his overall conclusions do not differ much from those of his predecessor. Simonton (1984b) cites many more recent studies of special groups (classical composers – Simonton, 1984b; psychologists – Lehman, 1966, Lynn, 1968, Zusne, 1976; mathematicians – Dennis, 1966, Cole, 1979) which confirm the shape of the regression curve. Quantity of production declines much less than quality (Dennis, 1966; Cole, 1979), a finding that agrees well with Lehman's own. Superficially obvious, such a result is not easy to account for, particularly as different people are involved in the comparisons in question. What is needed are studies looking at the ratio of outstanding work over total work published by selected scientists or artists over the course of their lives.

Such studies have been reported by Simonton (1977b) for 10 outstanding classical composers, Simonton (1983a) for 10 eminent psychologists, and Cole (1979) for scientists in six fields. All these studies show a significant correlation between quantity and quality, ranging roughly from .40 to .75. Simonton (1984b) interprets these findings as supporting the constant-probability-of-success model (Dennis, 1966), which maintains that the proportion of top-quality achievements remain relatively constant during a creator's career; the more good works are produced, the more bad works will be produced (Simonton, 1984b). As Dennis (1954) put it, 'Other things being equal, the greater the number of researches, the greater the likelihood of making an important discovery that will make the finder famous' (p. 182).

Simonton (1984b) argues that this model is theoretically compatible with Campbell's (1960) blind-variation and selective-retention model of creative thought which we have already encountered. If ideas are randomly produced, finding a good one will be a function of the quantity of ideas – the more the better. Thus the success of an individual creator is dependent upon his ability to generate big ideas, each representing a certain permutation of small ideas. 'The more permutations that are generated, the higher the odds that a particular formulation will survive the winnowing process imposed by posterity' (p. 83). The less prolific the creator, the less likelihood of leaving behind ideas surviving this selection process. Thus this model would predict that quality is a probabilistic consequence of quantity, giving a proper foundation for the empirical findings.

The conclusion may be correct, but it may also be premature. The correlations from Simonton's own work show less than 25% of the variance for quality to be accounted for by quantity; this is weak support for the theory, leaving 75% unexplained. Again, alternative hypotheses are possible. In science (and plausibly in art too) a novel idea is usually followed by several studies designed to carry the original idea further, defend it against attack, and apply it to various settings – all works of much lower originality. If these supporting studies are lacking, leaving us with a 'one off' product, this is much less likely to be honoured, and is likely to be overlooked. The theory has merit, but needs much more support than it has received so far.

While the general regression patterns of productivity on age is universal, there are differences in the age of peak production between different type of achievement. Poets and mathematicians achieve success earliest, scientists later, scholars later still. Even within one discipline, variations occur. Thus in the field of classical music, the peak for composing instrumental works occurred between 25 and 29, for symphonies between 30 and 34, for chamber music between 35 and 39, and for opera, light opera and musical comedy between 40 and 44 (Lehman, 1953; Dennis, 1955, 1966).

Can we explain the empirical observations psychologically? There are two major attempts. Beard (1874) postulated that creativity was a function of two basic variables, enthusiasm and experience. Neither can produce high-quality creative work without the other. Now Beard asserts that enthusiasm peaks early and then declines; experience continues to grow over time. The product of the two would give a peak around 40. This is not a bad theory; but it doesn't explain why 'enthusiasm' (motivation) should decline after 40; it is fairly obvious why experience goes on growing. If we substitute 'intelligence' for experience we might have a better model, with IQ declining with age. Simonton (1984b) rejects this theory quite vigorously;

> one potential explanation of the age curve can be discounted; any waxing and waning of productivity or influence with age cannot be reasonably be attributed to gains and losses in intelligence . . . the agewise fluctuations in an individual's intelligence do not decline rapidly enough to account for the agewise changes in cultural or political contributions (p. 107).

The position is nowhere as clear as Simonton suggests.

There are three major models of age developments in IQ (Dixon, Kramer and Baltes, 1985). The classical cross-cultural studies of the 50s and 60s show that average scores (e.g. on the WAIS – the Wechsler Adult Intelligence Scale) rise until about age 20 and then gradually decline throughout the adult years; verbal tests showed less (or no) decline than non-verbal tests (Botwinnick, 1977). Criticisms of cross-sectional studies brought the advent of longitudinal studies (Schare, 1983), which generally showed much less evidence of decline. And finally we have the Horn and Cattell (1967) theory which postulates a *decline* over age for *fluid* intelligence (usually measured by non-verbal tests), and a lack of decline, or even increase, for *crystallized* ability (usually measured by verbal tests.) My own preference would be for a variant on the Horn–Cattell theory, but the matter has not been definitely settled; and it would be premature to argue too forcibly on that basis. Nevertheless, the distinction between fluid intelligence, which is a *dispositional* trait (ability to learn), and crystallized ability (which is an index of past learning) is important and it is not unreasonable to associate creativity with fluid ability, Beard's 'experience' with crystallized ability. Hence the decline over age of fluid ability parallels the decline in creativity quite well, and Simonton would need to buttress his refusal to allow changes in IQ over age any relevance to the debate by much more extensive arguments than he has done so far.

Simonton's own theory is interesting but lacks any conceivable psychological support. He calls his theory a 'two-step' model; he postulates that each individual creator begins with a certain 'creative potential' defined by the total number of contributions the creator would be capable of producing in an unrestricted life span. (Rather like a woman's supply of ova!) There are presumably individual differences in this initial creative potential, which Simonton hardly mentions in the development of his theory. Now each creator is supposed to use up his supply of creative potential by transforming potential into actual contributions. (There is an obvious parallel here with potential energy in physics.) This translation of creative potential into actual creative products implies two steps. The first involves the conversion of creative potential into creative ideation, in the second step these ideas are worked into actual creative contributions in a form that can be appreciated publicly (elaboration). It is further assumed that the rate at which ideas are produced is proportional to the creative potential at a given time, and that the rate of elaboration if 'proportional to the number of ideas in the works' (Simonton, 1984b; p. 110). Simonton turns these ideas into a formula which generates a curve which gives a correlation between predicted and observed values in the upper 90s (Simonton, 1983b).

The theory is inviting, but essentially untestable – how would we measure the 'creative potential' which presumably is entirely innate, and should exist from birth? How could we measure ideation, or elaboration, independently of external events? Of course the curve fits observations beautifully, but then all the constants are chosen to make sure of such a fit! Given the general shape of the curve (inverted U), many formulae could be produced to give such a fit. Unless we are shown ways of independently measuring the variables involved, no proper test of any underlying psychological theory exists.

Indeed, there is no psychological theory, by which I would mean a theory built upon well researched concepts, which can be used to explain the observed phenomena. Superficially Beard's theory might seem to suggest such a theory – 'enthusiasm' certainly is a term with psychological overtones. However, there is no theory of enthusiasm which would predict its rise early in life and then its fall. Both features are assumed precisely in order to fit the data, and of course they do so – inevitably! It would be easy to write a formula incorporating enthusiasm and experience in just the right proportions, and with suitable weights, to reproduce the observed values as closely as Simonton's formula. Similarly, Simonton assumes that his 'creative potential' is translated into creative ideation at a rate proportional to the creative potential at a given time – why? Because that assumption enables him to fit a proper curve which is similar to the one actually discovered. But there is no reason why the translation of potentiation into ideation should take this particular form. We might just as well assume that the situation is modelled on the production of ova by females – there, too, we have a given stock, but it is depleted in a regular fashion. The choice is completely arbitrary, and made only because it enables us to approximate the observed curve.

What we need is a psychological concept which predicts, on the basis of laboratory experiments and theoretical considerations, that a most unlikely event will take place. To return to 'enthusiasm', or motivation as we would now call it, we have to explain why this is strong at the beginning (when there is little reinforcement for the young genius), but declines just when he is most successful, heaped with rewards and praise, and given far more money and facilities than ever! Fortunately such a concept does exist, derives from a well known theoretical system, and has good experimental backing. This concept is labelled 'conditioned inhibition', and it derives from the Hull–Spence learning theory (Kimble, 1961). Hullian theorizing has of course been heavily criticized, and as a general theory has been all but abandoned (Koch, 1954); however, many of his specific hypotheses are still of great value, and conditioned inhibition ($_{S}I_{R}$) is certainly one of them (Hull, 1943, 1951, 1952).

What does the theory say? We must begin with the concept of reactive inhibition (I_R). Hull assumes that every response of an organism, *whether reinforced or not*, leaves in the organism an increment of reactive inhibition, the magnitude of which is an increasing function of the *rate of response elicitation* and the *effortfulness of the response*. I_R is a primary negative drive resembling fatigue which leads to a cessation of the response which produces it, and like fatigue, it decays with rest. Formally, I_R is a performance variable which *subtracts* from the combined effects of learning and motivation ($_{S}E_{R}$) and, as a consequence, lowers response strength. (The expressions in brackets are Hull's attempt to develop a proper set of terms and relations to account for behaviour. Thus S stands for stimulus, R for response, I for inhibition, and so on.)

Clearly I_R is not the concept we are looking for; it decays with rest and is thus a short-term variable; we are looking for a more permanent effectiveness. Hull argued for such a concept along the following lines. Since I_R is a (negative) drive, the reduction of I_R is a reinforcement and, as such, is capable of strengthening any response which precedes it closely in time. In the theory, a particular response, i.e. the cessation of activity following I_R, was assumed to occupy such a position. Thus we have here a mechanism for the conditioning of a resting response to a certain stimuli; in the learning situation, it is the *resting response* evoked by I_R and reinforced by I_R reduction. In other words, as learning proceeds, fatigue (I_R) sets in and produces a resting response. This response is reinforcing (resting after hard mental or physical work is pleasant), and thus generates a conditioned response favouring rest/inhibition, called $_{S}I_{R}$ or conditioned inhibition. Conditioned responses are *additive*, and do not decay, hence there is a slow build-up of $_{S}I_{R}$ to repeated evocation of I_R in identical or similar situations, which ultimately enforces complete inhibition of the activity in question.

This theory has been heavily criticized on methodological grounds (Koch, 1954), and is inadequate in the face of certain empirical phenomena (Gleitman, Nachmius and Neisser, 1954). Because of the potential importance of the inhibition concept in personality theory (Eysenck, 1957), clinical psychology (Eysenck, 1973b) and experimental psychology (Eysenck and Frith, 1977), I set up a small research group to look into the problem theoretically (Kendrick,

1958a, b, 1960). The outcome was a reformulation of the Hullian equation which was found to be logically faulty (Iwahara, 1957; Jones, 1958), and in need of change. Instead of writing, as Hull would have done; ${}_SE_R = f(D \times {}_SH_R) - (I_R + {}_SI_R)$, in which D = drive, H = habit, and E = reaction potential, Jones suggested: ${}_SE_R = f[(D - I_R)\ ({}_SH_R - {}_SI_R)]$, thus making the net reaction potential a multiplicative function of the *resultant* drive state $(D - I_R)$ and the *resultant* habit state $({}_SH_R - {}_SI_R)$. This conception too has been criticized (Reid, 1960), but it overcomes some of the logical inconsistencies of Hull's formulation.

Kendrick (1958a,b) carried out an experiment specifically to test the intuitively unlikely deduction from this theory that 23-hour thirsty rats, trained to run a 12-ft, long runway to obtain water, would build up sufficient ${}_SI_R$ after many exposures to *refuse to run to the water in spite of their thirst*! There were 30 runs each day, reinforced by $\frac{1}{4}$ oz of water each time. Non-running (extinction) was defined in terms of the following criterion. On three consecutive days the rat must refuse to run for water twice consecutively within the first five trials. Thirty-six days after the last extinction day, each rat was re-run under the same conditions to see if there was any spontaneous recovery of the habit – as there should not have been according to the theory. It was found that all rats met the criterion for no-running, and there was no spontaneous recovery. There were important individual differences in the behaviour of the rats, some of whom came from an emotional strain, the others from a non-emotional strain (Eysenck, 1964). Emotional (neurotic?) rats show a lengthy growth of ${}_SI_R$ by its depressing effect on running times at the beginning of a session; in the non-emotional rats, cessation of running occurs relatively abruptly.

This is not the place to enter into a lengthy discussion of many theoretical points arising, or the important question of the degree to which animal studies can be used to explain human behaviour. As Kimble (1961) points out, in spite of imperfections and anomalies, 'the *kind of reasoning* employed by Hull in the development of the concept ${}_SI_R$, may represent the form which a more complete extinction theory might take' (p. 306). Obviously in the case of humans, the concepts of drive (enthusiasm) and habit (experience) are much more complex, and special experimentation is required to answer the question of applicability of the system. Our discussion is not concerned with *proving* the suggestion that the decline in productivity of creative persons is a function of ${}_SI_R$; the concern is with *suggesting* that psychologists, in considering problems of this kind, should consider related concepts which have some explanatory credentials, rather than appeal to common-sense concepts like 'enthusiasm', into which we can build any properties we like, as long as they may serve to 'explain' the observed phenomenon, or construct formulae based in non-testable hypotheses equally constrained to agree with observation.

I have to apologize to readers who are not intimately familiar with Hullian conceptions of learning theory (which would include many psychologists who have obtained their doctorates since 1960 or 1970); the matter is inevitably highly technical, but perhaps I can make amends by putting it in recognizable English, even though precision is lost thereby. Concentrated intellectual work causes fatigue; rest allows fatigue to dissipate, and produces a feeling of well-being. But this feeling of well-being acts as a reward or reinforcement for what has immediately preceded it, i.e. the state of rest, or not working. Such reinforcement produces a conditioned response of inhibition (resting or not working) which is reinforced every time a person rests after working hard. Conditioned responses do not dissipate over time, as fatigue does; hence the amount of inhibition felt during strenuous intellectual work grows and finally interferes with such work.

There is one further advantage in considering the Hullian theory. Much has been written about the unconscious elaboration of theories which suddenly result in some 'Aha!' experience; this unconscious elaboration has never found a theoretical explanation. I suggest in another chapter that we may find an explanation in the concept of I_R (reactive inhibition), and the dissipation of I_R during rest. Again, the suggestion links an observed phenomenon with an established theory which may (or may not) be used to explain the phenomenon, and explore the conditions under which it occurs. This, to my way of thinking, is the proper function of theory, and to use it in this fashion may guide research into fruitful channels. The rather pessimistic stance taken by the authors of the *Handbook of Creativity* (Glover *et al.*, 1989) to the theoretical and empirical advances in this field may be due largely to the absence of any guiding theories of this kind.

It might be argued that scientists and artists do not usually completely *cease* to produce, but rather decline slowly in their productivity. Actually it is a frequently observed phenomenon that even highly creative academics retreat into administrative positions as they advance in age, and cease to be creative, even by proxy. Others take early retirement, and cease to be active. There has not been any systematic study of this phenomenon, but I have seen many examples, even among my own students, and feel that such a study would be useful, particularly if guided by theory. But even when production does not cease, it is slowed down, and this slowing down is precisely what Kendrick found in his emotional rats. The finding, suggesting experiments contrasting high-neuroticism and low-neuroticism creators, demonstrates again the research-generating powers of a good theory. In the absence of such relevant research no more can be said about this particular attempt to explain the observed phenomena of aging.

Life span

Allied to the function of age is the concept of life span. Is a genius more likely to achieve a high position if he dies at an advanced age? It seems likely that the greater amount of work done by the long-lived should give them a considerable advantage; if you die at 30 or 40, you only had 10 or 12 years of possible achievement, but if you die at 70 or 80, you have 50 to 60 years of work behind you – should the longer-lived not have four or five times the chance of fame, as compared with those who died young? Simonton (1976a) found only a very small linear correlation between eminence and creative longevity among 656 classical composers; looking at the Cox (1926) sample of 301 geniuses he found a very marked curvilinear U-shaped relation for creators. The most eminent had died at 30 or at 90; the least eminent had died at 60! As he says about the latter, 'death at such a middling age comes too late to provoke a sense of tragic loss, and yet too early to allow the creator to become a living legend of artistic or scientific greatness' (p. 190). Leaders show the effect even more strongly; they are accorded more fame by posterity if they die young.

According to Simonton, the major lesson to be learned is that the attainment of eminence involves much more than the accumulation of accomplishments. 'Not only productivity but also the content of productivity determines ultimate historical influence' (p. 90). A person's right to be considered a 'genius' by posterity is influenced by quite irrelevant factors; he is more likely to be called a 'genius' if he behaves oddly, dies young or very old, has episodes of madness, is psychopathic in his relations with others, disregards normal restrictions, and generally lives up to the stereotype of genius people have in their minds. If genius, as we have agreed, is a social creation, we should not wonder if quite irrelevant criteria slip in!

Season of birth

A curious and at present inexplicable condition of eminence appears to be the season of birth. Huntington (1938) and Kaulins (1979) found that eminent people (defined as being listed in the *Encyclopaedia Brittanica*) showed a strong tendency to be born in the months between the winter solstice and the spring equinox; at the peak, in February, 36 eminent persons were born per day as compared with 27 at the trough! (The study encompassed over 11 000 people, and the trend is significant beyond doubt, whatever the reasons.) The explanation does not lie in IQ; if anything, the opposite trend applies to IQ data, although it is much weaker (Pinter and Forlano, 1943). The same trend as for eminence is observed with respect to psychosis (Hare, 1987a,b); psychotics too show a strong tendency to be born in February! It is not clear why either eminent people, presumably high on creativity, or schizophrenics should be conceived with very much greater frequency in May–June, as compared with ordinary people. Until this question is answered the observed close correlation between eminence and schizophrenia remains a puzzle, although it does of course support the relation between creativity and psychopathology we have noted several times. (One possible explanation may be that season-dependent virus infections produce changes in the cortex of the foetus which lead to creativity/psychosis depending on the amount of change produced (O'Callaghan, Gham & Takis, 1991).)

Motivation and the unconquerable will

It is commonplace that genius is accompanied by strong motivation; but such an observation does not get us very far; we infer the strong motivation from the fact that genius requires long-continued hard work ('ninety-nine per cent perspiration . . . '!) which surely would not be undertaken in the absence of such motivation. But we know very little about the kind of motivation involved, the way it expresses itself, or the possibility of increasing it. There is a good deal of work to suggest that 'intrinsic satisfaction' rather than 'extrinsic satisfaction' is the drive behind creative endeavour (Amabile, 1983a, b, 1985), but the empirical work involved in laboratory studies is rarely convincing. In the first place participation is usually constrained; in the second place achievements are minimal and hardly to be compared with works of genius; and in the third place, rewards are negligible (praise, toys, sweets, etc.). Few researchers would take up a new line of research if offered 100 dollars; they might think again if offered a million dollars! Proper experimentation is difficult if not impossible; Ochse (1990) has given a good survey of what is known, but this does not amount to much.

The English mathematician, Hardy, is perhaps a prime example of inner-directedness – he maintained that his profound hope was that nothing he had ever done would be of the slightest practical use. Pasteur, on the other hand, stated that 'nothing is more agreeable to a man who has made science his career than to increase the number of discoveries, but his cup of joy is full when the result of his observations is put to immediate practical use' (Vallery-Radot, 1937, p. 150). In fact, while Hardy's motivation was very much determined by internal sources of satisfaction, Pasteur's motivation was very much determined by external sources; as Dubos (1950) points out, most of his discoveries lay in topics commissioned by outside sources! Indeed, one may find it difficult to differentiate the two; success gives both intrinsic and extrinsic satisfaction (honours, advancement), and who is to say what is the major driving force? Many scientists have quite explicitly set out to become Nobel Laureates; this surely is an extrinsic type of motivation individually linked with intrinsic motivation. The concept of 'functional autonomy' (Allport, 1961) may be relevant here; early reward for certain activities may become internalized and thus independent of the normal source of reward. We may do a certain job because we are paid to do, gradually find it absorbing enough to develop intrinsic motivation for it.

I do not want here to discus these problems at any length, if only because such evidence as is available is not very compelling. I want rather to discuss a topic closely related to motivation, but very clearly demonstrating the importance of intrinsic motivation, coupled with certain personality traits which as we shall see, are closely related to psychoticism. I am referring to situations frequently encountered by geniuses, and quite generally by creative persons, in which the mediocre peer culture generally rejects the original products, so that the extrinsic motivation is negative. If under such circumstances the creative person keeps pressing on, we may be certain that his motivation is intrinsic. He receives nothing but rebuffs, criticism, ridicule and worse, yet he keeps on. The situation is not unusual, quite the opposite. As Einstein said: 'Great spirits always encounter violent opposition from mediocre minds'. What the genius needs is 'the unconquerable will . . . . and courage never to submit or yield' – something Milton recognized from his own experience.

There are three major sources of this resistance to novelty in science, and *pari passu* they can also be found in art. The first is religious; every one knows the fate of Galileo and Giordano Bruno (who was burned at the stake for holding the heliocentric view). Similarly, Copernicus was wise enough to postpone publication of his *magnum opus* until after his death, and even then it had to be published with a foreword pretending the theory had no real application to observable phenomena. Darwin (Gruber and Barrett, 1974) programmed his publication on evolution very carefully, but did not escape a torrent of criticism on religious grounds.

In more recent times social intolerance seems to have supplanted religious inquisition. Einstein's work was banned in Hitler's Germany because it was 'Jewish', and in Stalin's Russia because it was 'bourgeois'. The IQ was equally banned in both countries, for the same reasons; liberals (so-called) attempted to ban it in the USA, with some success, because it was not 'politically correct' – Pearson (1991). The sufferings of geneticists who refused to follow Lysenko in his anti-scientific and wholly absurd theories have been amply documented (Medvedev, 1969, 1971); their deaths were the direct consequence of this modern social inquisition.

Less often remarked, but possibly even more insidious, is the resistance by scientists to 'scientific discovery', as Barker (1961) has named this phenomenon. As he point out, in two systematic analyses of the social process of scientific discovery and invention, analyses which tried to be as inclusive of empirical facts and theoretical problems as possible, there was only one passing reference to such resistance in the one instance and none at all in the second (Gilfillan, 1935; Barker, 1952). This contrasts markedly with the attention paid to the resistance to scientific discovery on the part of economic, technological, religious ideological elements and groups outside science itself (Frank, 1957; Rossman, 1931; Shyrock, 1936; Stamp, 1937). This neglect is probably based on the erroneous notion embodied in the title of Oppenheimer's (1955) book *The Open Mind*; we assume all too readily that objectivity is the characteristic of the scientist, and that he will impartially consider all the available facts and theories. Polanyi (1958, 1966) has emphasized the importance of the personality of the scientist, and no one familiar with the history of science can doubt that individual scientists are as emotional, jealous, quirky, self-centred, excitable, temperamental, ardent, enthusiastic, fervent, impassioned, zealous and hostile to competition as anyone else. The incredibly bellicose, malevolent and rancorous behaviour of scientists engaged in disputes about priority illustrates the truth of this statement. The treatment handed out to Halton Arp (1987), who dared to doubt the cosmological postulate about the meaning and interpretation of the red-shift is well worth pondering (Flanders, 1993). Objectivity flies out of the window when self-interest enters (Hagstrom, 1974).

The most famous example of a priority dispute is that between Newton and Leibnitz, concerning the invention of the calculus (Manuel, 1968). The two protagonists did not engage in the debate personally, but used proxies, hangers-on who would use their vituperative talents to the utmost in the service of their masters. Newton in particular abused his powers as President of the Royal Society in a completely unethical manner. He nominated his friends and supporters to a theoretically neutral commission of the Royal Society to consider the dispute; he wrote the report himself, carefully keeping his own name out of it, and he personally persecuted Leibnitz beyond the grave, insisting that he had plagiarized Newton's discovery – which clearly was untrue, as posterity has found. Neither scientist emerges with any credit from the Machiavellian controversy, marred by constant untruths, innuendos of a personal nature, insults, and outrageous abuse which completely obscured the facts of the case. Newton behaved similarly towards Robert Hooke, Locke, Flamsted and many others; as Manuel (1968) says, 'Newton was aware of the mighty anger that smouldered within him all his life, eternally seeking objects. . . . many were the times when (his censor) was overwhelmed and the rage could not be contained' (p. 343). 'Even if allowances are made for the general truculence of scientists and learned men, he remains one of the most ferocious practitioners of the art of scientific controversy. Genteel concepts of fair play are conspicuously absent, and he never gave any quarter' (p. 345). So much for scientific objectivity!

Planck's experience with other leading physicists was no different. 'None of my professors at the University had any understanding of its contents', he said, after submitting his Ph.D. thesis in 1879 at the University of Munich. 'I found no interest, let alone approval, even among the very physicists who were clearly connected with the topic. Kirchoff expressly disapproved. I did not succeed in reaching Clausius. He did not answer my letters, and I did not find him at home when I tried to see him in person at Bonn. I carried on a correspondence with Carl Neumann, of Leipzig, but it remained totally fruitless' (Planck, 1949, p. 18). As he states, 'this experience gave me also an opportunity to learn a new fact – a remarkable one in my opinion: A new scientific truth does not triumph by convincing its opponents and making them see the light, but rather because its opponents eventually die, and a new generation grows up that is familiar with it.'

Once a theory has been widely accepted, it is difficult to displace, even though the evidence against it may be overwhelming. Kuhn (1957) points out that even after the publication of *De Revolutionibus* most astronomers retained their belief in the central position of the earth; even Brahe (Thoren, 1990) whose observations were accurate enough to enable Kepler (Caspar, 1959) to determine that the Mars orbit around the sun was elliptical, not circular, could not bring himself to accept the heliocentric view. Thomas Young proposed a wave theory of light on the basis of good experimental evidence, but because of the prestige of Newton, who of course favoured a corpuscular view, no-one accepted Young's theory (Gillespie, 1960). Indeed, Young was afraid to publish the theory under his own name, in case his medical practice might suffer from his opposition to the god-like Newton! Similarly, William Harvey's theory of the circulation of the blood was poorly received, in spite of his prestigious position as the King's physician, and harmed his career (Keele, 1965). Pasteur too was hounded because his discovery of the biological character of the fermentation process was found unacceptable. Liebig and many others defended the chemical theory of these processes long after the evidence in favour of Pasteur was conclusive (Dubos, 1950). Equally his micro-organism theory of disease caused endless strife and criticism. Lister's theory of antisepsis (Fisher, 1977) was also long argued over, and considered absurd; so were the contributions of Koch (Brock, 1988) and Erlich (Marquardt, 1949). Priestley (Gibbs, 1965) retained his views of phlogiston as the active principle in burning, and together with many others opposed the modern theories of Lavoisier, with considerable violence. Alexander Maconochie's very successful elaboration and application of what would now be called 'Skinnerian principle' to the reclamation of convicted criminals in Australia, led to his dismissal (Barry, 1958).

Another good example is Wegener's continental drift theory, which was given short shrift when he first announced it (Wegener, 1915), but which is now universally accepted. Coming from an outsider to geology, the theory was so revolutionary that most geologists rejected it out of hand. Many of them refused to take it seriously and simply ignored it. Opposition centred on two major pillars of the establishment, Hans Schille and Hans Cloos, who persecuted Wegener mercilessly. The criticisms were not all undeserved; the explanation Wegener gave of the mechanisms which caused the continents to drift were 'hopelessly inadequate', as Arthur Holmes pointed out. He himself postulated slow convection currents in the viscous magma of the mantle as the driving force, and thus indicated the path that plate tectonics would later follow. There are two features in Wegener's work which recur again and again. The first is that the advance is suggested by an outsider – Dalton, Pasteur and Young are other examples. (Wegener was a meteorologist by profession.) The second is that there are no causal theories to support the discovery, and without some such support traditional scientists are slow to believe in the existence and physical being of the new discovery.

The list is truly endless, and is continued in Barker's (1961) article. Here I will rather cite in a more detailed manner a particularly interesting case, that of Ignaz Philipp Semmelweis (Slaughter, 1950), who lived from 1818 to 1865. Born in Buda, he was educated at the universities of Pest (the other half of Budapest) and Vienna, where he took his M. D. and was appointed assistant in the first obstetric clinic under Johann Klein. His main concern was with puerperal infection, the scourge of maternity hospitals throughout Europe, which afflicted largely women who gave birth in hospital, because of poverty, illegitimacy or obstetrical complication. Mortality rates for these women varied widely from time to time, and from place to place, but they reached values of 25% and at times even higher. Women dreaded to be confined in hospitals. The cause of puerperal fever was unknown, and irrelevant factors such as overcrowding, poor ventilation, the onset of lactation or simple 'miasmas' were blamed, although John Burton in 1751 had advocated the view that the fever was brought to the patient by an outside agency, and that the carelessness of the midwife or the attendants might be the cause, and Charles White 22 years later published his book: *A Treatise on the Management of Pregnant and Lying-in Women, and the Means of Curing, More especially of Preventing, the Original Disorders to Which They are Liable*, in which he advocated a strict regime of cleanliness, fresh air, and hygiene – he claimed never to have lost a single patient from puerperal fever. Alexander Gordon of Aberdeen published his *Treatise of the Epidemic Puerperal Fever* in 1795, one of the first physicians to give a clear-cut statement on the contagious nature of childbed fever.

Others, such as Robert Callius in Dublin, were able through proper hygiene to reduce severe mortality in his hospital to 0.53%, without a single case of childbed fever. Oliver Wendell Holmes in America adopted a similar belief in his famous 1843 article 'The Contagiousness of Puerperal Fever'; he said that 'the disease known as Puerperal Fever is so far contagious as to be frequently carried from patient to patient by physicians and nurses'. Holmes, because of the very positiveness of his opinions, as well as his radical departure from the then accepted pattern of thinking produced an immediate storm of abuse from professors of obstetrics all over America.

Semmelweis soon noted that of the two divisions of the hospital, over twice as many mothers died of childbed fever in the first division as in the second; although admission was random. The difference was that students were taught in the first, midwives in the second. Semmelweis argued that perhaps the students carried something lethal from one patient to another, communicated when the patients were examined during labour. The students often came from an examination of cadavers, and did so without washing their hands; midwives of course did not carry out such an examination. Semmelweis was encouraged in his belief by the death of his friend Jakob Kolletschka, a pathologist, from a wound infection incurred during the examination of a woman who had died of puerperal infection. Semmelweis concluded that students who came directly from the dissecting room to the maternity ward carried infection from mothers who had died to healthy mothers they examined during labour. He promptly instituted a regime of handwashing, using soap and water, or later chlorinated lime, prior to any examination of a woman in labour. Semmelweis later described the effect of this measure:

> In 1846, when the chlorine washes were not in use, there died 459 puerperae out of 4010 in the First Clinic, or 11.4%. In the Second Division during 1846, out of 3754 there died 105 or 2.7%. In 1847, when about the middle of May the chlorine-washings were introduced, there had died in the First Division 176 out of 3490 puerperae, or 0.5%. In the Second Division, 32 died out of 3306 or 0.9%. In 1848, when the chlorine washings were used assiduously throughout the year, 45 puerperae died out of 3556, or 1.27%. In the Second Division during this year, 43 died out of the 3219 delivered, or 1.33%.
>
> I have assumed that the cadaveric material adhering to the examining hand of the accoucheur is the cause of the greater mortality in the First Obstetric Clinic; I have eliminated this factor by the introduction of the chlorine-washings. The result was, that the mortality in the First Clinic was confined within the limits of that in the Second, as the above cited figures show. The conclusion therefore, that the cadaveric particles adhering to the hand had in reality caused the preponderant mortality in the First Clinic, was also a correct one.
>
> Since the chlorine-washings were brought into use with such striking results, there was not the slightest change made in the conditions in the First Clinic, to which could be ascribed a share in the diminution of the mortality.

An almost ten-fold reduction in mortality might have been expected to provoke praise, interest and imitation. Nothing of the kind. The students did not like the inconvenience of constant handwashing, and Professor Klein, his boss, driven by jealousy, ignorance and vanity, put all sorts of obstacles in Semmelweis's way, underhandedly prevented his promotion, and finally drove him from Vienna.

Working as an obstetric physician in the maternity department at Pest, he promptly reduced the high maternal mortality rate to 0.85%. Some doctors followed Semmelweis and succeeded in reducing mortality rates in a similar fashion, but orthodoxy prevailed and tens of thousands of pregnant women died because of this stubborn blindness. Semmelweis became a successful physician in private practice, but never got over the failure of his profession to accept his doctrine, and thereby save the lives of countless young women; it literally 'drove him crazy', and he finally died from a wound in his right hand, a victim of the very disease which he had fought so manfully.

Like so many others whose original contributions raised a storm of criticism, Semmelweis was not the perfect knight in shining armour. He was stubborn, lacked diplomacy, and did odd things (such as suddenly leaving Vienna when he might have won his fight had he stayed). He refused unaccountably to write up his findings and theories in book form until it was much too late; *Die Aetiologie, der Begriff und die Prophylaxis des Kindbett-fiebers*, published in 1861, was badly written, too much concerned with personal quarrels, and not likely to win over opponents or even neutrals. Like so many geniuses, he was his own worst enemy (Slaughter, 1950).

Now, of course his contribution is widely recognized, and the University of Budapest bears his name. As Joseph Lister, the father of modern antiseptic surgery, wrote: 'Without Semmelweis my achievements would be nothing. To this great son of Hungary, Surgery owes most'. Even bearing in mind all his faults of character and behaviour, one must marvel at the blind stupidity, the absurd pretentiousness, the repulsive submission to false authority that characterizes his enemies, who, like Galileo's Aristotelian critics, refused to look through the telescope and see for themselves Jupiter's moons.

Another victim of mindless medical orthodoxy was the great Andreas Vesalius, who pioneered modern anatomy 450 years ago. The publication of his masterpiece on human anatomy, at the age of 28, made him the founder of a new science, but also involved him in harmful battles with the establishment. His critics remained faithful to the ancient authorities, like Galen, and refused to believe the evidence of their own eyes. Embittered by the harsh condemnation of his work, Vesalius gave up scientific work, burnt his notes, and became the personal physician to first, Charles V, and later Philip II, Kings of Spain. The Spanish Inquisition charged him with sinning against God and man, because of a rumour that he had once carried out an autopsy, and upon opening the chest had found the heart still beating! He would have been hanged had the King not intervened. As a penance Vesalius was made to undertake a pilgrimage to Jerusalem. On his return he was shipwrecked and perished.

The story is characteristic in many ways, but it would be quite wrong to imagine that this is the sort of thing that happened in ancient, far-off days, and that nowadays scientists behave in a different manner. Nothing has changed, and I have elsewhere described the fates of modern Lochinvars who fought against orthodoxy and were made to suffer mercilessly (Eysenck, 1990a). The battle against orthodoxy is endless, and there is no chivalry; if power corrupts (as it surely does!), the absolute power of the orthodoxy in science corrupts absolutely (well, almost!). It is odd that books on genius seldom if ever mention this terrible battle that originality so often has when confronting orthodoxy. This fact certainly accounts for some of the personality traits so often found in genius, or even the unusually creative non-genius. The mute, inglorious Milton is a contradiction in terms, an oxymoron; your typical genius is a fighter, and the term 'genius' is by definition accorded the creative spirit who ultimately (often long after his death) wins through. An unrecognized genius is meaningless; success socially defined is a necessary ingredient. Recognition may of course be long delayed; the contribution of Green (Connell, 1993) is a good example.

Two points deserve further discussion. The first relates to the problem:

When is a theory in science accepted as proven? Were Semmelweis's and Wegener's opponents correct in dismissing the assertions of these men as unproven? The philosopher might reply that you can never *prove* a theory to be correct, and that ultimately all theories are faulty (Popper, 1959, 1979). This is not helpful, although true enough in a sense. Newton's theory of gravitation was not true in any general sense, but it was true under certain limiting conditions; science would certainly not have benefited by rejecting it on philosophical grounds. Semmelweis and Wegener were certainly right in their major contentions, even though they failed to buttress their findings by causal theories. Why were their findings not accepted, while others received much less hostility when they were first announced? This is an important question which has not to my knowledge been treated seriously by philosophers and historians of science.

It may be suggested that one major factor is the existence of a felt lack, anomaly or lacuna in our knowledge. Einstein was accepted fairly quickly because of an accumulation of anomalies which had scientists looking for an alternative to Newton's theory. The Michelson–Morley experiment, the precession of the perihelion of Mercury and other observations and experiments made the Newtonian model unacceptable, and this perceived gap caused many attempts to fill it; Poincaré aud Lorentz are obvious examples. The mathematician who solves Fermat's theorem will be honoured because both the gap and the criteria for closing it are widely recognized. Few felt a gap in the areas pioneered by Semmelweis and Wegener, and methods for defining a proper solution were not universally agreed; this surely is a major factor deciding acceptance versus rejection of an original thesis.

The position of the innovator within science is another important guide. Semmelweis was a junior doctor, without influence or power, Wegener was not even a geologist, and powerless outsiders are not usually accepted readily when they preach new sermons. Where Poincaré had opened the doors, even a lowly patents officer in Zurich might be listened to; where a Ph. D. student like Planck attempted to get a hearing in a new field not so prepared, no-one would listen.

A third condition for acceptance is probably the diplomatic talents of the scientist. I have mentioned some of the awful errors committed by Semmelweis in trying to get his view accepted; had he been more sensible and tactful, he might have prevailed – and probably would have. These three factors, it may be suggested, play at least some part in deciding which novelties shall be accepted, which rejected.

The other point to be made is that that the moment at which new ideas are accepted is to some extent arbitrary. Pasteur's theories were fought by Liebig long after they had been widely accepted. The atomic theory was rejected by physicists (Dumas, Mach) long after it had been accepted by chemists. How many replications are needed to prove a point? There is also the problem of Fisher's Type One and Two errors – accepting a theory although it is false, rejecting it although it is true. In the case of Wegener and Semmelweis scientists made a Type Two error, but there are also many examples of Type One errors. Freudian psychoanalysis is a good example. It was embraced enthusiastically almost from the beginning in spite of the complete lack of evidence in its favour (Freud claimed the opposite, but historians have demonstrated that he and his followers were very economical with the truth – Eysenck (1990b).) Experimental studies have been almost universally negative (Kline, 1981; Eysenck and Wilson, 1973), and so have clinical investigations (Eysenck, 1990b); yet Freud still flourishes – admittedly only among the ignorant. A more recent example is the work of the health activist (Manhattan Institute, 1991), i.e. epidemiologists, oncologists, health specialists and others who have persuaded the community that almost everything we eat, drink or smoke causes cancer and/or heart disease. The evidence solidly contradicts the thesis (Coleman, 1988; Effron, 1984; Eysenck, 1991a) but the Type One error has been accepted, and reigns supreme.

Probably the balance between Type One and Type Two error in science is reasonable, but glaring errors occur in both directions. Genius, precisely because of the originality and creativity which characterize it, is more subject to Type Two error, and hence has to develop the attitude of the street-fighter – or go under. As we shall see, the personality traits which are needed here – aggression, Machiavellianism, ego-strength, dominance, self-reliance – are precisely those which research has shown to be characteristic of genius and the creative person generally – and the person high on the personality trait of psychoticism. Evolution triumphs again!

Genius and fraud

There is one point, however, which emerges from our discussion and which is even more likely to be omitted from books on genius; that is the frequency of fraud in science (Broad and Wade, 1982; Miller and Hersen, 1992). If orthodoxy treats the budding genius as it treated Semmelweis, we should not be surprised if the genius hits back, and if the weapons used against him are unfair, arbitrary, inequitable, dishonest, unscrupulous and unprincipled, then we should expect the genius to retaliate in kind. This appreciation that a genius may be flawed as a human being often comes as a shock to many people; they put Kepler, or Newton, on a pedestal, only to discover the feet of clay belatedly. Let us consider a few examples.

The Broad and Wade (1982) book names many famous scientists as having deviated significantly from the paths of righteousness, including Ptolemy, Galileo, Newton, Dalton, Mendel, and Millikan, the American Nobel Laureate who was the first to measure the electric charge on the electron. There seems to be little doubt about certain irregularities in their reports, but of course there are degrees of fraud, going from outright plagiarism and the invention of data to the omission of data contrary to one's theories and the adjustment of constants to suit one's case. The motivation is usually to give the impression of greater precision, closer correspondence of theory and experiment, and better fit to expectation.

Modern times also have their share of fraud, as Broad and Wade testify, but their account is somewhat selective. Thus they deal at some length with the Burt affair (Hearnshaw, 1979), i.e. with Burt's alleged invention of certain twin data, but do not mention the much more frequent unlikely claims of certain environmentalists to have succeeded in raising IQ by educational means (Spitz, 1986). Broad and Wade take it for granted that Burt was guilty, but the evidence is unequivocal; Joynson (1989) and Fletcher (1991) have published detailed accounts refuting the allegation. I have no wish here to adjudicate between the two sides, merely to point out that a doubtful case concerning an advocate of the importance of genetic factors in the genesis of intelligence is mentioned, but the numerous undoubted cases of fraud by environmentalists are ignored. Ideology may cause fraud, and it also seems to influence how the accusation is treated (Snyderman and Rothman, 1987).

Let us consider the case of Newton (Westfall, 1973), where the evidence seems to allow no doubt. Newton was involved in a mortal battle with continental physicists, opposing his quantitative, mathematical conception to their mechanistic views. He followed such men as Kepler and Galileo in this, but was much more concerned to establish a paradigm of *Philosophiae Naturalis Principia Mathematica*; 'universal precision' had replaced the world of more or less, as Alexandre Hayre put it. Clearly, in such an endeavour the successful demonstration of precision was vital, and this caused Newton to fudge his data. One example must suffice, and I have chosen that of velocity of sound. Newton had carried out the first successful analysis of what we now call simple harmonic motion, and undertook to extend that type of analysis to include the propagation of sound, thus inaugurating a new brand of theoretical physics. His demonstration rested on his understanding of the dynamics of the pendulum, which he extended to an analysis of waves on the surface of the water (Westfall, 1980). He demonstrated that the velocity of pulses (compressive waves propagated through air) varies as the square root of the elastic force divided by the density of water to that of air, and arrived at a preliminary velocity of sound of 979 feet per second. To this figure he made two corrections. The calculations had assumed a medium of pointlike particles, but of course the real particles are of finite dimensions in comparison with the spaces between them, a property which he called the 'crassitude' of the particles, and this he used as a factor in the corrections. Another correction had to be made for vapour, which does not vibrate with the air and thus causes an increase in the velocity proportional to the square root of the amount of air that the vapour displaces. What with one thing and another, Newton's calculation finally arrived at a figure agreeing precisely with the measured velocity of sound. Unfortunately, as Westfall (1973) demonstrates, 'any number of things were wrong with the demonstration' (p. 753).

Newton assumed a precise value for the velocity of sound which he took from the average of a large number or measurements varying over a wide range. His assumption that air contains vapour in the ratio of 10 parts to one, and that vapour does not participate in the sound vibration were completely arbitrary, without any empirical support. 'And his use of the "crassitude" of the air particles to raise the calculated velocity by more than 10% was nothing short of deliberate fraud' (p. 753). Newton's adjustment assumed that particles of water were completely solid; yet he believed that they contained a bare suggestion of solid matter in a vast preponderance of void! His calculation of the velocity of sound simply disregarded his often-varied and quite fundamental theory of matter in order to adjust the calculated speed of sound upwards by 109 feet per second!

One can sympathize with Newton's predicament. Having nailed his colours to the mast of precision, he was left in his calculations with an uncovered discrepancy of 20%, giving priceless ammunition to the hostile continental physicist jeering at him! 'The very flagrancy of his adjustment in this case becomes evidence for the compulsion behind the pretence of precision in the other case'! (p. 754), i.e. the acceleration of gravity, and the precession of the equinoxes, where he similarly fudged the data.

Kepler, too, presents a clear case of fraud (Donahue, 1988). His book, *New Astronomy*, published in 1609, used the results of triangulation in discovering the shape of the orbit of Mars, but only as a guide to his theorizing (Wilson, 1968). Wilson argues that Kepler could not have used triangulation to determine the orbit, because the procedure of triangulation was too imprecise. Kepler's presentation of his data is muddled and contradictory. The promised triangulation is reported in Chapter 51, but instead of proceeding directly to the comparison of the circular and elliptical theories, Kepler returns to the same question in Chapter 53, in a curious hybrid procedure difficult to follow. Donahue has this to say about Chapter 53:

> A closer look at Chapter 53, far from answering this question, only serves to increase the complexity. After telling us what he is going to do, Kepler proceeds to give two purported examples, which actually show quite a different procedure. Then, excusing himself from further presentation of computational details on the grounds that "it would be tedious", he sets out the table mentioned above. But, we are startled to note, the numbers in the table are dramatically different from the numbers in the computations from which they were supposedly derived. And that is not all: at the beginning of the next chapter, Kepler refers to a fifteen-minute correction in the mean anomalies, as an astonishingly large adjustment that he claimed was introduced in Chapter 53. Nothing of the kind is to be found in that chapter, however. Clearly there is more here than meets the eye (p. 217).

It is certainly startling to find an absence of essential computational details because 'taediesum esset' to give them. But worse is to follow. Donahue makes it clear that Kepler presented *theoretical deduction* as *computations based upon observation*. He appears to have argued that induction does not suffice to generate true theories, and to have substituted for actual observations figures deduced from the theory. This is historically interesting in throwing much light on the origins of scientific theories, but is certainly not a procedure recommended to experimental psychologists by their teachers!

Many people have difficulties in understanding how a scientist can fraudulently 'fudge' his data in this fashion. The line of descent seems fairly clear. Scientists have extremely high motivation to succeed in discovering the truth; their finest and most original discoveries are rejected by the vulgar mediocrities filling the ranks of orthodoxy. They are convinced that they have found the right answer; Newton believed it had been vouchsaved him by God, who explicitly wanted him to preach the gospel of divine truth. The figures don't quite fit, so why not fudge them a little bit to confound the infidels and unbelievers? Usually the genius is right, of course, and we may in retrospect excuse his childish games, but clearly this cannot be regarded as a licence for non-geniuses to foist their absurd beliefs on us. Freud is a good example of someone who improved his clinical findings with little regard for facts (Eysenck, 1990b), as many historians have demonstrated, *Quod licet Jovi non licet bovi* – what is permitted to Jupiter is not allowed the cow!

One further point. Scientists, as we shall see, tend to be introverted, and introverts show a particular pattern of level of aspiration (Eysenck, 1947) – it tends to be high and rigid. That means a strong reluctance to give up, to relinquish a theory, to acknowledge defeat. That, of course, is precisely the pattern shown by so many geniuses, fighting against external foes and internal problems. If they are right, they are persistent; if wrong, obstinate. As usual the final result sanctifies the whole operation (fudging included); it is the winners who write the history books!

The historical examples would seem to establish the importance of motivational and volitional factors, leading to persistence in opposition against a hostile world, and sometimes even to fraud when all else fails. Those whom the establishment refuses to recognize appropriately fight back as best they can; they should not be judged entirely by the standards of the mediocre!

Periodic variations

One of the most obvious features of the history of science and the arts is the fact that certain periods stand out as times when genius was relatively plentiful (fifth century BC in Athens; the Renaissance; golden ages in Sumerian and

Chinese cultures), others when for centuries mediocrity seemed to reign (medieval Europe). The Roman historian Velleius Pateraulus already noted that geniuses appeared in clusters (Arieti, 1976), and now the recognition of 'golden ages' and 'dark ages' is commonplace. William James (1890) suggested that such clusterings of genius were determined by chance, but this is unlikely and can be shown to be untrue. Ertel (1991) concentrated on science and plotted data taken from historical records, using 21 reference sources for important discoveries and the dates when they were made. He discovered both short-term and long-term cyclic variations. The raw data were delta-transformed and detrended, there being of course a very notable upward trend from 1700 to 1920. After detrending his data, and performing a special analysis, Ertel found that 'two notable power deviations in the spectrum indicated significant cyclic oscillations of intellectual achievements within a period of six to seven years' (p. 112).

Long-term changes were similarly plotted from 1620 to 1820. Again there is a general upward trend, simply indicating an increasing level of scientific activity. In addition, however, there was an exceptionally flourishing period from 1650 to 1675; this may be identified with the Renaissance, occurring rather later in science than the arts. Ertel checked this finding by using individual scientists rather than individual discoveries as his units, and found a similar clustering in his time series. He also showed that the 'golden age' reflected quality rather than quantity; 'the breakthrough period is prominent only for the most eminent and not for the less eminent scientists' (p. 116). Many others have found evidence for the clustering of geniuses and certain periods (e.g. Bullough, Bullough and Munro, 1981; Gray, 1966; Kroeber, 1917, 1944; Simonton, 1984b; Sorokin, 1951). Chance may safely be ruled out as a factor.

What are the answers for the flourishing of geniuses in this cyclic fashion (short-term) and this clustering (long-term)? Causes may be internal, produced by human interactions, e.g. by such causes as the ending of the Thirty Years War (1618–48), a peace which released much rational effort which had been devoted to war-like aims. Causes may also be external, physical events independent of human interaction. I will discuss such causes presently.

Genetic factors as proposed by Galton (1869), Ellis (1904) and Cattell (1971) are unlikely to account for short-term cycles, or even 'golden ages' extending over a few dozen years; genetic changes occur over centuries or millenia, not such short periods. Other genetic suggestions might be worthy of more detailed examination than they have received. Thus Ellis, in his examination of pictures in the National Gallery found that blonde-haired, blue-eyed Nordic types were frequently involved with mathematical and scientific achievements, while dark-haired, dark-eyed Celtic types were more famous for their verbal-social skills. Similarly Cattell argued that there was in Europe a preponderance of mechanical and scientific inventions in the northern parts, and of artistic and religious achievements in the Mediterranean areas. Jews, of course, are an exception; originally a Mediterranean nation, they have become famous for their highly creative scientific work (Storfer, 1990, pp. 321–30). Thus Sarton (1927–48) identified 626 'outstanding' scientists living between 1150 and 1300 AD, of whom 195, or more than 15%, were Jewish! This is roughly 30 times as many as would be expected on a chance basis (Patai, 1971). Similar accomplishments have been characteristic of Jews in the field of music (Storfer, 1990), but not in painting, architecture, and the more observational sciences. In chess, almost half of the 15 world champions since 1851 have been able to claim Jewish ancestry (Cranberg and Albert, 1988, p. 158).

Differences between northern and Mediterranean groups cannot be due to differences in IQ; they do not exist. There is a possibility that northerners may be more introverted, southerners more extraverted, but the data do not give much support to such an hypothesis (Barrett and Eysenck, 1984). Possibly a somewhat higher neuroticism score for southerners makes them more artistic, and somewhat greater stability may make northerners more scientific (Barrett and Eysenck, 1984); the data are not sufficient to answer this question. But as far as Jews are concerned, there is no question that they score very highly on IQ tests as already noted. Storfer (1990) has summarized American studies to this effect; two British studies, a Canadian one, and several Israeli ones all concur that the Jewish IQ is very significantly above that of other Caucasian or Mongoloid groups. Whether these differences are genetic, in part or in whole, or whether they are mainly due to infant care and environmental stimulation, as Storfer suggests, cannot be answered with any degree of confidence at the moment; most likely both sets of variables are involved.

A more likely approach to cyclic phenomena is that of Kroeber (1944), who attempted to show that creativity in a given society waxes and wanes as a particular scientific or artistic approach or style becomes exhausted; habituation sets in. In science, a similar process exists (Kuhn, 1976), due to which scientific revolutions or 'paradigm shifts' occur when 'ordinary science' has found insuperable anomalies in old-established theories. Martindale (1990), in a path-breaking series of studies, put forward a dual theory of creative change, particularly in poetry and the arts. Artists and scientists inevitably search for novelty; what has been done once cannot be done again. A given style in art finally collapses, and a new style (say impressionism) arrives because the old style has nothing new to contribute. Artistic production is judged in terms of its 'arousal potential', i.e. in terms of what Berlyne (1971) called 'collative properties', such as complexity, surprisingness, incongruity, ambiguity, and variability. Over the period from 1290 to 1949, Martindale showed that as required by his theory arousal potential increased at an accelerating rate in texts from 170 British poets.

Collative properties are finally exhausted, and a new style takes over. Within a given style, arousal potential is produced in particular by what Martindale calls 'primordial content', i.e. by what Kris (1952) called 'primary process' thinking. This is contrasted with 'secondary process' thinking, which is abstract, logical, and reality oriented (left-hemisphere thinking, as often defined, although the relation to the cerebral hemisphere is much more complex than ordinarily assumed – Efron (1990)), while primary process thinking is free-associative, concrete, irrational and artistic. Concepts like Nietzsche's (1927) Apollonian versus Dionysian thinking, Berlyne's (1965) autistic versus directed thinking Wundt's (1896) associonistic versus intellectual thought, or Peter McKellar's (1957) A-thinking versus R-thinking all refer to the same differentiation.

Primordial content, like arousal potential, can be measured, as Martindale demonstrates. His prediction for the contribution of primordial content over time is that it should be rising overall, but show quasi-periodic oscillations. The reason for these oscillations would be found in the occurrence of stylistic changes. Within a given style primordial content increases in order to increase arousal potential. When change to a new style occurs, this is sufficiently arousing, enabling new ways of expression to arrive, so that primordial content can take a back seat for a while. But gradually these novel contributions become exhausted, and primordial content is again required to increase in order to prevent habituation, and produce increased arousal potential. Fig. 4.1 shows this quasi periodic oscillation quote clearly; the occurrence of the oscillation is obvious. Martindale attempts to support his thesis by anlayses of modern French poetry, American short stories, classic Greek vases, Japanese prints and Gothic cathedrals; he certainly succeeds brilliantly in his efforts to show how the internal processes of artistic creation produce periods of great creativity (when a new style is born), and rather more barren periods when an old style runs out of steam.

Analogous consideration may apply to scientific changes. In the nineteenth century Newtonian physics seemed to run out of steam; physicists assumed that he had discovered the laws according to which the universe was run, and all that could be done was to dot the *i*s and cross the *t*s, and perhaps explain the reasons for the few remaining anomalies. Physics was getting boring, and it needed Einstein and Planck to change the 'style' and restore interest. How far Martindale's evocation of the concept of 'arousal' is applicable to the scientific field remains to be seen; it would need an extensive search for relevant evidence before anything very definite can be said.

Another suggestion by Velleius Pateraulus to explain the clustering of geniuses he had observed was that exceptional role-models might produce a whole group of highly gifted pupils whose genius would be kindled by their teacher. This hypothesis has been adopted by Kroeber (1944); great masters set examples and set standards by virtue of their own outstanding contributions. Simonton (1978) carried out a historiometric study to show that the amount of creative work produced by a given generation was significantly related to the amount of creative productions in the *two preceding* generations, suggesting that the development of creative ability in a rising generation is in

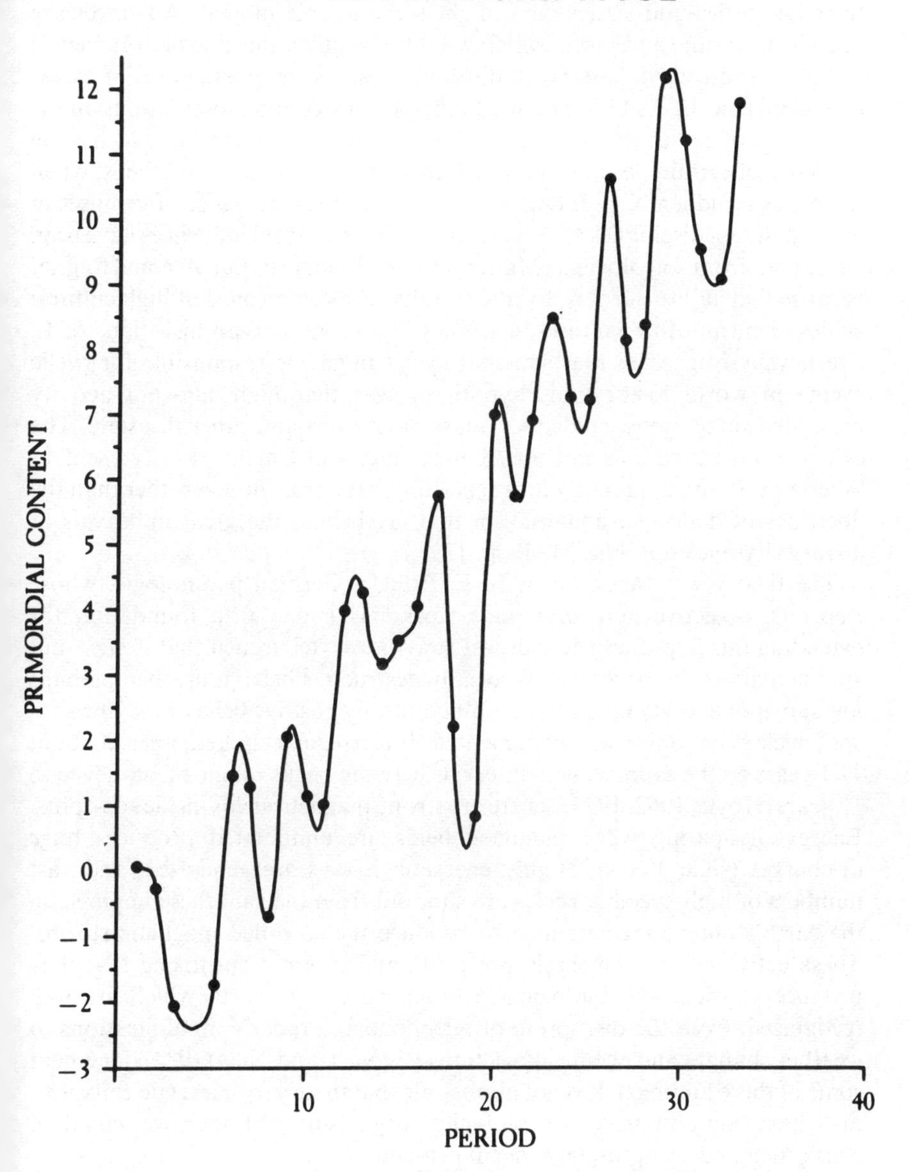

Fig. 4.1 Growth of primordial content in poetic works over time, with periodic stylistic changes (Martindale, 1990). From *The Clockwork Muse* by Colin Martindale, copyright © 1990 by Colin Martindale. Reprinted by permission of Basic Books, a division of HarperCollins Publishers, Inc.

part determined by the number of contemporary creators and patriarchs in their late-fifties and sixties who might serve as role models. Advances are specific as to time and place, which would strengthen this line of argument.

An alternative, or possibly additional cause, is the postulation of extra-terrestrial factors. As I have pointed out, there are certain observations on the emergence of genius which are difficult to explain, not the least of which is the existence of certain periods, like the Renaissance, or Pericles's Athens, when there was a sudden blossoming of culture, with the emergence of geniuses in many different disciplines of art and science. Many causal influences have been suggested, from sociological (Marxist) to psychological, but all come to grief because they fail to account for the simultaneous emergence of high cultural achievement in other cultures, e.g. the Chinese. A Russian historian, A. L. Chizhevsky, suggested that sun-spot cycles might be responsible for cyclic events in world history; his hypothesis was that high sun-spot activity promoted revolutions, epidemics, mass migrations and other disasters. The evidence presented was not very convincing, and Chizhevsky was sent to Siberia by Stalin, apparently for suggesting that it was the sun rather than the doctrines of dialectical materialism that lay behind the great upheavals of history (Eysenck and Nias, 1982, p. 132).

The theory was taken up by S. E. Ertel, a German psychologist whose rigorous, objective and large-scale work has laid a firm foundation for extending this hypothesis to cultural activities. Ertel argued that if *high* sun-spot activity really triggered off socially destructive behaviour, then perhaps *low* sun-spot activity might trigger off culturally positive behaviour. The sun-spot cycle is of course well documented. It is irregular, but emerges at about 11.1 years on the average, with intervals between peaks ranging from seven to 17 years (Hoyle, 1962, 1975). There are strong magnetic fields inside sun-spots. Energy dissipating when magnetic fields are annihilated produces huge discharges (solar flares). Highly energetic X-rays are generated, and vast numbers of high-speed particles are shot out from the sun; these impinge on the earth's outer magnetic field to produce the so-called magnetic storms. These activities are immensely powerful, and it is not surprising that they produce physical and biological phenomena on the earth which are well recognized, from the disruption of telephone and radar communications to weather changes and chemical reactions (Eysenck and Nias (1982) document some of these findings). It is not impossible that these very energetic emissions may have some influence on biological organisms, although we would of course demand exceptionally cogent evidence.

Ertel used recorded sun-spot activity going back to 1600 or so, and before that by analysis of the radiocarbon isotope C14, whose productions as recorded in trees, which give an accurate picture of sun-spot activity. Plotted in relation to sun-spot activity were historical events, either wars, revolutions, etc. or specific achievements in painting, drama, poetry, science and philosophy. Note that Ertel's investigations resemble a 'double blind' paradigm, in

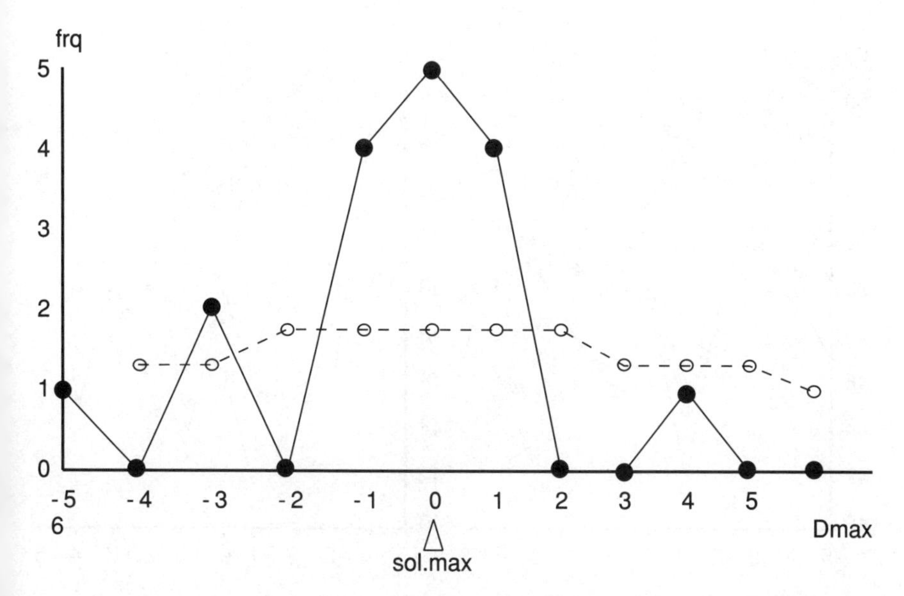

Fig. 4.2 Frequency of occurrence of 17 major conflicts, in relation to solar activity (sun-spots). (Ertel, unpublished, by permission).

that the people who determined the solar cycle, and those who judged the merits of the artists and scientists in question, were ignorant of the purpose to which Ertel would put their data, and did not know anything about the theories involved. Hence the procedure is completely objective, and owes nothing to Ertel's views, which in any case were highly critical of Chizhevsky's ideas at the beginning.

The irregularities of the solar cycle present difficulties to the investigator, but also, as we shall see, great advantages. One way around this problem was suggested by Ertel; it consists of looking at each cycle separately; maximum solar activity (sol. max.) is denoted 0, and the years preceding or succeeding 0 are marked −1, −2, −3 etc., or +1, +2, +3 etc. Fig. 4.2 shows the occurrence of 17 conflicts between socialist states from 1945 to 1982, taken from a table published by the historian Bebeler, i.e. chosen in ignorance of the theory. In the figure the solid circles denote the actual distribution of events, the empty circles the expected distribution on a chance basis. Agreement with theory is obvious, 13 of the 17 events occurring between −1 and +1 solar maximum (Ertel, 1992a,b).

Actually Ertel bases himself on a much broader historical perspective, having amassed 1756 revolutionary events from all around the world, collected from 22 historical compendia covering the times from 1700 to the present. There appears good evidence in favour of Chizhevsky's original hypothesis. However, in this book we are more concerned with Ertel's extension to cultural events, i.e. the view that *art and science prosper most when*

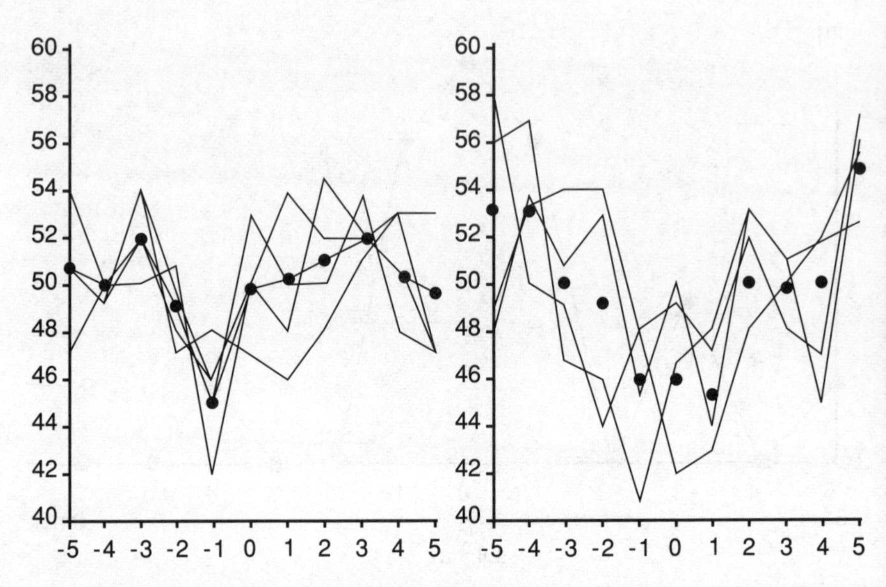

Fig. 4.3 Scientific discoveries in four different sciences, as related to solar activity, with minimum achievement.

solar activity is at a minimum. Following his procedure regarding revolutionary events, Ertel built up a data bank concerned with scientific discoveries. Fig. 4.3 shows the outcome; the solid lines show the relation between four scientific disciplines and solar activity, while the black dots represent the means of the four scientific disciplines. However, as Ertel argues, the solar cycle may be shorter or longer than 11 years, and this possibility can be corrected by suitable statistical manipulation; the results of such manipulation, which essentially records strength of solar activity regardless of total duration of cycle, are shown on the right. It will be clear that with or without correction for duration of the solar cycle, there is a very marked U-shaped correlation with this activity, with an average minimum of scientific productivity at points -1, 0 and $+1$, as demanded by the theory.

It also appears that the corrected data give a slightly better correlation than the uncorrected ones; this is important in answering a possible criticism that the 11-year cycle of sun-spots may not be in a causal relation to scientific discovery, but that there may be another, independent cycle which accidentally runs alongside the solar one. If that were so, scientific productivity would not follow closely the variability of the solar cycle, as it apparently does. Thus this variability, instead of being a nuisance, is an actual advantage; it enables us to form a crucial calculation which disproves a potent alternative explanation of the findings. The observed variability (for which no good theory exists in physics or astronomy) may also be used in another direction, namely predictions deriving from certain long-term failures of solar activity, as indexed by sun-spots. As originally suggested by the German astronomer, Sporer, and the British astronomer, Maunder, there was a minimum amount of solar activity from around 1620 to 1710, the so-called 'Maunder Minimum', and if Ertel's theory is correct, there should have been a special flourishing of the arts and sciences during that period, not only in Europe, but world-wide. This is an important deduction from Ertel's theory; what are the facts?

Fig, 4.4 shows scientific and philosophical activity in Europe, and a combination of the two in China, before, during and after the Maunder Minimum; there is no question that a marked burst of activity emerged during that period for all three groups. Similarly for the arts; curves showing abnormal creativity for painting, poetry and a combination of science and philosophy are given by Ertel; the agreement is astonishing.

Fig. 4.5 shows a comparison of literary productivity in the world as a whole, and specifically in European, Persian, Osmanic, Arabic, Chinese and Japanese language groups. The congruence or peak activity in all these groups during the Maunder Minimum is unmistakable. Fig. 4.6 compares peak activity in painting, poetry, literature and science for European and non-European countries. In addition to the overall congruence, it should be noted that for both groups peak activity in painting *precedes* that in poetry, poetry precedes literature, and literature precedes science; the straight dotted line indicates this precession.

Ertel finally presents a much larger parallel set of curves, based on radio carbon C14 dating, of Chinese and European literature, painting and science, from AD 600 to 1700, i.e. antedating the periods already covered. Discontinuities are obvious, and it is apparent that these discontinuities show cultural synchronicity, and are also synchronous with the radiocarbon C14 curves. Also, indicated are several minima other than the Maunder, namely the Schwabe, Oort, Wolf and Sporer minima; notable is the correspondence with cultural productivity, particularly the Chinese – before the thirteenth century there was as little artistic or scientific productivity in Europe, which was still locked into a medieval sleep (Ertel, 1989).

A final attempt to prove his theory led Ertel to look at the C14 phenomena in the seventh to eighth century BC, when an unparalleled outburst of creativity occurred in ancient Greece, in India, in China and in the Near East; here too agreement with (lack of) solar activity was found, as demanded by the theory.

Not too much should of course be made of these findings prior to their being published in detail, and subject to formal criticism by experts in history, astronomy and historiometrics. I was privileged to have access to all of Ertel's writings, including a projected book, unpublished lectures, etc., and with his permission have tried to give a brief overview of the huge mass of material covered. If found acceptable, this undoubtedly constitutes a major contribution to an understanding of genius, the relation between the occurrence of periods like the Renaissance when all cultural activities seem to receive an enormous boost, and external determinants whose mode of action can in

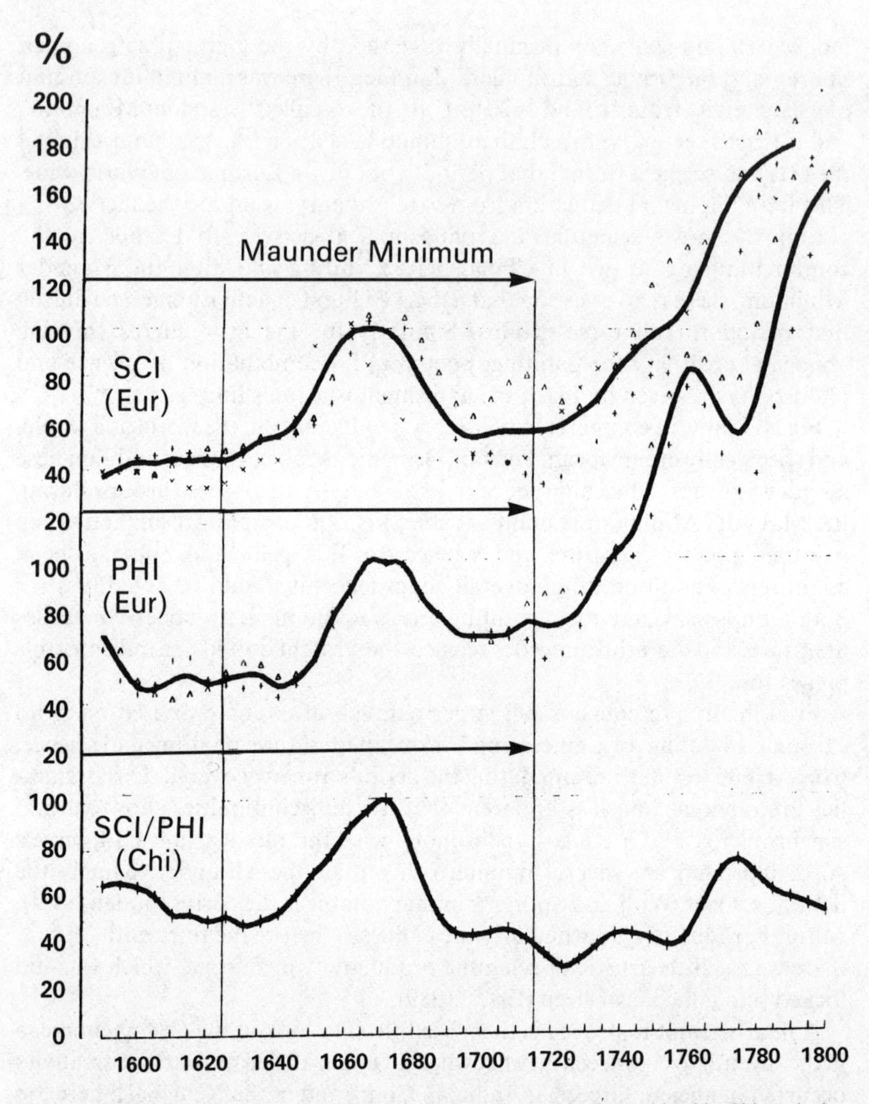

Fig. 4.4 Flourishing of science and philosophy in Europe, and both combined in China, during the Maunder Minimum (long continued lack of solar activity). (Ertel, unpublished, by permission).

principle be understood and studied along orthodox scientific lines (which distinguishes it most markedly from astrology with which it has of course no connection at all). If further historiometric studies should support Ertel's theory, then clearly it will be up to biologists to look into the physiological–hormonal link between solar activity and terrestrial magnetism, on the one hand, and human creativity on the other. If support is found, observationally

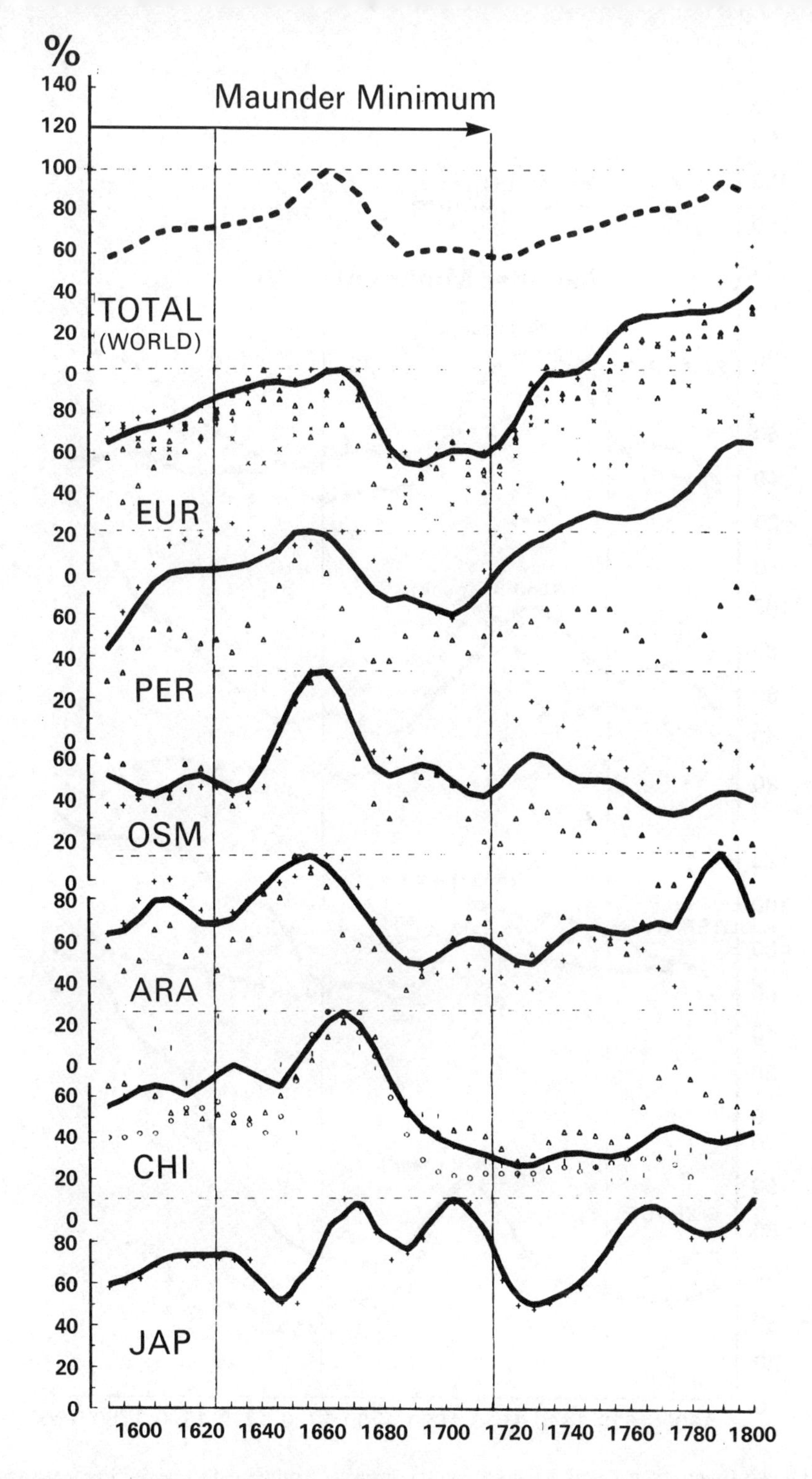

Fig. 4.5 Comparison of literary productivity in the world, and specifically in Europe, Persia, Osman Europe, Arabia, China and Japan, during Maunder Minimum. (Ertel, unpublished, by permission.)

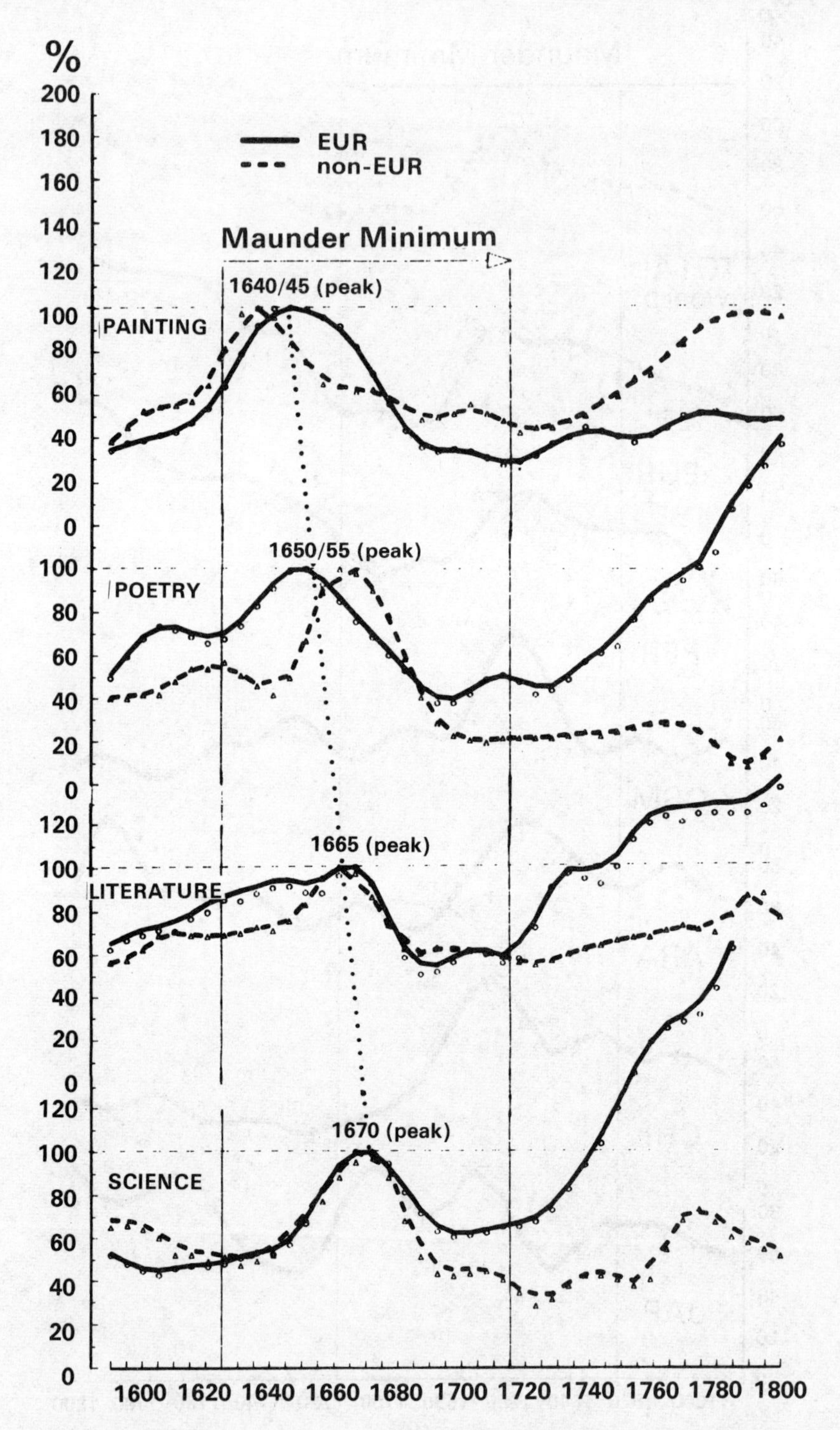

Fig. 4.6 Comparison of peak activity in painting, poetry, literature and science for European and non-European countries during Maunder Minimum. (Ertel, unpublished, by permission.)

and experimentally, for the observed correlations and their causal substratum, we may claim to have discovered an important outside factor determining to some extent creative achievement.

There is one other important observation that is very relevant to the distinction between *genius* and *talent*. Ertel found a marked difference between production by the *eminent* and by the *ordinary* practitioners in painting and science. During the Maunder Minimum the *eminent* showed a very significant increase, while the talented painters and scientists did not, or at least not to anything like the same extent. This finding is very clear, but difficult to explain; it might suggest that there is some kind of *qualitative* difference between genius and talent, while I have argued for a continuum. Until we have a better understanding of the underlying physical, physiological and psychological connections involved, it will not be possible to arrive at any proper interpretation of all these data. Nevertheless, they are so clear and relevant that it is to be hoped that physiologists, psychologists and physicists will try and elucidate the complex causal web that is involved.

Do the data here presented enable us to paint a rough portrait of the budding genius? Clearly he should be male, of middle or upper-middle parentage, and preferably come from a Jewish background. He should receive intellectual stimulation at home, but ought to lose one parent before the age of 10. He should preferably be born in February, and die at 30 or 90, but on no account at 60! He should so arrange it that at the time of his maximum creative powers (between 35 and 45, or even younger for mathematicians and poets) there should be minimum of solar activity. If he also inherited a very high IQ, a high-degree of psychoticism, and great ego-strength, plus great creativity and high special talents (verbal, numerical, musical, visio-spatial), how could he go wrong? Perhaps he could take care to be born into an advanced society which has an artistic and/or scientific tradition, at peace with other societies and not in a state of violent revolution. He might of course succeed against all the odds, like Washington Carver, but chances are much better if he fits into the picture outlined here. Perhaps the picture would be more complete still if we added gout to the background of the genius; it has been found that there is a higher then average ratio of gout in the eminent men Ellis (1926) selected for his *A Study of British Genius*! (Gout is associated with high levels of uric acid in the blood, and uric acid has been found to be quite highly correlated with achievement and productivity, though only slightly with IQ – Jensen and Sirka (1993). The molecular structure of uric acid is closely similar to that of caffeine, and hence acts as a brain stimulant, raising cortical arousal and increasing mental activity.) The role of cortical arousal will be discussed in a later chapter, as will that of other plausible biological factors in producing creative achievement.

5 Intuition and the unconscious

Thou shalt not muzzle the ox that treadeth the corn

Bible

Stages of problem-solving

Two notions, ideas, concepts – call them what you like – have always been attached to the problem of creativity. Is has been widely surmised that the creative genius generates his major ideas by way of *intuition*, rather than rational thinking; reason can *test* and prove or disprove the insights achieved by intuition, but cannot *produce* them. Furthermore, the *process* by means of which intuition works is unconscious; the *Unconscious*, whether with or without a capital 'U', is the cradle of creativity. I will spell out some of the data on which this commonsense view rests (mainly anecdotal), and then go on to *define* the concepts experimentally.

Many famous scientists have identified *stages* of problem-solving; these nearly always include a central stage during which unconscious cerebration, mentation, or whatever predominates. Thus Helmholtz (1896) postulated an initial *investigatory* stage, during which he became saturated with the large body of relevant facts. This was followed by a stage of *rest and recovery*, during which he gave no conscious attention to the problem, and finally a stage of *illumination*, when he had sudden insight into the solution – the 'aha!' experience. Similarly descriptions are given by Dewey (1910), Rossman (1931), Osborn (1953), Johnson (1955), Cattell (1971), and Mansfield and Busse (1981). Wallas (1926) put various introspective accounts together in his frequently cited four-stage scheme: (1) Preparation; (2) Incubation; (3) Illumination; and (4) Verification. To these has recently been added a stage of Problem-finding, which precedes Wallas's stage one (Arlin, 1975; Bunge, 1967; Dillon, 1982; Chand and Runco, 1993).

These stages are often presented as psychological discoveries, and the scientists and psychologists putting them forward are regarded as empirical investigators. However, as Smetslund (1984) has pointed out, many allegedly contingent propositions in psychology, supported by experimental evidence, are really non-contingent and thus necessarily true, as opposed to contingent propositions which are empirical. Contingent propositions could be wrong;

non-contingent propositions are true *a priori*. In a similar vein, Bradley and Swartz (1979) have drawn a clear distinction between *a priori* and *empirical* knowledge.

Consider problem-solving. Obviously you cannot solve a problem unless you *have* a problem; there must be a stage of *problem finding*. To find a problem, you must have some *general knowledge* in the field in which you hope to find your problem. Having found your problem, you must obviously investigate it, i.e. read around it, discuss it with experts, do some experiments. Having done that, or while you are doing it, you must obviously think about the problem; it will not solve itself! Hopefully you will find an answer, the stage of illumination. But your answer may be wrong, so you need to verify your answer – or disprove it, as the case may be! All these steps are non-contingent; they are *inherent* in the notion of problem-solving. Empirical investigation can clarify the nature of the mental processes involved at each stage – i.e. are unconscious processes involved in the incubation stage? They can also clarify other questions – i.e. are the stages clearly separated, or do they overlap? And they can tell us whether the answer comes suddenly – the 'aha!' experience, and as a kind of intuition, or slowly and in degrees, as a consequence of logical thinking. But the nature of the stages and their sequence, is given logically *a priori*; it could not be different!

All these theories have been lumped together under the rubric 'invention as heroic quest' (Perkins, 1992), who opposes to it his 'smart foraging model' (p. 239). His 'typography of invention' is probably more related to applied science, where the *aim* is often prescribed by the practical needs of industry, and where the importance of a search well-tuned to the topography of ideas involved may be exceptionally important (Weber & Perkins, 1992). Invention and patenting would seem to incorporate features additional to those treated in this book, but not on the whole differing in principle from the views of creativity here put forward (Usher, 1954; Barsalla, 1988). Certainly there are important differences in the context of invention, i.e. technology, which differentiates it from traditional discovery (Houndshell & Smith, 1988), but the essential act of creativity does not seem to differ.

The stages in the model are usually given as distinct and sequential, but in reality this is almost never true. Problems of any importance are usually complex, different parts interrelated and each requiring a solution which in turn changes the complexion of other, related problems. Thus solution of one sub-problem may lead to the reformulation of other sub-problems; this may necessitate surveys of new information, subsequent periods of incubation, new 'aha!' experiences, leading to a reformulation of the problem, and so on in rich profusion. The complexity of the pathways involved can be enormous, but it is not my intention to go into that question here; I am concerned with the *incubation* stage, and the question of involvement of the unconscious, and with the resultant *illumination* stage, and the role of intuition.

The importance of unconscious incubation and problem-solving (whether in science or not) is well illustrated in autobiographical tales told by famous scientists, writers and artists (Ghiselin, 1952). These have been repeated so often that I will restrict myself to just one or two; these prove nothing, but suggest relevant concepts and problems to attack with more sophisticated methods. Kipling (1918) once put it in typical verse form (typical in the sense of unmistakable):

> *This is the doom of the Makers – their Daemon lives in their pen.*
> *If he be absent or sleeping, they are even as other men.*
> *But if he be utterly present, and they swerve not from his behest,*
> *The word that he gives shall continue, whether in earnest or jest.*

In other words, genius relies on the Daemon slumbering in the unconscious; if he tries to go it alone, even genius becomes ordinary.

The way Daemon works is well illustrated in Coleridge's (1895) famous prefatory note to *Kubla Khan*. He writes:

> In the summer of the year 1797, the Author, then in ill-health, had returned to a lonely farmhouse between Porlock and Linton, on the Exmoor confines of Somerset and Devonshire. In consequence of a slight indisposition, an anodyne had been prescribed, from the effects of which he fell asleep in his chair at the moment that he was reading the following sentence, or words of the same substance, in "Purchas's Pilgrimate": 'Here the Khan Kubla commanded a palace to be built, and a stately garden thereunto. And thus ten miles of fertile ground were inclosed with a wall.' The Author continued for about three hours in a profound sleep, at least of the external senses, during which time he has the most vivid confidence, that he could not have composed less than from two or three hundred lines; if that indeed can be called composition in which all the images rose up before him as *things*, with a parallel production of the correspondent expression, without any sensation or consciousness of effort. On awaking he appeared to himself to have a distinct recollection of the whole, and taking his pen, ink, and paper, instantly and eagerly wrote down the lines that are here preserved. At this moment he was unfortunately called out by a person on business from Porlock, and detained by him above an hour, and on his return to his room, found, to his no small surprise and mortification, that though he still retained some vague and dim recollection of the general purport of the vision, yet, with the exception of some eight or ten scattered lines and images, all the rest had passed away like the images on the surface of a stream into which a stone has been cast, but alas! without the after restoration of the latter!

As we know only too well, the pool never became a mirror; what the Daemon had wrought in the unconscious, the conscious mind could not restore.

Of course, Coleridge is not known for his veracity; he often added details to twice-told stories which would make them more interesting – as indeed many people are wont to do! Did the legendary 'person from Porlock' in fact exist, or was he conjured up to explain away Coleridge's failure of imagination to complete the poem? This is the weakness of the anecdotal method, it demands reliance on self-motivated introspection which experiment teaches us to distrust.

Poincaré on creativity

The one account of creativity which above all others has been quoted again and again was composed by Henri Poincaré, a mathematician of genius whose word is perhaps to be trusted more than that of an opium-crazed poet, who was often economical with the truth. I will quote parts of his story which are relevant to our quest. Note the similarity of his account to the Campbell–Simonton–Fourneaux theory mentioned previously.

> What is mathematical creation? It does not consist in making new combinations with mathematical entities already known. Any one could do that, but the combinations so made would be infinite in number and most of them absolutely without interest. To create consists precisely in not making useless combinations and in making those which are useful and which are only a small minority. Invention is discernment, choice.
>
> How to make this choice? The mathematical facts worthy of being studied are those which, by their analogy with other facts, are capable of leading us to the knowledge of a mathematical law just as experimental facts lead us to the knowledge of a physical law. There are those which reveal to us unexpected kinship between other facts, long known, but wrongly believed to be strangers to one another.
>
> Among chosen combinations the most fertile will often be those formed of elements drawn from domains which are far apart. Not that I mean as sufficing for invention the bringing together of objects as disparate as possible; most combinations so formed would be entirely sterile. But certain among them, very rare, are the most fruitful of all.
>
> To invent, I have said, is to choose; but the word is perhaps not wholly exact. It makes one think of a purchaser before whom are displayed a large number of samples, and who examines them, one after the other, to make a choice. Here the samples would be so numerous that a whole lifetime would not suffice to examine them. This is not the actual state of things. The sterile combinations do not even present themselves to the mind of the inventor. Never in the field of his unconsciousness do combinations appear that are not really useful, except some that he rejects but which have to some extent the characteristics of useful combinations. All goes on as if the inventor were an examiner for the second degree who would only have to question the candidates who had passed a previous examination.
>
> But what I have hitherto said is what may be observed or inferred in reading the writings of the geometers, reading reflectively.
>
> It is time to penetrate deeper and to see what goes on in the very soul of the mathematician. For this, I believe, I can do best by recalling memories of my own, but I shall limit myself to telling how I wrote my first memoir on Fuchsian functions. I beg the reader's pardon; I am about to use some technical expressions, but they need not frighten him, for he is not obliged to understand them. I shall say, for example, that I have found the demonstration of such a theorem under such circumstances. This theorem will have

a barbarous name, unfamiliar to many, but that is unimportant; what is of interest for the psychologist is not the theorem but the circumstances.

For fifteen days I strove to prove that there could not be any functions like those I have since called Fuchsian functions. I was then very ignorant; every day I seated myself at my work table, stayed an hour or two, tried a great number of combinations and reached no results. One evening, contrary to my custom, I drank black coffee and could not sleep. Ideas rose in crowds; I felt them collide until pairs interlocked, so to speak, making a stable combination. By the next morning I had established the existence of a class of Fuchsian functions; those which come from the hypergeometric series; I had only to write out the results, which took but a few hours.

Then I wanted to represent these functions by the quotient of two series; this idea was perfectly conscious and deliberate, the analogy with elliptic functions guided me. I asked myself what properties these series must have if they existed, and I succeeded without difficulty in forming the series I have called theta-Fuchsian.

Just at this time I left Caen, where I was then living, to go on a geologic excursion under the auspices of the School of Mines. The changes of travel made me forget my mathematical work. Having reached Coutances, we entered an omnibus to go to some place or other. At the moment when I put my foot in the step the idea came to me, without anything in my former thoughts seeming to have paved the way for it, that the transformations I had used to define the Fuchsian functions were identical with those of non-Euclidean geometry. I did not verify the idea; I should not have had time, as, upon taking my seat in the omnibus, I went on with the conversation already commenced, but I felt a perfect certainty. On my return to Caen, for conscience' sake I verified the results at my leisure.

Then I turned my attention to the study of some arithmetical questions apparently without much success and without a suspicion of any connection with my preceding researches. Disgusted with my failure, I went to spend a few days at the seaside, and thought of something else. One morning, walking on the bluff, the idea came to me, with just the same characteristics of brevity, suddenness and immediate certainty, that the arithmetic transformations of indeterminate ternary quadratic forms were identical with those of non-Euclidean geometry.

Returned to Caen, I meditated on this result and deduced the consequences. The example of quadratic forms showed me that there were Fuchsian groups other than those corresponding to the hypergeometric series; I saw that I could apply to them the theory of theta-Fuchsian series and that consequently there existed Fuchsian functions other than those from the hypergeometric series, the ones I then knew. Naturally I set myself to form all these functions. I made a systematic attack upon them and carried all the outworks, one after another. There was one however that still held out, whose fall would involve that of the whole place. But all my efforts only served at first the better to show me the difficulty, which indeed was something. All this work was perfectly conscious.

Thereupon I left for Mont-Valerian, where I was to go through my military service; so I was very differently occupied. One day, going along the street, the solution of the difficulty which had stopped me suddenly appeared to me. I did not try to go deep into it immediately, and only after my service did I again take up the question. I had all the elements and had only to arrange them and put them together. So I wrote out my final memoir at a single stroke, and without difficulty.

I shall limit myself to this single example; it is useless to multiply them. In regard to my other researches I would have to say analogous things. Most striking at first is this appearance of sudden illumination, a manifest sign of long, unconscious prior work. The role of this unconscious work in mathematical invention appears to me incontestable, and traces of it would be found in other cases where it is less evident. Often when one works at a hard question, nothing good is accomplished at the first attack. Then one takes a rest, longer or shorter, and sits down anew to the work. During the first half-hour, as before, nothing is found, and then all of a sudden the decisive idea presents itself to the mind. It might be said that the conscious work has been more fruitful because it has been interrupted and the rest has given back to the mind its force and freshness. But it is more probable that the rest has been filled out with unconscious work and that the result of this work has afterwards revealed itself to the geometer just as in the cases I have cited; the revelation, instead of coming during a walk or a journey, has happened during a period of conscious work, but independently of this work which plays at most a role of excitant, as if it were the goad stimulating the results already reached during rest, but remaining unconscious, to assume the conscious form.

There is another remark to be made about the conditions of this unconscious work; it is possible, and of a certainty it is only fruitful, if it is on the one hand preceded and on the other hand followed by a period of conscious work. These sudden inspirations (and the examples already cited sufficiently prove this) never happen except after some days of voluntary effort which has appeared absolutely fruitless and whence nothing good seems to have come, where the way taken seems totally astray. These efforts then have not been as sterile as one thinks; they have set agoing the unconscious machine and without them it would not have moved and would have produced nothing.

The need for the second period of conscious work, after the inspiration, is still easier to understand. It is necessary to put in shape the results of this inspiration, to deduce from them the immediate consequences, to arrange them, to word the demonstrations, but above all verification is necessary. I have spoken of the feeling of absolute certitude accompanying the inspiration; in the cases cited this feeling was no deceiver, nor is it usually. But do not think this is a rule without exception; often this feeling deceives us without being any the less vivid, and we only find it out when we seek to put on foot the demonstration. I have especially noticed this fact in regard to ideas coming to me in the morning or evening in bed while in a semi-hypnogogic state.

Incubation and inspiration

This long quotation is here given because it well describes what countless mathematicians, scientists, writers, artists, and composers have described somewhat less clearly. (Many other examples are given by Hadamard (1959).) There is the preliminary labour; the incubation period; the sudden integration, owing its existence to inspiration rather than conscious logical thought, and finally the verification or proof, perfectly conscious. There is the description of the conditions under which incubation and inspiration occur; quiet, low cortical arousal conditions, not occupied with mental work, or any conscious consideration of the problem whose solution is sought. Such is the description of the *problems* the psychologist has to find a solution for; the description does not pretend to give us an answer. Poincaré's description preceded the theories of Campbell–Simonton and Furneaux quoted in a previous chapter, but Poincaré clearly disagrees (as I have done) with the notion that the associations formed during the incubation period are *random*. Indeed, he gives some of the same reasons as I have done why this is impossible.

Poincaré's (1908) very important observations are discussed at length by his biographer (Toulouse, 1910), and his mode of thinking compared with Einstein's by Miller (1992). He was one of the most noteworthy of the early pioneers to contribute introspectively to the discussion of creativity, and his book is still worth reading.

Poincaré's description tells us that no solution arrives during the preliminary unconscious work on the problem; it occurs after what psychologists call a rest period. But why? What happens during the rest period? Poincaré suggests two possibilities. One is that the rest 'has given back to the mind its force and freshness'. The other is that the rest 'has been filled out with unconscious work'. Experimental studies of 'reminiscence', as this improvement of performance during a rest period has been called, have always demonstrated a very robust effect, almost regardless of the activity involved (Eysenck and Frith, 1977). The major theory trying to account for this improvement was put forward by Kraepelin already in the 1890s; it corresponds to Poincaré's first alternative. This is usually phrased in terms of *dissipation of inhibition during rest*. The inhibition is built up during the work period preceding the rest, and dissipated during the rest, thus allowing a sharp rise in output to occur from the beginning of the rest period to its end.

An alternative explanation offered by Eysenck and Frith (1977) is in terms of *consolidation of the memory trace*. Memory traces acquired during the preliminary work period have to be consolidated to be available for future improvement, but such consolidation can only occur during rest – continued work does not allow consolidation to happen. Thus simply allowing work to continue is useless, it does not improve performance and output. Such consolidation is of course unconscious, and may be similar to Poincaré's second alternative.

A third possibility, not entirely different to that just mentioned, suggests that if during the conscious work period wrong ideas have been developed they prevent correct associations from emerging; it needs a period of quiet, during which the wrong ideas may weaken, to make possible alternative attempts at solution. These erroneous ideas, having most work done on them, require a lengthy period of conscious disregard to make possible the emergence into consciousness of the correct ideas, suppressed until then by the more strongly learned wrong ideas. There is good experimental evidence to show that high cortical arousal, typical of conscious problem solving, narrows the associative field and supresses the emergence of remote associations; a lower degree of cortical arousal allows these remote and unusual associations to emerge (Eysenck, 1973a; Easterbrook, 1959).

Much work has been done on reminiscence, and we know many of the laws according to which it operates; unfortunately little if any of this work has been directly concerned with our present problem. Usually the experiments have been concerned with the learning of certain motor skills and mental activities, like *pursuit rotor learning* (where the subject has to keep a metal stylus in contact with a small metal disc affixed to the rim of a rotating flat circular plate) or path learning (where the subject has to learn to trace with a pencil the path from a circle marked A to one marked B, and from there to C, etc., with the circles dotted irregularly around a page). Here would seem to lie one of the most inviting possibilities of being able to test specific hypotheses concerning what actually goes on during the course of incubation.

The two other variables suggested by Poincaré (and of course many others!) are inseparably joined. One is unconscious mental activity, the other is intuition, or the arrival at a solution to a problem on a basis other than conscious logical thought. (Intuition is defined by the dictionary as 'knowledge or perception not gained by reasoning and intelligence; instinctive knowledge or insight.') There is one important proviso: insight or intuitive knowledge is not necessarily correct; as we shall see, the intuitive insights gained by Ramanujan, one of the greatest mathematicians of our century, were right for the most part when checked by Hardy, another (non-intuitive) mathematician of outstanding fame, but quite a few were in fact mistaken. To link intuition with exactitude, correctness, precision or truth would be a mistake; intuition can be as wrong, treacherous, mistaken and erroneous as logical thinking; it is no guarantee of truth. The adjective 'intuitive' refers to the method of arriving at a conclusion; not any property of the conclusion itself. Intuition is the basis of the 'aha!' experience, but how many 'ahas' have had to be retracted?

Creativity and the unconscious

Our concern, then, is with unconscious mentation (an oxymoron to many readers!) and the (sometimes) resulting intuitive solution of problems, leading to the 'aha!' experience. Before turning to a detailed discussion of what experimental science has to say on these issues, I want to make it quite clear that nothing in this chapter is in any way connected with the theories of the man often believed to have been the creator and originator of the concept of the unconscious, namely S. Freud. As Whyte (1962) has shown so clearly in his book, *The Unconscious Before Freud*, there were over 200 philosophers, psychologists, medical men, historians, poets, men of letters, psychiatrists and scientists who postulated the existence and importance of unconscious processes of cerebration, motivation and perception before Freud. Freud appropriated (without acknowledgement usually!) many of the results of all this work, and added what to many critics seemed absurd elaborations which made the Freudian unconscious unacceptable and useless for serious theoretical work – or for experimental testing, for that matter (Eysenck, 1990a).

In a recent symposium on academic studies of unconscious processes, Greenwald (1992) emphasizes that 'psychoanalytic conceptions of the unconscious cognition lack empirical confirmation' (p. 166), and adds that 'academic psychologists have sometimes gone beyond empirical scepticism to suggest that the concept of *unconscious cognition* has no place in psychology' (ibid.). Neither extreme view, i.e. that of the 1930s to 1950s, which attempted to explain all human behaviour in terms of 'dynamic' concepts based on the Freudian unconscious, or that just mentioned which dismisses all appeals to unconscious forces as unscientific, is tenable; we have to proceed on a rather more pragmatic and less ideological level. It might be, for instance, that we are conscious of mental *content*, not mental *processes* (Nisbett and Wilson, 1977; Mandler, 1985); I shall return to a consideration of the experimental literature presently.

Let me first pursue the reasons for rejecting the Freudian unconscious. It seems that here as elsewhere, what is new in Freud is not true, and what is true is not new. To the impressive evidence existing at the turn of the century for unconscious mentation, Freud added notions of infantile trauma and Oedipus complexes, repressed into the unconscious, actively determining behaviour ever after, and leading to neurotic (or even psychotic) symptoms. There is no factual support for this view (Kline, 1981; Eysenck and Wilson, 1973), and Freud's own work amply disproves it. Let us consider, if only briefly, Freud's theory of dreams – dreams being, in his own words, the royal road to the unconscious. Freud insists that dreams are always wish-fulfilments, the wishes relating to repressed infantile material. In his book on dream interpretation, he gives many examples of the way in which he interprets dreams, but astonishingly enough *not one of those dreams deals with repressed infantile material*! Richard M. Jones (1970), a well-known Freudian, writes:

> I have made a thorough search of 'The Interpretation of Dreams' and can only report that there is not one illustration of wish-fulfilment which meets the criterion of reference to a repressed, infantile wish. Every illustration posits a wish, but every wish . . . . . is either a wish of out-and-out conscious reflection, or is a suppressed wish of post-infantile origin.

Thus the royal road to the unconscious even when travelled by Freud himself, leads to a non-Freudian unconscious. It is curious that thousands of readers have assumed that the Freudian dream interpretations served to prove the correctness of his theories, when in fact they utterly disproved them!

Of course, saying this does not mean that some of Freud's views may not be useful – particularly when he merely rechristens processes often described by others. Thus, it has become acceptable to talk of Freud's (1900/1938) notion of primary process thinking as opposed to secondary process thinking, notions used by Kris (1952) to investigate creative thinking. This contrast is fundamental in describing the continuum along which states of consciousness and types of thought vary (Fromm, 1978). Primary process thinking is free-associative, concrete, irrational and autistic; it is the matter of dreams and reveries, and in its more extreme forms it is the thought of psychosis. Secondary process thinking is abstract, logical and reality-oriented; it is the thoughts of everyday, waking reality. As Martindale (1990) points out, there is nothing new in this postulation except the terms used; Nietzsche (1927/1972) called the same states apollonian vs. dionysian, Wundt (1896) called them associationistic vs. intellectual, and even the ancient Greeks had a word for it. Freud had a talent for coining new names for old ideas, and then claiming these ideas as his own. It may be useful to employ the new terms, but at the risk of being saddled with much surplus intellectual luggage!

It might be said that it is Jung, rather than Freud, who has had most influence on the study of creativity; does what I said about Freud's notion of the unconscious apply to Jung's conception also? All one can say is that if Freud's notions have failed to be supported by empirical research, there has been practically no attempt to even test Jung's ideas; indeed he would probably be the first to agree that his ideas were far removed from natural science testing (Naturwissenschaft). This being so, of course, there is no empirical support for his vague and essentially untestable theories.

A possible exception is the research that has been done with the Myers–Briggs Type Indicator (Thorne and Gough, 1991), already mentioned in a previous chapter. This is a questionnaire allegedly based on Jungian typological thinking, giving scores on extraversion–introversion, the intuitive vs. sensing dimension, thinking vs. feeling dimension, and the judging–perceiving dimension. This creates 16 personality types which are said to be similar to Jung's theoretical concepts. I have always found difficulties with this identification, which omits one half of Jung's theory (he had 32 types, by asserting that for every conscious combination of traits there was an opposite unconscious one). Obviously the latter half of his theory does not admit of questionnaire measurement, but to leave it out and pretend that the scales measure Jungian concepts is hardly fair to Jung. Nevertheless, the results achieved with the questionnaire have been discussed in an earlier chapter; I have taken up there the other problem with this questionnaire, namely its very weak psychometric underpinnings.

To return to the more civilized unconscious of non-Freudian origin, particularly as it is related to creativity and intuition. The Neoplatonists, and in particular Plotinus, were already convinced that 'feelings can be present without awareness of them', and that 'the absence of a conscious perception is no proof of the absence of mental activity'. Koestler (1964) lists many other preceding Descartes's dualistic definition of mind as awareness, which would make unconscious ideas an oxymoron. There are theologians like St. Thomas Aquinas, mystics like Jacob Brehme, physicians like Paracelus, astronomers like Kepler, writers and poets like Dante, Cervantes, Shakespeare and Montaigne; by all of these the importance of the unconscious was taken for granted. Many sided with Descartes, but others took a more realistic view. Leibnitz stated that 'our clear concepts are like islands which arise above the ocean of obscure ones' Kant agreed;

> The field of our sense-perceptions and sensations, of which we are not conscious, though we undoubtedly can infer that we possess them, that is, the dark ideas in man, is immeasurable. The clear ones in contrast cover infinitely few points which lie open to consciousness; so that in fact on the great map of our spirit only a few points are illuminated.

Koestler lists a number of German poets and writers as supporting the idea of an unconscious origin of creative ideas, among them Herder, Schelling, Hegel, Fichte, and Goethe, who said: 'Man cannot persist long in a conscious state, he must throw himself back into the Unconscious, for his roots live there . . . . Take for example a talented musician, composing an important score: consciousness and unconsciousness will be like warp and weft.' And Fichte postulated the existence of *pre-conscious* states as 'a middle-condition of the mind'.

More recently, scientists like Fechner and Wundt entered the arena, with Fechner's famous metaphor of the mind as an iceberg, with only a fraction of it above the surface of consciousness, moved by the winds of awareness, but mostly by hidden under-water currents. Wundt seemed to support the contents-conscious, processes-unconscious distinction when he wrote:

> Our mind is so fortunately equipped, that it brings up the most important bases for our thoughts without our having the least knowledge of this work of elaboration. Only the results of it become conscious. This unconscious mind is for us like an unknown being who creates and produces for us and finally throws the ripe fruits in our lap.

Many authors used analogies to clarify their view of the relation between the conscious and the unconscious. I have already cited Fechner's notion of awareness as 'the top of the iceberg', Kant's 'few illuminated points on the great map of our spirit', Leibnitz's 'islands which arise above the ocean'. Others have used the notion of a spotlight which attention throws upon the scene, illuminating one central part and leaving others dark. All are agreed that *conscious awareness is a matter of degree*; there is no absolute difference. It is this problem that has stymied much of the work done by experimental psychologists in an effort to clarify the situation. If the consciousness–unconsciousness dichotomy is to be studied, our main weapon is the verbal report of awareness, but this is unreliable. The concept of the *limen* (conscious above, unconscious below) was established in the perceptual field, but it turned out to be not a point, but a region. If you show a letter, or a number, varying the size (as on an optician's chart), or its luminosity, or its distance, there is no sudden point where it is seen or recognized no more; instead the probability of correctly identifying it decreases gradually until it approximates chance. As Fig. 5.1 indicates, there is no abrupt change, as in 5.1a, but a gradual one, as in 5.1b. Even worse, different people have different cut-off points. When a percept is correctly identified 50% of the time, one person will answer 'Yes' to the question of whether he can see it, while another will say 'No'. Thus our criterion is far from perfect, and statistical methods are needed to sort out accuracy of perception from decision-making parameters (Tanner and Swets, 1954).

What did our ancestors base their views on? There was the matter of sleep; clearly unconscious for the most part, but with occasional bursts of altered consciousness (dreams). But something mental seems to be happening during sleep; Tartini is reported to have composed the Devil's Trill Sonata while asleep. Many scientists have reported the solution of a problem appearing mysteriously after a good nights sleep; so have mathematicians; Hadamard reports:

> One phenomenon is certain and I can vouch for its absolute certainty; the sudden and immediate appearance of a solution of the very moment of sudden awakening. On being very abruptly awakened by an external noise, a solution long searched for appeared to me at once without the slightest instant of reflection on my part – the fact was remarkable enough to have struck me unforgettably – and in quite a different direction from any of those which I have previously tried to follow,

Another good example is the physicist and Nobel Laureate, Otto Loewi, quoted by Ochse (1990), who awoke one night to find that he had discovered in his sleep a brilliant solution to a long-standing problem. He made a few notes in a notebook beside his bed, but on waking the next morning discovered to his chagrin that the notes were utterly illegible, and the solution forgotten! He went to his laboratory to try and make sense of his scrawl while working on the problem, but in vain. However, the next night he awoke with the same flash of insight, and this time was careful to record the idea that led him to gain the Nobel Prize, in a more legible form!

Poets, too, have acknowledged the positive contributions of sleep (Ghiselin, 1952). I have already quoted Coleridge's famous *Kubla Khan* episode. The evidence is too voluminous to be discussed in more detail, but of course it may be doubted whether dreaming can really be regarded as evidence of 'unconscious' thinking; perhaps it is merely less directed thinking using non-verbal and symbolic forms (Hall, 1953)?

More impressive manifestations of the unconscious are the following.

(1) Memory; memories must subsist in some unconscious form because we are certainly only conscious of a minute part of all that we know.
(2) The tip-of-the-tongue phenomenon; we know that we know some-

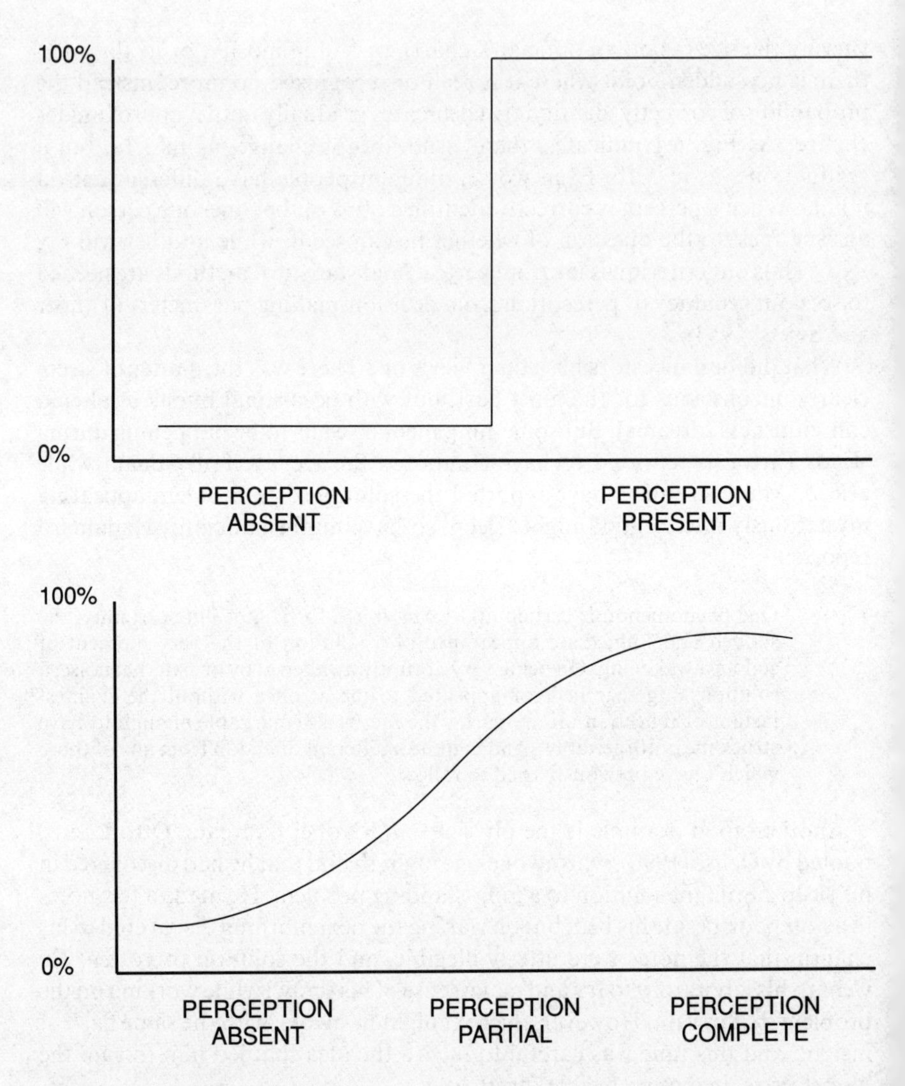

Fig. 5.1 Incorrect (5.1a) and correct (5.1b) notion of unconscious and conscious perception of a physical and mental stimuli (percepts).

thing, but temporarily cannot recall it; ergo it is at present unconscious.

(3) Hypnosis; under hypnosis we can be ordered to forget something we know very well, or not see or hear something clearly present. We are unconscious, in the sense that we cannot verbally describe or react to clearly present phenomena.

(4) We react to other people with like or dislike, but cannot consciously give a reason; we may later on discover the reason which at first was unconscious (i.e. unverbalizable).

All these and many other observations tell us that such cognitive material is temporarily unverbalizable, but it does not tell us whether the unconscious mind is simply a reservoir of memories, or whether it is in fact as active as the anecdotes quoted above may suggest. To answer that question we must look at the experimental evidence. The amount of such evidence is too large to be discussed in detail; a brief summary of work relevant to our main topic must suffice.

A good introduction is a much-quoted paper by Nisbett and Wilson (1977), entitled 'Telling more than we can know: verbal reports on mental processes.' As they point out, many cognitive psychologists have put forward the view that we may have no direct (introspective) access to higher order mental processes such as those involved in evaluation, judgment, problem-solving, and the initiation of behaviour (Mandler, 1975a,b; Miller, 1962; Neisser, 1967). As Miller (1962, p. 50) put it, 'It is the *result* of thinking, not the process of thinking, that appears spontaneously in consciousness.' Or, as Neisser (1967, p. 301) put it: 'The constructive processes (of encoding perceptual sensations) themselves never appear in consciousness'. And Mandler (1975a, p. 241) maintained that: 'The analysis of situations and appraisal of the environment . . . goes on mainly at the unconscious level'. Indeed, 'unconscious processes . . . include those that are not available to conscious experience, be they feature analyzers, deep syntactic structures, affective appraisals, conceptual processes, language production systems, action systems of many kinds' (p. 250).

Nisbett and Wilson (1977) survey a large amount of material to discover to what extent these suggestions (for they are not conclusions, not being based on experimental data for the most part) are in fact true. They deal with verbal reports on cognitive processes in dissonance and attribution studies, subliminal perception, reports of problem-solving processes, subjects' inability to report accurately on the effects of stimulation on responses, failure to report the influence of effective stimulus factors, erroneous reports about position effects on appraisal choice, erroneous reports about anchoring effects on prediction, reports on the influence of ineffective stimulus factors, erroneous reports about the emotional impact of literary passages, erroneous reports about the effects of distraction on reaction to films, etc.

They conclude that we are 'unaware of our unawareness', and imagine that consciousness covers a much larger ground than it actually does. We do indeed know a great deal about ourselves and others; we have a great storehouse of private knowledge (Jones and Nisbett, 1972); 'the only mystery is why people are so poor at telling the difference between private facts that can be known with near certainty and mental processes to which there may be no access at all'

(p. 255). In addition, 'we are often capable of describing intermediate results of a series of mental operations in such a way as to promote the feeling that we are describing the operations themselves' (p. 254). There are many other reasons for our mistaken notion that mental processes can be made conscious, and are capable of being studied introspectively; for the most part the evidence cited by Nisbett and Wilson seems to contradict that belief.

A more recent symposium entitled: 'Is the unconscious smart or dumb? (Loftus and Klinger, 1992) brings the issue up-to-date. There is universal agreement 'that the reality of unconscious processes is no longer questionable' (p. 764), although there is no uniform agreement about how sophisticated these processes are. Contributors to the symposium point out many of the problems inherent in the whole postulation of a clear-cut dichotomy between conscious and unconscious.

> To the clinician it is a helpful nomenclatural heuristic: patients do often give the impression of having no awareness of memories, scripts, or emotions that control their behaviour. Nevertheless, when laboratory psychologists try to discover the true divide between the subliminal (and the supraliminal), their efforts inevitably bog down in a tangle of methodological problems that in fact are covers for conceptual problems (Erdelyi, 1992).

However that may be, 'the psychological unconsciousness documented by latter-day scientific psychology is quite different from what Sigmund Freud and his psychoanalytical colleagues had in mind' (Kihlstrom, Barnhardt and Tataryn, 1992, p. 789). As they point out, his unconscious was hot and wet, seething with lust and anger; it was hallucinatory, primitive and irrational. Ours is more cold and dry, kinder and gentler than Freud's, and more reality-bound and rational. In other words, Freud's unconscious would be useless in accounting for any number of unconscious processes influencing creativity or genius; modern notions are much more relevant, even if there is still much argument (Kihlstrom, 1987; 1990).

Essentially, there is much agreement between anecdotal accounts of the importance of unconscious mental processes in creativity, intuition and the accomplishments of genius in science and the arts, and the experimental studies carried out by psychologists, usually without much thought about their possible relevance to creativity. The major agreements seem to be as follows.

(1) Unconscious mental processes exist and play an important role in human mentation.
(2) It is the *results* of thinking, not the *processes* of thinking which appear in consciousness.
(3) The unconscious of the experimenter is very different from that of the psychoanalyst.
(4) The former is concerned with rational cerebration, problem-solving,

and reality-bound, the latter with highly emotional reactions, lust and aggression.

(5) It may be the *low cortical arousal* which is typical of unconscious cerebration which enables it to use lower associative thresholds, wider associative horizons, and thus facilitates creative endeavour.

(6) The experience of intuition, so often associated with creative thought, is a function of unconscious processes. We must next turn to a discussion of this last point.

The measurement of intuition

That there is a clear distinction between *intuition* and logical, Aristotelian thinking has been an axiom of philosophy since the days of Plato. Westcott (1968) has given a succinct history of the development of philosophical schools and their use of these respective 'ways of knowing'. Thus Plato argued that ultimate reality may be known through a series of steps: induction, operating on the sensory world, yields *conceptions*; intuition, operating on conceptions, yields *ideas*, which are the ultimate reality. Descartes, Locke and Hume, on the other hand, argued that ultimate reality consisted of the fundamental fact of immediately present ideas and impressions which were known through intuition. Such knowledge-through-intuition is the only *certain* kind of knowledge, *pace* Bishop Berkeley's idealism and Kant's never-to-be-known *things-in-themselves*. These philosophical speculations are of interest only because they bestowed a continuing controversy on the newly-developing psychology of Wundt and his followers; this too is chronicled by Westcott. It has issued, among other things, in the controversy between the idiographic (understanding, intuitive) approach to personality, and the nomothetic (measuring, predictive) approach. This in turn led to the distinction between the clinical–intuitive and the statistical–measurement approach (Meehl, 1954; Kleinmuntz, 1990).

Knowledge gained through unconscious (unverbalizable) intuition may be of many kinds. Take an example from physics. Pakistan's bowlers have discovered and perfected a 'reverse swing' of the cricket ball; the bowler holds the ball as if to swing it one way, but it confounds the batsman by curving in the opposite direction. The bowlers who perfected this technique worked entirely by intuition, having no idea of the physics involved. Physicists have given the effect a good deal of thought, to assimilate it to our formal knowledge of physics. and discovered that the amount of swing imparted $= C_Y \times \frac{1}{4} \times p \times \mathrm{p}^2 \times \mathrm{s}$ divided by m, where C_Y is the sideforce coefficient, p is the air density, p is the distance the ball travels, s is the ball's surface area and m is the ball's mass. The *creative impulse* came from the bowlers, highly motivated to find a new

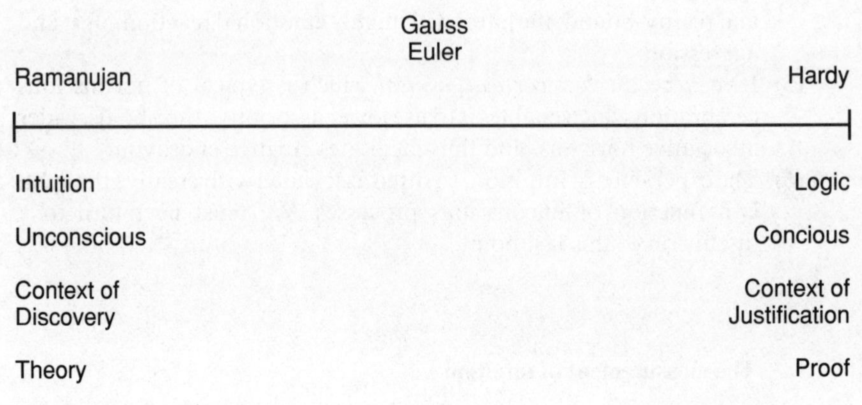

Fig. 5.2 Intuitive-logical dimensions, illustrated by famous mathematicians.

way to beat the bat; only after this breakthrough did orthodox science *explain* the intuitive discovery.

It was of course Jung (1926) who made *intuition* one of the four functions of his typology (in addition to thinking, feeling, and sensation). This directed attention from the *process* of intuition to intuition as a personality variable – we can talk about the intuitive type as opposed to the thinking type, the irrational as opposed to the rational. (Feeling, too, is rational, while sensation is irrational, i.e. does not involve a judgment.) Beyond this, Jung drifts off into the clouds peopled with archetypes and constituted of the 'collective unconscious', intuitions of which are held to be far more important than intuitions of the personal unconscious. Jung's theory is strictly untestable, but has been quite important historically in drawing attention to the 'intuitive person', or intuition as a personality trait.

Jung, like most philosophers, writers and psychologists, uses the contrast between 'intuition' and 'logic' as an absolute, a dichotomy of either – or. Yet when we consider the definitions and uses of the terms, we find that we are properly dealing with a continuum, with 'intuition' and 'logic' at opposite extremes, rather like the illustration in Fig. 5.2. In any problem-solving some varying degree of intuition is involved, and that may be large or small in amount. Similarly, as even Jung recognized, people are more or less intuitive; personalities are ranged along a continuum. It is often easier to talk *as if* we were dealing with dichotomies (tall vs. short; bright vs. dumb; intuitive vs. logical), but it is important to remember that this is not strictly correct; we always deal with continua.

The main problem with the usual treatment of 'intuition' is the impossibility of proof; whatever is said or postulated is itself merely intuitive, and hence in need of translation into testable hypotheses. Philosophical or even commonsense notions of intuition, sometimes based on experience as in the case of

Poincaré, may seem acceptable, but they suffer the fate of all introspection – they may present us with a problem, but do not offer a solution.

The same is true of psychoanalytic intuitions, which are discussed at some length by Westcott (1968); they are often contradictory, and there is no conceivable proof. Freud argued that therapeutic success, based on clinical intuition, could be used to demonstrate the truth of these intuitions, but the failure of achieving such therapeutic success must invalidate the intuition. (In the latest meta-analysis available, Swartberg and Stiles (1991) studied 19 experiments comparing the effects of short-term psychoanalysis – the most frequently used – with the effects of no treatment: there was no difference!) Thus certain types of 'intuition' cannot aspire to scientific status because they do not lend themselves to empirical testing.

Westcott (1968) appealed to another tradition when looking for a definition and conception of intuition that would be susceptible to empirical testing, and to scientific study and proof or disproof. As he points out, within the general context of contemporary cognitive theory the term 'intuition' is frequently used with specific reference to the process of rapid classification of a stimulus event, and the equally rapid attribution of the characteristics of its class to that event, even when these characteristics are not obviously manifest. This is an inferential event in which some of the premises are contained in the stimulus event and some in the coding system of the perceiver. Thus intuition deals essentially with the rapid (and incomplete) coding or categorizing of stimuli, and rapid extrapolation based on the class membership of the event categorized (Bruner, 1957; Bartlett, 1958; Levy, 1963). Such a definition lends itself easily to empirical study; it also clearly implies a *continuous*, not a categorical classification of problem-solving.

Westcott quotes with approval a definition, based on similar reasoning, given by Guiora *et al.* (1965):

> Intuition is the result of a process of making an apparently direct, immediate and accurate judgment and/or prediction – fashioned out of idiosyncratic associations reached through allo-logical principles – that has been set in motion by an amount of immediate external cues, normally inadequate for a logical judgment and/or prediction (Guiora *et al.*, 1965, pp. 9–10).

My own definition would run something like this: Intuition is a mode of cognitive functioning located at the opposite end of a contiuum from logical thinking, characterized by speed and suddenness of reactions (aha! experience), small number of relevant facts known or considered, feelings of certainty about the conclusion reached, reliance on unconscious (non-verbalizable) processes, not following the rule of Aristoletic logic, and relying on unusual associations and analogies.

This definition is not right or wrong; definitions try to embody the major lines of theoretical thinking and empirical study, and they do so more or less successfully. They are judged by their usefulness in bringing together known facts, and helping the discovery of new ones. This definition will serve to link together my general argument and such experimental work as has been done in this field, notably by Westcott; no more is claimed for it.

It may be noted that from this definition certain deductions follow which give rise to experimental paradigms. Thus by restricting the time allowed for thinking, we may force subjects in the direction of intuitive reactions (Farmer, 1961). We may use successful problem-solving accompanied by inability to verbalize the principle involved as an index of intuition (Bouthilet, 1948). We might score subjects on the basis of reaching conclusions founded on limited information (Westcott and Ranzoni, 1963). Only the Westcott studies have been pursued over a lengthy period, and with sufficient energy to give rise to a sufficiently large body of evidence to be cited in this chapter (Westcott, 1961, 1964, 1966). More recent studies have used lack of conscious (verbalizable) ability to describe the solution principles involved (Bowers *et al.*, 1990) and degrees of confidence (Griffin and Tversky, 1992), but there have been no studies recently looking at individual differences in the context of the intuitive personality in an experimental design.

Westcott (1968) has set down the requirements for measuring intuition according to his methodology. First, we need a series of problems which can be solved in the presence of varying amounts of information. Second, there must be a way of determining how much information a given individual requires, as compared with how much is ordinarily required. Finally, there must be consensus on what constitutes the correct conclusion. Four types of problems are in fact used by Wetscott (verbal series, verbal analogy, numerical series and numerical analogy problems.) The problems do not require any specialized subject-matter knowledge not familiar to all subjects. Each problem is presented in a sequence of steps, all of which are obscured from the subject, and the subject has to complete the problem after seeing as few or as many of the steps as desired. The analogy problems are constructed so that repeated examples of a relationship are obscured, but available, and the subject is to complete an analogy to the relationship after seeing as few or as many examples as desired. Each step in a series or each example in an analogy is called a 'clue'. Subjects are instructed to work out the solutions to each problem using as few clues as possible.

Performance was evaluated using four criteria.

(1) Clue use: the number of clues uncovered.
(2) Success: the number of problems solved correctly.
(3) Efficiency: the ratio of success/clue use.
(4) Confidence: the subject's rating of his degree of confidence in the correctness of his solutions.

Westcott (1968) gives a detailed account of all the many studies in which he used this paradigm; here I will only mention the major findings.

First, consider the question of *reliability*. Over many samples, clue use appears highly reliable (r = .80); in other words, *clue use is a highly consistent personality variable from one occasion or test to another*. Second, it is more reliable than success, which averages only about .5 to .6. Third, when verbal scores are correlated with numerical scores, clue use gives much higher values (.51) than success (.27) or efficiency (.27). Fourth, test–retest reliability for success and efficiency remains about the same as split-half reliability, but clue use drops from .80 to .50, i.e. shows much less temporal continuity. Fifth, there is practically no correlation between clue use and success, correlations for different samples varying from .24 to −.24. This is perhaps the most unexpected finding; one might have expected that the more clues a person has, the more likely he is to find the correct solution. Sixth, the intuitively obvious solution, namely that the brighter students are more successful and also need fewer clues is incorrect; success shows a moderate correlation with SAT ability scores (.12 verbal; .35 mathematical), but hardly any clue use (−.08 verbal; −.17 mathematical). Possibly because of the restriction of range in SAT scores in university student samples, ability is responsible for at most a very small proportion of the clue use variance, and little of the success variance (5% approximately). Seventh, academic achievement is correlated *neither* with clue use *nor* with success or efficiency (practically all correlations are insignificant statistically). As Wetscott summarized the evidence, 'it seems evident that neither academic aptitude nor academic success, is very concerned with intuitive thinking or with particularly efficient information-gathering and use' (p. 117.)

The correlations with *confidence* are of interest. Success, as might have been expected, correlates positively with confidence (around .45). Cue use correlates negatively with confidence, i.e. subjects who use fewer cues (are more intuitive) have higher confidence levels (−.25 approximately). Efficiency has the highest correlations with confidence, averaging .50. *Success based on minimal information is more likely to be associated with confidence.*

Personality measures of impulse expression, flexibility and manifest anxiety showed only very moderate correlation with the experimental variables, flexibility correlating *negatively* with cue use, anxiety correlating negatively with success, and efficiency correlating positively with flexibility and negatively with anxiety. The choice of personality variables was not optimal, and a much better choice could now be made. Using items rather than scales for analysis, and contrasting *extreme* groups, Westcott and Ranzoni (1963) found that low cue users were less conforming and socialized, while high cue users were more cautious, conservative and compliant. In these ways, the intuitive, low cue users seem to be very similar to the creative subjects considered in earlier chapters, the non-intuitive, high cue users to the non-creative.

Westcott and Ranzoni divided their groups into four: the low cue users who were successful, the low cue users who were unsuccessful, the high cue users who were successful, and the high cue users who were unsuccessful. They discovered that successful intuitive thinkers were unconventional, affectively involved, confident and comfortable. The 'wild guessers' were also unconventional and affectively involved, but desperate and anxious. The steady, successful problem solvers were cautious, orderly, and confident, while the careful but not unsuccessful problem solvers were cautious and compliant, but defensive and moralistic about themselves and the world.

Interviews, adjective checklists and the Allport – Vernon – Lindzey Study of Values were also administered. These very numerous data are given by Westcott (1968) in detail, but may be summarized as follows, and disclose traits and behaviours additional to those mentioned in the last paragraph. Thus the successful intuitive thinkers tend to be self-sufficient and comfortable in their unconventionality. As far as social interaction goes their skills and investment are relatively low; in social situations they show little affect; rather, they become emotionally involved in non-social affairs. Their investment is primarily related to abstract issues. In other words, they may be described as stable introverts.

Westcott goes on to say that these people had to explore uncertainties and entertain doubts, and live with these doubts and uncertainties without fear. In addition they enjoy taking risks, and are willing to expose themselves to criticism and challenge. They resist control and order imposed from without, but they do maintain a high sense of morality which is generated from within; they describe themselves as independent, foresighted, confident and spontaneous.

The wild guessers claim a high degree of social competence and involvement, although they are the least interested group of all as far as other people are concerned. Social interactions are characterized by considerable strife; they seem to be quite self-absorbed, and their affective investments seem to be directed toward themselves. This self-conscious and social conflict takes form, according to Westcott, through a driven and anxious unconventionality, coupled with strong and rigid opinions, and overlaid with cynicism. Risk and challenge are poorly managed, and do not culminate in creative production.

How about the non-intuitive? Those who are successful are distinguished by a very strong preference for order, certainty and control, combined with a high respect for authority. They are well socialized, with their stated interests and values well in the mainstream of their culture, but in the realm of interpersonal relations there is some social awkwardness and anxiety. They have difficulties in handling affect, and describe themselves as cautious, kind, modest and confident. The failures, too, describe themselves as cautious, kind and modest; they are concerned with religious matters, and the centralizing of authority and power. Everything is risky, and they are powerless to influence or control it, combined with a lack of self-confidence. They are quite conservative, generally passive, and don't like making waves.

Wescott (1968) also reports on a series of studies carried out with children, using a perceptual inference test in which a series of pictures is shown starting with a very rudimentary and fragmentary one to which details are gradually added in successive presentations; the crucial score is the stage at which children can correctly tell what the picture represents. The major finding, apart from the demonstration that the methods work as well with the children as did the earlier one with adults, was the discovery of a large difference in scores between children rated as creative vs. non-creative by their teachers. This reinforces belief that *creativity* and *intuition* are quite closely related. (These tests resemble those used to measure a factor called 'closure speed' in ability testing – Carroll (1993). There is good evidence for such a factor in the psychometric literature. There is also some similarity to the Frenkel-Brunswick (1949) measures of intolerance of ambiguity.)

The theory of intuition

Westcott's work is absolutely fundamental to an understanding of the concept of intuition, and remains unrivalled in that field. There have been sporadic efforts to measure intuition, but not in any systematic way, and not in reaction to creativity or personality (e.g. Metcalfe and Wiehe, 1987). The only major contribution has been a study by Simonton (1980a), which has been concerned with furnishing us with a predictive and exploratory model of intuitional analysis. Such a model is missing from the more experimental approach of Westcott, and as it makes an important contribution to the subject matter of this chapter, I will summarize Simonton's contribution in brief. He defines his use of the term by stating that by 'intuitive' he means behavioural adaptations to the environment which are unconscious, ineffable, and essentially probabilistic in character. (Ineffable is probably not a good term to use here; I interpret it to mean 'impossible to verbalize'.) Simonton interprets the term 'unconscious' along the lines of Miller's (1942) conception, and contrasts 'intuitive' with 'analytical', i.e. processes that are definitely conscious, capable of being communicated to others, and largely subject to discrete, logico-symbolic mediations. Simonton's major contribution is probably his discussion of the *probabilistic* character of intuitive mentation. This, in turn, is part of his general *associative* approach, which is probably the major approach to creativity, and the only one which has been subjected to experimental investigation, although it has been criticized as being over-simplistic and cumbersome (Anderson and Bower, 1973).

I am aware that during the 1950s and 1960s, associationist theory in experimental psychology was abandoned in favour of the cognitive approach, describing mental functions in terms of symbols and rules that operate on these symbols. More recently associationism has made a come-back in the form of connectionist or parallel distributed processing (PDP) models (Morris, 1989). This 'new associationism' differs from the old in postulating distributed representation of stimulus and response and a large number of parallel connections between input and output representations. But these differences, based as they are on computer simulation and neurobiological findings, cannot hide the essential similarities. The attempt is made to explain behaviour in terms of *simple units*, the *connection* between them, and some *general principles* concerning the origin of these connections. It would not be sensible here to argue the case for a remodelled associationist theory, or to translate Mednick and Simonton into PDP language. It would even be arguable that their theories can be rephrased in terms of cognitive models without loss. Fashions in theory come and go in psychology, but *plus ça change* . . . . .

Simonton begins by defining conditional probability, taking two events, X and Y, which often occur together. After a number of such coincidences, people acquire an expectation that whenever one event occurs, the other has a certain probability of occurring which can be calculated. Thus the conditional probability of X, given Y, is designated p (X/Y), and is defined by the equation: $p(X/Y) = N_{xy}/N_y$. The numerical value of any conditional probability must thus lie between 0 (no connection) and 1 (certainty).

This conditional probability function can be used to define more precisely the psychological concept of *association* or *conditioning* (Uttley, 1956). The *frequency* with which events occur together is one of the major predictors of the likelihood of an association between them being formed, or a conditional response arising. Thus, differences in associative strength would be reflected in the conditional probability function. Conditional probabilities can estimate perceptual expectancies and the strengths of habits (Atkinson, Bower and Crothers, 1965; Krantz *et al.*, 1974; Laming, 1973; Norman, 1972; Rosenblatt, 1958). Similarly, in classical Pavlovian conditioning the association between conditioned (CS) and unconditioned stimuli (UCS) would assume the value of p (CS/UCS).

Simonton goes on to argue that the psychological consequences of any given association will, in large part, depend on the strength of the conditional probability; the greater the strength, the more likely that action will follow. Simonton proposes that human information processing is sliced into 'layers' involving increasing conditional probabilities, and increasing action potentials appropriate to each layer. This gives rise to four distinct thresholds for different levels of behavioural and mental activity, each imposing higher probabilistic requirements before an association can enter the corresponding level of processing.

(1) *The threshold of attention.* (TA). This basically determines if the association between two events, ideas, or percepts has a high enough conditional probability to be worth further examination. There is wide recognition of the need for some selective attention of filtering prior to more extensive processing, leading to the ignoring of low probability associations (Broadbent, 1971).

(2) *The threshold of behaviour* (TB). This determines the point where conditional probability is large enough to have behavioural consequences. More specifically, this threshold represents the minimal

requirement for perceptual expectancies, classical and operant conditioning, and generalizations of discriminations free from linguistic or symbolic mediation.

(3) *The threshold of cognition* (TC). This determines if a conditional probabilistic relation is strong enough to become consciously embodied as a symbol. Even though below this threshold, 'a behavioural adaptation to the environment may occur without any corresponding ability to articulate the knowledge in verbal or symbolic form' (p. 9), only stimuli crossing the threshold of cognition make symbolization possible.

(4) *The threshold of habituation* (TH). This evaluates a particular conditional probability as being so high that it can be acted upon without further deliberation. Behaviours and cognition passing the threshold can be enacted with a minimum of attention.

At level 1, associations are very weak, and essentially *unconscious*; they have no phenomenological consequences. With higher conditional probability, and hence greater intensity, behavioural associations will be more phenomenologically prominent, but still weaker than associations which have surmounted the threshold of cognition; Simonton calls them *infraconscious*, in contrast to *fully conscious* associations which have passed the third threshold. Attention needs become reduced when behaviour becomes habitual, and Simonton labels habituated association *ultraconscious*.

Simonton locates the point of transition from *intuitive* to logical or *analytic* thinking between levels 2 (TB) and 3 (TC). Behavioural associations can continue to strengthen through practice, however, 'the status of the associations may change dramatically once the conditional probability passes the threshold of cognition' (p. 12). The association is now sufficiently strong to produce *symbolic representation* and enter consciousness, a moment often remembered as a flash of insight, or a 'Eureka' experience. This sudden change has several consequences. The first is that, because of the marked influence of verbal mediation on the level of cognition, *consciousness tends to be much more logical than infraconsciousness*. And second, *conscious processes often tend to become conservative rather than infraconscious* processes. Neisser (1963) has argued in more detail about these differences, and the Whorf–Sapir hypothesis (Whorf, 1960) documents how our thoughts may be shaped or even moulded by the syntactic and lexical limitations of our language, limitations from which level 2 associations are of course free.

Simonton goes on to postulate a personality typology based on the intuition–analytical dichotomy, but before turning to a discussion of this aspect of his theory we must return to the fundamental problem of conscious–unconscious mentation already considered to some extent earlier in this chapter. As we saw there, it has proved extremely difficult to mark a clear-cut difference, or sudden shift, from one to the other, such as Simonton postulates.

As Zuriff (1985) remarks, the debate about cognitive mediation in conditioning has remained unsolved. The question, he says, is plagued by methodological pitfalls. We are largely restricted here, as in the case of subliminal perceptions already discussed, to verbal reports which are responses to a variety of stimuli of which those entering the arguments are only some. Awareness *measurement* can itself create contingency *awareness*, as Baeyens, Eelen and Bergh (1990) have shown. Thus for conditioning, as for subliminal perceptions, the evidence for complete lack of awareness of the CS–UCS connection is doubtful.

There are suggestions that better evidence may be available from evaluative conditioning (Razran, 1954; Levey and Martin, 1975; Martin and Levey, 1978). A special issue of *Cognition and Emotion* (Watts, 1990) has been devoted to this topic, featuring particularly the work of Baeyens *et al.* (1990). Their work reinforces the evidence from numerous studies cited by them and by Levey and Martin (1990) to the effect that Pavlovian conditioning *can* take place in conditions of contingency unawareness, thus supporting Simonton's distinction between levels 2 and 3.

A rather different approach to answer the same question, and partly sidestep the problem of awareness measurement creating contingency awareness, was undertaken by Gidwani (1971). She was testing the hypothesis that extraverts would acquire verbal conditioned responses more quickly than introverts (Eysenck and Eysenck, 1985). However, this prediction only covers *unaware* conditions of acquisition; if contingencies are known, the extravert's higher degree of sociability might lead him to respond in the direction the experimenter might be thought to expect; under those conditions extraverts might learn the conditioned response more quickly. Awareness was tested after completion of the experiment, and it was found that when aware, introverts conditioned better; when unaware extraverts did. This suggests some degree of validity for the awareness assessment, and also seems to validate the assumption that Pavlovian conditioning is possible even when the subject is unaware of the contingencies. Thus the evidence *does not rule out* the existence of Pavlovian conditioning under conditions of lack of conscious awareness, but suggests that this is relatively rare, and that we are probably dealing with a continuum, as suggested in Fig. 5.1. Even so the distinction between levels 2 and 3 is a valid one, although it cannot be regarded as absolute.

We may now return to Simonton's typology. He postulates two orthogonal parameters, rather similar to Welsh's (1975) *origence* and *intellectence*. First he suggests there are important differences with respect to sheer numbers of associations; this would be related to a person's wealth of ideas and skills, i.e. his intellectence, or *g* (general intelligence). However, this is not enough; according to the threshold model, associations have different psychological consequences depending on their respective conditional probabilities. 'Hence there arises the need to hypothesize a second parameter: namely, the *probabili-*

ties belonging to the associations‘ (p. 15). This results in four characteristic types of associative conditional probability distributions. Types A and B have *large* numbers of associations available, Types C and D only much smaller ones (bright vs. dull). Types A and C have more infraconscious than conscious associations; their intuitive knowledge is much more extensive than their analytical knowledge. Types B and D on the other hand, tend to have far more habits and cognitions and far fewer intuitions. ‘Expressed differently, most of the knowledge of Type B persons is conscious rather than infraconscious so that their adaptive competence can be shared with other human beings through linguistic or symbolic communication’ (p. 17). Although presented as ‘types’, presumably Simonton would agree that both of these distinctions are based on *continuous* distributions, i.e. both the height of the distribution and its departure from normality are quantative variables. The ‘intuitive’ vs. ‘analytical’ distinction is of course closely related to Mednick's (1962) associative gradient, which brings us back to issues already discussed in an earlier chapter. He also links intuition with low cortical arousal, along the lines established by Easterbrook (1959) and Berlyne (1960), and already discussed. (I shall return to this notion in chapter 7.) The rest of the monograph is concerned with elaborating his theory, and suggesting, ‘empirical propositions’ which lay it open to experimental testing.

Simonton (1975) has linked his conceptions of intuition explicitly with creativity (as measured by the Barron–Welsh Scale) in an experiment involving tasks of high or low complexity. He started from the observation that individuals confronted with *complex* problems are more likely to solve these intuitively (Hull, 1920; Rees and Israel, 1935; Snapper, 1956) and added the hypothesis that intuitive thinking was more accessible to highly creative individuals. He found that the three variables involved were related very much as hypothesized: more creative subjects found *intuition* more effective for complex tasks, *analysis* for the simple tasks; this relation was reserved for the less creative subjects. It seems a pity that this highly promising experiment has not been replicated and followed-up; it seems to attack a fundamental question very successfully.

A case history of intuition: Ramanujan

So much for theoretical argument and factual demonstration concerning the intuitive and the analytical mind (always remembering that these two are the extremes of a continuum, with most people somewhere in the middle, using both intuition *and* analysis). It may be interesting (although of course proving nothing!) to consider briefly the intertwined lives of two great mathematicians who stood at opposite ends of this continuum – the Indian intuitionist *par excellence* Ramanujan and the rigorous analyst G. H. Hardy (Kanigel, 1991; Hardy, 1940). Mathematics lends itself to such a comparison particularly well because both intuition and rigour can be seen most clearly, and because there is a high degree of agreement – much greater than would be found in art, say.

Ramanujan was born in 1887, in Kumbakonam, a small Indian town on the sacred river Cauvery. He scarcely spoke during the first three years of his life; there were fears that he was dumb. He was an enormously self-willed child, and reacted very negatively to school when he was enrolled at the age of five. 'Even as a child he was so self-directed that, it was fair to say, unless he was ready to do something on his own, in his own time, he was scarcely capable of doing it at all' (Kanigel, p. 13); school to him meant shackles to throw off, rather than keys to knowledge. He was solitary, lacked all interest in sport, but was fond of asking questions of a vaguely scientific nature. His father was a lowly clerk, working all day in a shop; he was essentially brought up by an intensely religious mother.

To say that Ramanujan excelled at mathematics would be an understatement. He not only excelled his classmates but also his teachers. At a ceremony in 1904 Ramanujan was awarded the K. Rangenatha Rao prize for mathematics, the Headmaster stating that he deserved higher than the maximum possible awards – Ramanujan, he said, was off-scale. His genius took flight when a book on mathematics came into his possession, G. S. Carr's *Synopsis of Elementary Results in Pure and Applied Mathematics*. It contained some five thousand equations, and other mathematical facts – unsystematic, dealing with algebra, trigonometry, calculus, differential equations, analytical geometry – not a real synopsis, but rather a compendium. There was no *method* of arriving at formulas, or proving them in the book, so that Ramanujan had largely to fashion his own – which he did with delight.

Having discovered Carr, Ramanujan graduated from high school and entered Kumbakonam's Government College. He neglected all school subjects except mathematics, and thus he was left to do pretty well what he liked – it seems doubtful if his teachers could teach him anything. He failed English composition, and had his scholarship taken away from him. This began a long story of battling with the establishment, blind as always to true greatness, insisting on piddling rules made for mediocrity, and furious when faced with anything it could not understand. Ramanujan lost all his scholarships, he failed in school, he even lost students he was tutoring – his ever-original mind was not suited to rote-learning and traditional methods of doing things. Instead he started to keep notebooks in which he would record ideas, proofs, theorems – anything that engaged his interest and stimulated his original and inventive genius. He kept up the habit of entrusting his ideas and findings to his notebooks throughout his troubled youth and even later; they were the spur for many excellent mathematicians to follow in his footsteps and *prove* his intuitive ideas.

In any case, for five years Ramanujan was left alone to pursue mathematics, receiving no guidance, no stimulation, no money, except a few rupees earned from tutoring. College limited him, teachers could not evaluate his mathematics, and anything else would not do – one can see their problem. He tried to interest the leading professional mathematicians in his work, but failed for the most part. What he had to show them was too novel, too unfamiliar, and additionally presented in unusual ways; they could not be bothered. Kanigel describes in detail all the rejections, misapprehensions, and misconceptions Ramanujan encountered in his dealings with competent mathematicians; they simply could not deal with the novelty of his approach, the unusual methodology, and the exceptional fertility of his imagination. Above all, they failed to appreciate his intuition because of the lack of analytical proof; who was he to imagine he was right in his anticipation when there was no proof to show that his imagination was along the right lines (Mordell, 1941)?

Finally, in despair, Ramanujan wrote to G. H. Hardy, at the time our leading British mathematician, well-known for his attempts to introduce continental ideas of rigour and proof into the somewhat sloppy mathematics current in England. In many ways Hardy is the opposite of Ramanujan. Widely read where Ramanujan was unfamiliar with classical and modern methods and developments; insistent of proof where Ramanujan was only interested in intuition; omniscient in academic subjects where Ramanujan was entirely ignorant. At this point it may be useful to say a few words about intuition and proof in mathematics; the matter is discussed in more detail by Hardy (1979).

As Hardy points out, proof and its importance in mathematics is a very difficult subject. Most readers will probably imagine that proof is a central and all-important part of mathematical science, but it is a rather late arrival which was firmly established on the continent. Newton's and Leibnitz's notions of calculus, for instance, lacked all rigour; it was only Cauchy in his *Cours d'Analyse* who introduced rigour into the subject. Hardy points out that 'all physicists, and a good many quite respectable mathematicians, are contemptuous about proof' (p. 15). He quotes Eddington as maintaining that proof, as pure mathematics understand it, is really quite uninteresting and unimportant, and that no-one who is really certain that he has found something good should waste his time looking for a proof. Of course this is a gross exaggeration; Eddington suggested that there were exactly 136.2^{256} protons in the universe, but could not resist the temptation of proving it. (He was quite wrong, of course!)

Hardy states that there are some points concerning proof where nearly all mathematicians are agreed. The first point is that even though we may not understand exactly what proof is, we can, in ordinary analysis at any rate, recognize a proof when we see one (rather like the proverbial elephant!). Secondly, there are two different reasons for any presentation of a proof. The first is simply to secure conviction. The second is to exhibit the conclusions as the climax of a conventional pattern of propositions, a sequence of propositions whose truth is admitted and which are arranged in accordance with rules. Nearly always, the first reduces to the second. As he says, we may be able to recognize that 5, or even 17, is a prime, but nobody can convince himself that $2^{127}-1$ is a prime except by studying a proof. 'No one has ever had an imagination so vivid and comprehensive as that' (p. 16).

Now a mathematician usually discovers a theorem by an effort of intuition; the conclusion strikes him as plausible, and he sets to work to manufacture a proof. Sometimes this is a routine matter, 'but more often imagination is a very unreliable guide' (p. 16). This is particularly true in the analytical theory of numbers, where even Ramanujan's imagination led him very seriously astray. As an example, Hardy uses a false conjecture which was even endorsed by Gauss, and which took about 100 years to refute. It concerns the distribution of prime numbers, which is the central problem of the analytical theory of numbers.

How would we expect the very embodiment of proof and rigour to react to the outpourings of the untutored, intuitive, entirely original and incredibly creative Ramanujan? Hardy came from a lower middle-class family; his father was a shop-keeper, but the family emphasized intellect and learning. Hardy made his way to Winchester, one of England's leading public schools (for 'public' read 'private' of course!), and from there to Cambridge, through outstanding excellence and achievement. The letter he received, out of the blue, began as follows:

> Dear Sir,
> I beg to introduce myself to you as a clerk in the Accounts Department of the Port Trust Office at Madras on a salary of only £20 per annum. I am now about 23 years of age. I have had no University education, but I have undergone the ordinary school course. After leaving school I have been employing the spare time at my disposal to work at Mathematics. I have not trodden through the conventional regular course which is followed in a University course, but I am striking out a new path for myself. I have made a special investigation of divergent series in general and the results I get are termed by the local mathematicians as 'startling'.

In the second paragraph, he insisted that he could give meaning to negative values of the gamma function. In the third he disputed an assertion contained in a mathematical pamphlet Hardy had written three years earlier. Ramanujan enclosed a number of pages with some of his intuitive findings.

For Hardy, as Kanigel says, Ramanujan's pages of theorems were like an alien forest whose trees were familiar enough to call trees, yet so strange they seemed to come from another planet. Indeed, it was the strangeness of Ramanujan's theorems, not their brilliance, that struck Hardy first. Surely this was yet another crank, he thought, and put the letter aside. However, what he had read gnawed at his imagination all day, and finally he decided to take the letter to Littlewood, a mathematical prodigy and friend of his. The whole story is brilliantly (and touchingly) told by Kanigel; fraud or genius, they asked themselves, and decided that genius was the only possible answer. All honour to Hardy and Littlewood for recognizing genius, even under the colourful disguise of this exotic Indian plant; other Cambridge mathematicians, like Baker and Hobson, had failed to respond to similar letters. Indeed, as Kanigel says, 'it is not just that he discerned genius in Ramanujan that stands to his credit today; it is that he battered down his own wall of skepticism to do so' (p. 171).

The rest of his short life (he died at 33) Ramanujan was to spend in Cambridge, working together with Hardy who tried to educate him in more rigorous ways and spent much time in attempting to prove (or disprove!) his theorems, and generally see to it that his genius was tethered to the advancement of modern mathematics. Ramanujan's tragic early death left a truly enormous amount of mathematical knowledge in the form of unproven theorems of the highest value, which were to provide many outstanding mathematicians with enough material for a life's work to prove, integrate with what was already known, and generally give it form and shape acceptable to orthodoxy. Ramanujan's standing may be illustrated by an informal scale of natural mathematical ability constructed by Hardy, on which he gave himself a 25 and Littlewood a 30. To David Hilbert, the most eminent mathematician of his day, he gave an 80. To Ramanujan he gave 100! Yet, as Hardy said,

> the limitations of his knowledge were as startling as its profundity. Here was man who could work out modular equations and theorems of complex multiplication, to orders unheard of, whose mastery of continued fractions was, on the formal side at any rate, beyond that of any mathematician in the world, who had found for himself the functional equation of the Zeta-function, and the dominant terms of many of the most famous problems in the analytical theory of numbers; and he had never heard of a doubly periodic function or of Cauchy's theorem, and had indeed but the vaguest idea of what a function of a complex variable was. His ideas as to what constituted a mathematical proof were of the most shadowy description. All his results, new or old, right or wrong, had been arrived at by a process of mingled arguments, intuition, and induction, of which he was entirely unable to give any coherent account (p. 714).

Ramanujan's life throws some light on the old question of the 'village Hampden' and 'mute inglorious Milton'; does genius always win through, or may the potential genius languish unrecognized and undiscovered? In one sense the argument entails a tautology: if genius is *defined* in terms of social recognition, an unrecognized genius is of course a *contradicto in adjecto*. But if we mean, can a man who is a potential genius be prevented from demonstrating his abilities?, then the answer must surely be in the affirmative. Ramanujan was saved from such a fate by a million-to-one accident. All his endeavours to have his genius recognized in India had come to nothing; his attempts to interest Baker and Hobson in Cambridge came to nothing; his efforts to appeal to Hardy almost came to nothing. He was saved by a most unlikely accident. Had Hardy not reconsidered his first decision, and consulted Littlewood, it is unlikely that we would ever have heard of Ramanujan! How many mute inglorious Miltons (and Newtons, Einsteins and Mendels) there may be we can never know, but we may perhaps try and arrange things in such a way that their recognition is less likely to be obstructed by bureaucracy, academic bumbledom and professional envy. In my experience, the most creative of my students and colleagues have had the most difficulty in finding recognition, acceptance, and research opportunities; they do not fit in, their very desire to devote their lives to research is regarded with suspicion, and their achievements inspire envy and hatred.

Bronowski (1956) once wrote a paean to scientists, which is as naïve as it is stereotyped:

> By the worldly standards of public life, all scholars in their work are of course oddly virtuous. They do not make wild claims, they do not cheat, they do not try to persuade at any cost, they appeal neither to prejudice nor to authority, they are often frank about their ignorance, their disputes are fairly decorous, they do not confuse what is being argued with race, politics, sex or age, they listen patiently to the young and to the old who both know everything. These are the general virtues of scholarship, and they are peculiarly the virtues of science.

In my experience scientists are no different from other human beings. Many of them do make wild claims, cheat, appeal to prejudice and authority, claim omniscience, fight like Manx cats, and are jealous of their peers. To help potential young rivals, more creative and more original in their thinking, would certainly go against the grain. There are exceptions, but to deify scientists is factually quite wrong and historically incorrect; Newton (Westfall, 1980) is a much more likely role model – cheating, constantly engaged in underhand battles about priority, insanely jealous of others' achievement, trying to do rivals down, seeking authoritative office – a poor human being, yet one of the greatest scientists ever! If the budding genius had to depend on older peers, he would be in a sorry state! Potential geniuses can fail, particularly in our own bureaucratized society where governments seek to treat science as a milchcow, and organize universities like factories. The free spirit of the genius will find it more and more difficult to find a niche in such a world!

Together, Hardy and Ramanujan achieved what neither could have done alone; they made an entry in the history of mathematics that will endure over time. But, as Bollobas (1988) observed, while Hardy furnished the technical skills needed to attack the problem, 'I believe Hardy was not the only mathematician who could have done it. Probably Mordell could have done it. Polya could have done it. I'm sure there are quite a few people who could have played Hardy's role. But Ramanujan's role in that particular partnership I don't think could have been played at the time by anybody else' (p. 78).

The greatest mathematicians – Gauss, Euler – of course combine the gifts of intuition and analysis; what makes Hardy and Ramanujan so interesting is precisely the one-sidedness of their gifts. Ramanujan without a doubt was a genius, but in the words of Mark Kac (1985), a Polish émigré mathematician, he was a 'magician' rather than an 'ordinary genius'.

> An ordinary genius is a fellow that you and I would be just as good as, if we were only many times better. There is no mystery as to how his mind works. Once we understand what he has done, we feel certain that we, too, could have done it. It is different with the magicians. They are, to use mathematical jargon, in the orthogonal complement of where we are and the working of their minds is for all intents and purposes incomprehensible. Even after we understand what they have done, the process by which they have done it is completely dark.

This is another way to characterize the *intuitive* ('magician') as opposed to the analytical ('ordinary genius') scientists or mathematician (and probably, *ceteris paribus*, artist as well); we see the intuitive worker as intrinsically more 'creative' just because the origins of his creativity are hidden in the unfamiliar cliffs and caves of the unconscious. It is interesting to note that Hardy did not enjoy Hadamard's explanation of creativity in terms of unconscious activity; he 'did not find it convincing', and thought that his distaste for all forms of mysticism might be prejudicing him unduly. Even when forced to agree with Hadamard's description of the nature of the unconscious activity, he 'did not *like* this kind of language'. To him, it verged on nonsense. (Perhaps he had been frightened off by reading about psychoanalysis!)

Kanigel's book on Ramanujan and Hardy should be read by anyone interested in genius, and its varieties; it gives sum and substance to more abstract discussions and arguments. It also shows just how much distance there is between that which we have to explain, and our puny achievements in explaining it. To contemplate this abyss stretching between reach and grasp may be a salutory experience; as Browning said: 'Ah, but a man's reach should exceed his grasp, or what is heaven for?' Galileo's rolling wooden balls down inclined planes was a true predecessor of our landing a man on the moon; perhaps in a few hundred years we shall see our many efforts to measure intuition as the beginnings of a true understanding of genius!

Possibly we will then also be able to integrate our ideas of 'creativity' with the findings of quantum mechanics (Goswami, 1993). According to the quantem model, the mind-brain consists of classical and quantum functions and structures. It is in the latter that we would look for creative ideation in the form of 'unlearned coherent superpositions or states of simultaneous multiple possibilities' (McCarthy, 1993, p. 201). It is suggested that 'the relationship among attention, self-awareness, context, ambiguity, uncertainty, and conscious and unconscious perception in creative ideation are proposed to be experienced discontinuously due to an indeterminate element inherent in nature' (*ibid*, p. 201). Whether deductions from such a position are really testable at the moment, as McCarthy (1993) suggests, is perhaps doubtful, but the curious inclusion of human consciousness in the esoteric formulations of quantum mechanics (Edelman, 1992) suggests that we cannot reject such unlikely speculations as completely unreasonable. Only time will tell whether quantum theory can really help us in our quest.

6 The nature of psychopathology: psychoticism[1]

No great genius has ever been without some madness.

Aristotle.

Psychosis and psychoticism

In previous chapters we have seen that while psychopathology was clearly enmeshed with genius and creativity, the usual psychiatric concepts, e.g. psychosis, were clearly counter-indicated; psychosis (schizophrenia, manic-depressive disorder) was rarely found in geniuses, equally rarely, in creative writers, mathematicians, scientists, or architects, and in any case its diagnosis was so unrealiable, even in live patients studied intensively, that diagnosis on the basis of accounts written centuries ago would be pretty worthless. An even more serious objection would be that the orthodox psychiatric system of nosology is not in line with modern discoveries, and entails many serious defects. The major defect, which is absolutely fundamental, concerns the notion that psychiatric disorders differ *qualitatively* from normality, just as tuberculosis, or malaria, or mad cow's disease is differentiated qualitatively from healthy normality. Such a categorical differentiation was already criticized by Kretschmer (1936, 1948) and many others, and does not accord with reality (Claridge, 1985). This point will be discussed in more detail presently.

A very relevant weakness, often noted in psychiatric nosology, is the *reliability* of diagnoses; different psychiatrists very frequently apply different diagnostic labels to a given patient. It is frequently claimed that the arrival of DSM-III, the Diagnostic and Statistical Manual of Mental Diseases, published by the American Psychiatric Association, has overcome this difficulty, and that now all is plain sailing. A recent book by Kirk and Kutchins (1992) has shown clearly that these claims, often made by official sources, are quite unjustified; *parturient montes nascetur ridiculus mus*! (The mountains will be in labour, to produce a ridiculous little mouse! Horace.) Reliability is still as low as ever, and the notion that long-deceased geniuses can be reliably diagnosed is no nearer to reality. The reasons for the failure of DSM-III are not far to seek; its authors did not look at scientific evidence, but relied on committees and

[1] Parts of this chapter are taken from Eysenck (1992a).

backroom politicking, not the best way of constructing a scientific taxonomy (Eysenck, Wakefield and Friedman, 1983)!

Another weakness of modern psychiatric nosology is the whole-hearted acceptance of Kraepelin's (1897) distinction between schizophrenia and manic-depressive disorder (to use the modern terms). As we shall see, these are not separate diseases, separated categorically and qualitatively as would be physical disorders like cancer and heart disease. This notion also is not true; the evidence against this view of 'separate diseases' is quite strong. Thus the basis on which most of the papers concerning the psychopathology of genius have been grounded is so insecure that the results require drastic re-interpretation (Eysenck, 1992a). The major aspect of this re-interpretation is the postulation of a fundamental dimension of personality, namely *psychoticism*, which is a dispositional variable or trait predisposing people to functional psychotic disorders of all types. It is the major purpose of this chapter to introduce this variable, define it, discuss its measurement, and suggest its relation to creativity and genius. I will start with a brief historical introduction.

In 1952, I suggested that in addition to neuroticism (N) and introversion–extraversion (E) there existed a third major dimension of personality, called psychoticism (P) which was orthogonal to N and E (Eysenck, 1952a, b). Work testing various deductions from this hypothesis has been published periodically (Eysenck, Granger and Brengelmann, 1957; Eysenck, 1970b; Eysenck and Eysenck, 1976; Eaves *et al.*, 1989), with largely positive results; reviews of the many studies generated by the original theory have been published by Claridge (1981, 1983) and Zuckerman (1989b; Zuckerman, Kuhlman and Camac, 1988). A recent paper (Eysenck, 1991b) has attempted to consider the relation between the three dimensions postulated in the P–E–N model, and other typologies, such as Cattell's 16 PF (Cattell, Eber and Tatsuoka, 1970) and the 'Big Five' (John, 1990). It is suggested that the P–E–N model constitutes a paradigm (in the Kuhnian sense) in personality research (Eysenck, 1983b), and fulfills the stringent criteria suggested for acceptance of a paradigm (Eysenck, 1991b).

In this chapter I shall be concerned with certain substantial criticisms of the P dimension, i.e. criticisms which are concerned not with purely psychometric issues, but with issues of great theoretical importance. To understand these issues, it may be useful to introduce in some detail the model I am putting forward. Fig. 6.1 illustrates its major features. The abscissa constitutes a dispositional personality trait, psychoticism, which extends from the left (low P – high empathy, socialization, co-operativeness) to the psychotic characteristics and syndromes shown on the right (traits characteristic of high P are shown in Fig. 6.2). The distribution of P is more or less normal; in actual fact it has usually been skewed to the right (Eysenck and Eysenck, 1975), and although recent improvements in the scale have ameliorated this tendency, it has not been abolished (Eysenck, Eysenck and Barrett, 1985), but still persists. Whether this feature is the product of psychometric faults in

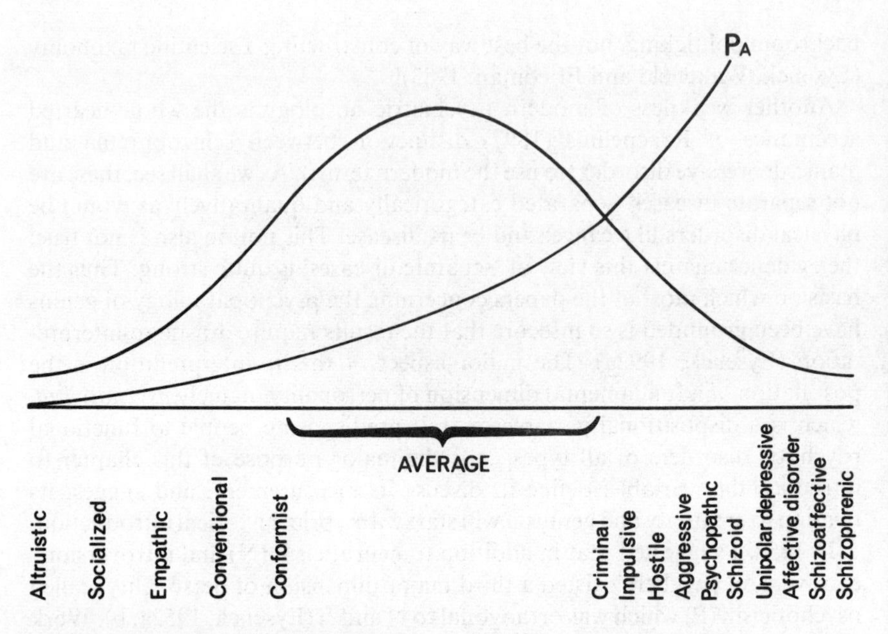

Fig. 6.1 Psychoticism as a personality variable P_A: probability of a person at a given period on the abscissa developing a psychotic disorder.

questionnaire construction, or inherent in the 'true' distribution of P, is not known; for our present purpose the answer is irrelevant. It may be noted in passing, however, that J-shaped distributions are quite common in psychology (Walbert *et al.*, 1984), and that Allport (1934) many years ago demonstrated the applicability of his J-curve hypothesis is conforming behaviour, which in many ways is the obverse of psychoticism. Conformity is frequent, psychoticism relatively rare – thank goodness!

Psychosis (schizophrenia, manic-depressive illness) is postulated to occur under environmental stress with a probability P_A, which is a monotonic function of psychoticism, as shown in the figure. Psychosis is not regarded as a category qualitatively different from normality, a point I have tried to establish by reference to criterion analysis (Eysenck, 1950, 1952a, b), and similarly different psychoses are not regarded as categorically different from each other; in both cases we are dealing with *continua* of one kind or another. Close to psychosis at the right of the diagram are behaviours variably diagnosed as schizoid, 'spectrum' or psychopathic, with 'personality disorder' a more recent synonym. Fig. 6.2 shows some of the traits the observed intercorrelations between which provide the ultimate justification for the postulation of P as a dimension of personality (Eysenck and Eysenck, 1976).

This model clearly violates a number of psychiatric assumptions, and it is important to answer possible objections if the model is to prove acceptable. In

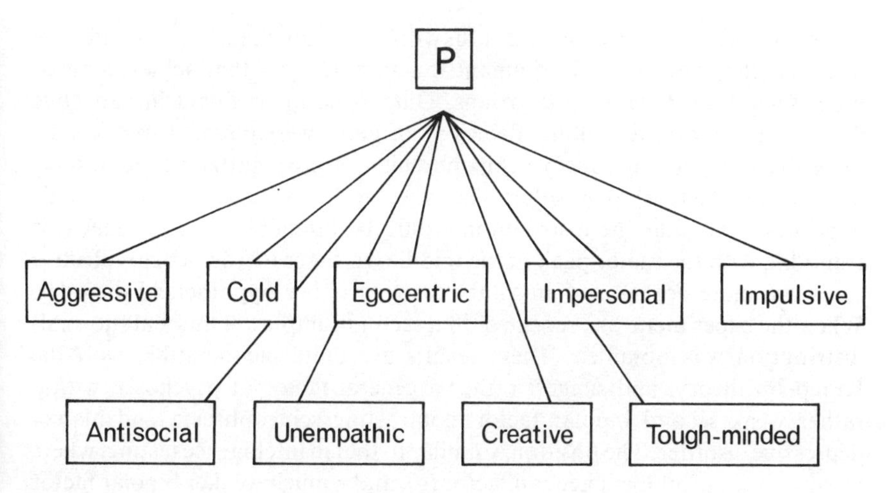

Fig. 6.2 Traits characteristic of high psychoticism.

addition, there have been some psychological criticisms which also demand an answer. I shall try to answer these criticisms in detail.

'Psychosis' or different psychoses

The first problem to be tackled is that of whether we are dealing with 'Psychosis' or different psychoses. During the last century it used to be assumed that there was a common feature to all functional psychoses (the theory of the *Einheitspsychose*); Griesinger (1861), Guislain (1833), Neumann (1859) and Zeller (1837) may be quoted in support of such a view. Kraepelin (1897) separated manic-depressive insanity from dementia praecox (schizophrenia), conceiving of them as unrelated diseases; this conception is prevalent in most if not all textbooks of psychiatry and clinical psychology. (See Berrios (1987) for an historical introduction.) Kraepelin (1920) himself pointed out some of the difficulties raised by this bifurcation; as he explained:

> No experienced psychiatrist will deny that there is an alarmingly large number of cases in which it seems impossible, in spite of the most careful observation to make a firm diagnosis . . . it is becoming increasingly clear that we cannot distinguish satisfactorily between these two illnesses and this raises the suspicion that our formulation of the problem may be incorrect.

The well-known difficulty of obtaining acceptable reliabilities in the diagnosis of psychotic disorders bears ample testimony to this problem.

Few studies, oddly enough, have been devoted explicitly to the solution of this very fundamental problem. Kendall and Gourlay (1978) constitute an important exception to this criticism. Selecting 146 patients with a group diagnosis of schizophrenia and 146 with a group diagnosis of affective disorder, they devised a discriminant function analysis that achieved maximum separation between the groups. Diagnostic items favouring affective illness (early morning waking, delusions of guilt) were counted to one side, items favouring a diagnosis of schizophrenia (affective flattening, Schneiderian first-rank factors) to the other. The expectation on the basis of Kraepelin's hypothesis was that the distribution would be bimodal; in actual fact it is trimodal, with the major peak in the middle. In other words, schizo-affective cases are more common than 'pure' cases of either hypothetical disorder. When the experiment was repeated on a second sample a unimodal (normal) distribution was obtained. These results are quite incompatible with the Kraepelin theory, and suggest rather a general factor of psychosis, with a rather weak second bipolar factor, contrasting schizophrenia and manic-depressive disorder. The position is similar to that in intelligence testing, where we also find a prominent general factor (*g*), and a much weaker bipolar factor dividing verbal from non-verbal tests (Eysenck, 1979).

Cloninger *et al.* (1985) have presented some contrary evidence to suggest that there does exist a 'point of rarity' between the symptom complex of schizophrenia and that of other psychiatric disorders. They used the self-report ratings of a series of 500 psychiatric outpatients and 1249 of their first-degree relatives, and derived a discriminant function distinguishing between the symptoms of schizophrenia and those of other conditions from the ratings of half their subjects; they then obtained a bimodal distribution of scores on this function when it was cross-validated using the ratings generated by their remaining subjects. However, their data are much less objective than those used by Kendell and Gourley.

It seems, as Farmer, McGuffin and Bebbington (1988) point out, that 'in general, criteria that incorporate longitudinal variables such as duration of illness in their definition . . . fare better (with respect to the prediction of short and long-term outcome) than those relying purely on cross-sectional psychopathology' (p. 43), but this, combined with the varying nature of such criteria, leaves us in the curious position where, as Brockington, Kendell and Leff (1978) have put it, the previous state of inarticulate confusion in the diagnosis of schizophrenia has been replaced by a babble of precise but differing formulations of the same concept. This uncertainty must add considerably to the problem of deciding between a dimensional and a categorical basis of classification.

Kendell and Brockington (1980) tested another deduction from the Kraepelinian model, namely that discontinuity between diagnostic groups could be discovered by analysing the relation between diagnostic score and some outcome variable, such as time in hospital, occupational record, social outcome, etc. Non-linear regression would be indicative of a genuine discontinuity, but in testing eight outcome criteria against the schizophrenic-affective continuum, Kendell and Brockington failed to discover any such lack of linearity. As they say, 'the results of this further analysis do not lend support to the view that schizophrenic and affective psychoses are distinct entities' (p. 266).

Quite generally, it appears that the likelihood of schizo-affective patients getting better is intermediate between that of schizophrenics (worst prognosis) and affective (best prognosis) (Crougham, Welner and Robinson, 1974; Brockington, Kendell and Wainwright, 1980). Bimodality of outcome is not usually found, but continuity seems to be the rule (Crow, 1986).

When Kasanin (1933) introduced the concept of 'schizo-affective' psychoses, these were supposed to be relatively rare. Clearly they are not, and neither do they constitute a third type of psychosis; most psychotics seem to fall into this category, and there are no clear boundaries between them, schizophrenics and manic-depressives. Kendell and Brockington (1980) developed a method for establishing a non-linear relationship between symptomatology and outcome, but failed to find any such relationship in a sample of 127 unselected psychotic and 105 schizo-affective patients; as they say, 'it has to be noted that yet another attempt to demonstrate discontinuity has failed'. Dimensional rather than categorical taxonomic thinking is clearly indicated.

It may be useful to look at one further type of evidence, namely that furnished by factor analysis and cluster analysis. The data were collected and originally analysed by Everitt, Gourlay and Kendell (1971), who rated 146 schizophrenics and an equal number of affective psychotics on 44 variables which were section scores on the Mental State Schedule used in the US–UK Diagnostic Project (Cooper *et al.*, 1972). Means and standard deviations on each variable are given by Maxwell (1972), who factor analysed the correlations between the items separately for each group. If each group was suffering from a completely different illness, and if some items were relevant to one type of illness, others to the other, then one would not expect similar covariance matrices, or similar factor patterns. In actual fact there is considerable similarity. Equally, on the hypothesis of different diseases one would expect a quite different distribution and frequency of symptoms for the two samples. Maxwell's Table 1 shows that this is not so; the two distributions are very highly correlated, as our Fig. 6.3 shows very clearly. There are 44 symptoms listed on the abscissa, with the percentage incidence indicated on the ordinate.

There are a few symptoms indicated in the figure where slight differences do appear, and they are pretty much where one would expect them. Affectives have higher scores on worry, muscular tension, secret thoughts, depressed mood, signs of depression, somatic symptoms, fading interests and lack of concentration; schizophrenics on frequency of voices, subjective thought disorder, delusions of persecution, blunting and incomprehensibility, but the differences are nowhere absolute, but only relative. It should be added that the items used were but a small section of those employed in the original study, concentrating on those which best discriminated the two groups; had another sample of items been employed, the similarities would have been much greater.

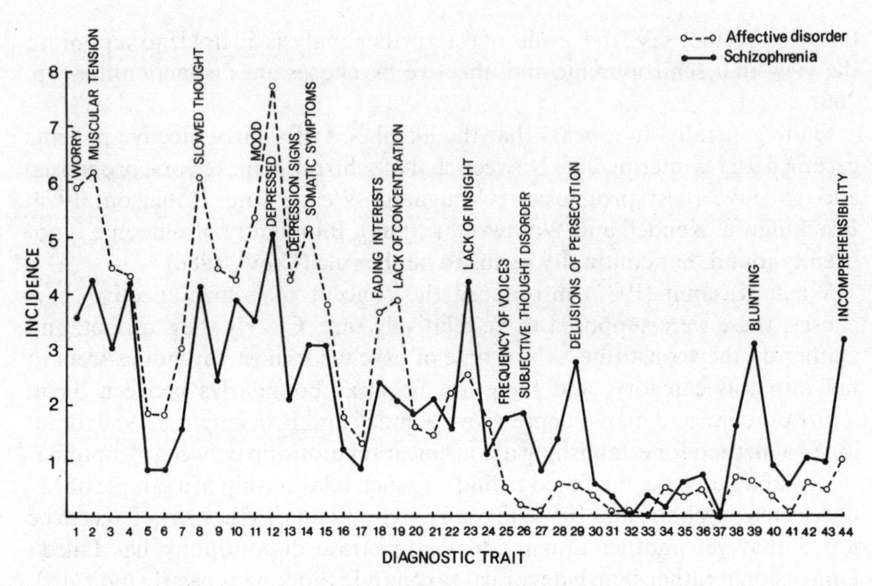

Fig. 6.3 Forty-four diagnostic traits of groups of patients diagnosed schizophrenic or affective. (Drawn using figures given by Maxwell (1972).)

These results are not compatible with a rigid form of the Kraepelinian dichotomy, but fit very well into a dimensional view, considering a continuum of psychoticism, together with a small bipolar factor contrasting schizophrenic and affective disorders. Such a view of course requires other types of confirmation, and these are discussed below.

It might be thought that the prophylactic value of medication might show sufficient specificity to mark a clear discontinuity, lithium preventing relapse in affective illness, and neuroleptic medication in schizophrenia. However, some schizophrenic illnesses respond to lithium (Biederman, Lerner and Belmaker, 1974; Delva and Letemendia, 1982, 1986), and in addition to their effectiveness in mania, neuroleptics may be of value in depression (Hollister *et al.*, 1967). There has never been any doubt that mania responds to phenothiazines, as do some depressions (Klerman and Cole, 1965). It is also well-documented that many schizophrenics respond well to ECT (Brandon *et al.*, 1985; Taylor, 1980). As Crow (1986) sums up, 'no unequivocal demarcation of the functional psychoses can be made on the basis of symptoms, outcome or response to treatment' (p. 421). Response to treatment is perhaps the weakest of these three sets of arguments, partly because of the poor reliability of psychiatric diagnoses, partly because it is not unlikely that such diagnoses are often influenced by the responses of the patients to treatment. The major problem, of course, is an ethical one; as long as we *believe* that certain drugs are better for treating certain types of psychosis, whether such beliefs are true or false, so long will it be ethically impossible to mount decisive experiments incorporating random allocation of patients to drug treatments.

Kendell (1987) makes another important point. As he says, 'it is disconcerting how frequently the biological abnormalities reported in schizophrenia and assumed to be of aetiological significance are subsequently found in affective disorders also' (p. 501). He mentions the enlargement of the lateral ventricles (Dolan, Calloway and Mann, 1985), abnormal smooth pursuit eye movement (Iacono, Peloquin, Lumry, Valentine and Tuason, 1982), the role of 'high expressed emotion' in precipitating relapse (Horley, Orley and Teasdale, 1986), and even the season of birth (Hare, 1987a, b). These similarities certainly speak against a rigid adoption of the Kraepelinian system, although they do not necessarily support Crow's apparent denial of *any* distinction between the different functional psychoses.

Another source of evidence is epidemiology. As Hare (1987a) has pointed out,

> comparison of the findings in schizophrenia and affective psychosis shows the two groups to be similar in sex ratio, age-incidence, risk of suicide, and seasonal variations in onset and birth: and to be different in personality type, premorbid impairment, and age of onset by sex. Differences in prognosis and fertility are less marked now than formerly (p. 514).

Thus here also there are factors favouring a dimensional, rather than a categorical approach.

The data so far considered are certainly not in accord with what one might have expected to find if Kraepelin's hypothesis had been correct; as far as they go the data are rather in accord with a dimensional theory which would suggest that differential diagnosis of psychotic patients grades them along a continuum of severity, rather than classifying them in terms of non-overlapping categorical disease entities, with schizophrenia the most severe, followed by schizo-affective disorder, bipolar affective illness, and finally unipolar affective illness. What does genetic research have to say on this topic (Crow, 1987; McGuffin, Murray and Reveley, 1987; McGuffin and Murray, 1991)?

The genetic approach

First, let us look at the genetic approach which has also been used in studies of genetically identical individuals. McGuffin, Revely and Holland (1982) reported on a set of monozygotic triplets, two of whom had received a hospital diagnosis of schizophrenia, while the third was considered to be a manic depressive. Re-evaluation and the use of 'blind' raters suggested that the discordance was not simply due to misdiagnosis or differing diagnostic bias. Nor is this finding an isolated instance; Dalby, Morgan and Lee (1986) and

Farmer, McGuffin and Gottesman (1987) have also reported similar discrepancies for twins. These data illustrate 'some of the shortcomings of a strictly applied Kraepelinian dichotomy' (McGuffin *et al.*, 1987), and support the existence of a general psychoticism factor.

Following a rather different line of argument, Decina, Lucan and Linder (1989) concluded a survey of parent–child pairs who both required hospital admission for a psychotic illness, by saying that 'while no patient with affective disorder was found amongst the children of schizophrenic parents, 50% of children of parents with affective disorder presented with schizophrenia'. These results are difficult to understand genetically. If we agree that on the psychoticism continuum schizophrenia is further removed from normality than affective disorder, the law of regression to the mean would lead us to expect that the parent–child trend would be from schizophrenia to affective disorder; instead we get an increase in severity of pathology from one generation to another.

These studies are very relevant to the question of whether schizophrenic and affective psychosis are genetically related. Gershon and Rieder (1980) answer the question in the negative: 'evidence from twins and family studies suggests that bipolar manic depressive illness and chronic schizophrenia are distinct entities'. (See also Gershon *et al.*, 1976, 1982.) Similarly, Shields, Heston and Gottesman (1975) declare that 'the genetic diathesis for affective disorders is independent of that for other psychiatric disorders'. Crow (1986) disagrees: 'affective illness in one generation may predispose to schizophrenia in the next' (p. 421), and again (Crow, 1990): 'attempts to draw a line of genetic demarcation between schizophrenic and affective illnesses have failed. It must be assumed that these diseases are genetically related' (p. 788). In view of such diametrically opposed opinions, a detailed look at the evidence may be in order.

Rosenthal (1970) already noted that, in five studies, there was an excess of schizophrenia in children of parents with affective disorder; there was a mean incidence of 2.3%, as compared with the 0.8% lifetime prevalence in the general population. This would not be expected on the 'separate diseases' hypothesis. Two studies of psychotic parents with psychotic children amplify these early findings (Penrose, 1968; Powell *et al.*, 1973). In 621 such pairs, Penrose found that among children of parents with affective disorder, schizophrenia was almost as common as affective disorder (205 vs. 232). Similarly, in the Powell *et al.* study the number of schizophrenic children was actually greater than that of manic-depressive children (15 vs. 10). Of the children of schizophrenic parents, the majority was schizophrenic (150 vs. 34 for Penrose; 9 vs. 0 for Powell *et al.*). In a methodologically rather less satisfactory study, Cammer (1970) found that 26.6% of 533 children of 273 manic-depressive parents were schizophrenic. In a smaller way, Elsasser (1952) found that of 169 children of 2 parents with non-schizophrenic psychoses, 18 suffered from definite affective illness and 6 from definite schizophrenia. Schulz (1940), in a small-scale study, found that of 25 children of 2 manic-depressive parents, 7 suffered from the same disorder, but 3 suffered from schizophrenia.

Equally interesting is the evidence from Pollock and Malzberg (1940), who collected family histories of psychosis over three generations, and found an excess of affective illness (15 cases) over schizophrenia (11 cases) in relatives of preceding generations who had been diagnosed as suffering from schizophrenia. Slater (1953) found a ratio of 4:3 (affective disorder to schizophrenia) in the parents of schizophrenic patients. The corresponding ratio for siblings was 3:5, also showing an unexpectedly large number of affective psychosis. He did not find a similar excess of schizophrenia in the parents or siblings of patients with affective disorders. These findings support an earlier one by Slater (1936), to the effect that in a study of manic-depressive illness there occurred a surprisingly large number of schizophrenics among the children; in 10 of 15 such cases he was unable to find schizophrenia in other members of the patient's family or that of the husband or wife.

An interesting study by Kant (1942) suggested that the proportion of schizophrenic to manic-depressive cases of relatives of deteriorated schizophrenics was 5:1, whereas, in relatives of recovered schizophrenics, the ratio was 1:5. Thus, in milder cases of schizophrenia, i.e. cases closer to affective disorder on our continuum, relatives actually are much more likely to suffer from affective than schizophrenic disorders, a finding difficult to explain on Kraepelinian lines. Similarly, Weinberg and Lobstein (1943) found in a study of 199 schizophrenic personalities that there was a higher percentage of affective disturbances in the ancestry of remitting schizophrenics than in that of steadily deteriorating ones. Again, Vaillant (1962) compared the heredity of 30 recovered schizophrenics with a non-recovered group, and found that at least 50% of the recovered group had heredity positive for an affective, mostly depressed psychosis, as contrasted with 7% in the non-recovered group.

Particularly impressive is a large-scale study by Tsuang, Winokur and Crowe (1980) of 1587 first-degree relatives of schizophrenics, manics, depressives, and controls who were personally interviewed without knowledge of the proband's diagnosis. Schizophrenia in first-degree relatives carried a morbidity risk of 0.6 in the controls; for schizophrenia it was over five times higher, for mania over three times higher, for depression over twice as high, but not significantly different. For affective disorder in first-degree relatives morbidity risks or depression was twice as high as in controls; mania twice as high, and schizophrenia just a little higher. Thus depression is twice as high in first-degree relatives of schizophrenics as in controls, and also twice as high in first-degree relatives of affective disorder patients as in controls. (I have combined the findings from two forms of data collection, personal interview and records used by the author, because they give very similar results.) Bipolar affective disorder was if anything more common in relatives of schizophrenics than of depressives, while unipolar disorder was much less so. The authors emphasize that 'schizophrenia and affective disorder were different and support heterogeneity of major functional psychoses' (p. 500); they rather underplay the equally clear indications of overlap, linking bipolar affective disorder rather more closely with schizophrenic than unipolar disorders. These two disorders may then be conceived as respectively mild and severe forms of the same disorder (Tsuang, Faraone and Fleming, 1985), with the more severe disorder more closely related to schizophrenia.

Equally impressive is a more recent paper by Angst and Scharfetter (1990), looking at over 250 probands of variously diagnosed psychotics, and examining the ratio of schizophrenic to affective illnesses in the first-degree relatives. This ratio rose from 0.30 in unipolar to 0.47 in bipolar affective disorders, to 0.92 in predominantly affective and 2.99 in predominantly schizophrenic schizo-affective disorder, and finally 5.05 in schizophrenia. 'As the formal illness in the proband changes from affective to schizophrenic, the ratio of schizophrenia to affective illness in the first-degree relatives increases. There is no discontinuity such as would allow one to assert that there are two quite separate genetic components' (Crow, 1990, p. 791).

On such a continuum model, where would we expect schizo-affective disorder to go? Clearly it should appear between schizophrenia and affective disorders, giving us a continuum ranging from normal through unipolar and bipolar affective disorders to schizo-affective disorder and finally schizophrenia. The evidence supports such a view of schizo-affective disorder. Angst, Felder and Lohmeyer (1979) found that the risk of schizophrenia and affective disorder was approximately equal in first-degree relatives of schizo-affective probands, while that of schizo-affective illness was appreciably less. Tsuang *et al.* (1977) and Baron *et al.* (1982) found schizo-affective illness to be more closely related to affective illness than to schizophrenia; they agreed with Tsuang (1979) that schizo-affective illness was not a genetic entity.

Gerston *et al.* (1982), in a study of 1254 relatives of patients with major affective disorder, concluded that

> these data were compatible with the different affective disorders representing thresholds on a continuum of underlying multifactorial vulnerability. In this model schizo-affective illness represents greatest vulnerability, followed by bipolar . . . then unipolar (affective) illness.

They also found that there was a highly significant ($p<0.001$) excess of schizophrenics among the relatives of patients with schizo-affective illness in comparison with those having other types of affective illness. Curiously, Gerston *et al.* do not extend their continuum concept to include schizophrenia!

It is true that the morbid risks of psychosis in the relatives of probands with schizo-affective disorders is particularly high, but the suggestion that these disorders 'breed true' (Perris, 1974) is unwarranted. As McGuffin *et al.* (1987) point out, 'a number of studies have now failed to support the concept of schizo-affective pychosis as a distinct genetic entity. Although schizo-affective disorder does occur in the families of probands with schizo-affective disorder, these relatives are also at increased risk of both schizophrenia and affective disorder' (p. 550). They also suggest that schizo-affective disorder could be heterogeneous with some cases part of a phenomenological spectrum attributable to mainly 'schizophrenic' genes and other cases a severe variant of affective psychosis, but this seems far-fetched and much less likely than the dimensional hypothesis outlined above.

Even when the schizophrenia is narrowly defined, the risk of schizo-affective disorder is significantly increased in relatives of schizophrenics so diagnosed (Kendler, Gruenberg and Tsuang, 1989). Crow (1986) concludes his survey of the evidence relating to schizo-affective disorder by saying that 'parsimony requires the conclusion that schizo-affective disorder is but the bridge between affective disorders and schizophrenia. The psychoses constituted a genetic continuum rather than two unrelated diatheses' (p. 424). This conclusion has been elaborated by Crow and Cooper (1986) along genetic lines. Crow (1986) enumerates several additional studies which tend to support the thesis of a continuum (see also Menninger *et al.*, 1958; Karlsson, 1974; Rennert, 1982; Flor-Henry, 1983, in support). Such a conclusion does not oblige one to agree to the specific genetic theories developed by Crow (1990), and it does not rule out the existence of specific genes responsible for different manifestations of the underlying vulnerability concept (Eysenck, 1972b).

Kendell (1987) has expressed the position admirably.

> Four main groups of functional psychoses have been recognised since the early years of the century: schizophrenia and affective psychoses, acute psychoses of good prognisis; and chronic paranoid psychoses. The air of permanence and stability is misleading. None of these four groupings, or of the individual psychoses included within them, has been clearly demonstrated to be a disease entity. All are still defined by their clinical syndromes and these syndromes appear to merge inevitably into one another and into other syndromes in the domain of neurotic illness and personality disorder. As a result it is not clear where the boundaries should be drawn (p. 499).

This is a clear statement of the facts which demand a form of *dimensional* rather than categorical description, even though some categorical specificity cannot be ruled out as an additional descriptive variable.

Problems of diagnosis

All the data so far surveyed have been based on psychiatric classification, and it is well known that diagnoses in this field are far from reliable or accurate (Beck, 1962; Hunt, Wittson and Hunt, 1953; Kreitman, 1961; Kreitman *et al.*, 1961; Ley, 1972; Norris, 1959; Sandifer *et al.*, 1968; Schmidt and Fonda, 1956). These studies have been chosen because of the number of psychiatrists involved, and the number of categories used (Ley, 1972). Other references are to Ask (1949) and Shepherd *et al.* (1968). (It might be thought that more recent data involving DSM-III would be more appropriate, but of course the sources cited are more relevant because they reflect practices during the time when most of the studies here mentioned were planned and executed.)

The amount of agreement between different psychiatrists is not high. The actual figures for percentage agreement are 55% for Schmidt and Fonda, 58% for Norris, 63% for Kreitman *et al.*, i.e. averaging around 59% for two observers. For the Sandifer *et al.* study the figure is 34% (4 raters) and 10% (6–10 raters). These figures are clearly unsatisfactory, but not far removed from those obtained for medical diagnoses of physical diseases generally (Eysenck, 1991a).

The relativity of psychiatric diagnosis is most clearly shown by the huge national differences which have been found. Thus Kramer (1961) showed that in some age/sex categories 10- or even 20-fold differences are involved when comparing British and American diagnoses. Cooper *et al.* (1972) have shown that these differences arise from a much more inclusive conception of schizophrenia, leading to a much higher frequency of this diagnosis in the USA than in the UK, with an opposite tendency as regards the diagnosis of manic-depressive disorder. The recent adaptation of DSM-III criteria has of course considerably altered the picture, but the new standards are more relevant to future studies.

Granted that reliability of psychiatric diagnoses is poor, and often owes more to the nationality of the psychiatrist making the diagnosis than to the actual mental state and behaviour of the patients (Cooper *et al.*, 1972), it is worth nothing that diagnoses of major psychotic illnesses show little overlap with neurotic disorders of various kinds, and have somewhat higher rates of agreement. It does not seem likely that random errors (Ley, 1972) can explain the observed relationships. This is the 'observer agreement' model of reliability; more interesting from our point of view is the consistency or stability model, i.e. comparing diagnoses assigned to patients on successive admissions, or at other widely separated points in time.

An early study by Masserman and Carmichael (1938) looked at a series of 100 patients admitted to a university psychiatric clinic in Chicago and found that a 'major revision' of diagnosis was necessary in over 40% only 12 months later. In a much larger study, Babigian *et al.* (1965) compared the diagnoses assigned to 1215 patients on more than one occasion and found that diagnoses of schizophrenia were more reliable than of affective disorder. However, the actual time interview between diagnoses was only a few days or weeks, so there was little room for change of symptomatology over time. Odegaard (1966), in a comparison of first and last diagnoses of patients first admitted in 1950–4, and then readmitted at the end of 1963, found that many patients originally diagnosed as having a reactive psychosis were subsequently diagnosed as having either schizophrenia or a manic-depressive illness – but of course these are patients readmitted, and thus probably atypical of all first cases so diagnosed. Cooper (1967) studied diagnoses of 200 patients on four different occasions, finding that only 54% were allocated to the same broad category on all four occasions. However, many changes were due to changes in the doctor rather than to changes in the patient's symptoms. Such changes were found in 16% (32 cases), the commonest being for schizophrenic symptoms to develop in patients who had previously had purely depressive symptoms, but with some paranoid features. Here, as in the transgenerational shift from affective (parent) to schizophrenic (child), the direction is toward a more severe disorder.

Of particular interest for our purpose is a study by Kendell (1974) who followed up 1913 patients originally diagnosed in 1964 and readmitted at least once before the end of 1969. It was found that 58% of patients did not undergo any significant change in diagnosis. Unchanged at final diagnosis were 69% of all depressive illness, 75% of all schizophrenics, and 35% of all personality disorders. A change from schizophrenia to depressive illness was found in 10.9%, from depressive illness to schizophrenia in 7.4%. From mania to schizophrenia there was a change in 17.0%, but only 2.3% in the reverse direction. (Depressive illness here included reactive and endogenous; there appears to have been little stability to the more refined diagnosis.) It is interesting that there was a significant interchange of diagnosis for personality disorder and depressive illness, as usually personality disorder is considered to belong with the schizophrenic *Erbkreis*.

The data so far considered are incompatible with a purist interpretation of Kraepelin's theory, but they only partly support Crow's view. It is true that the major two psychotic disorders do not 'breed true', but equally there is no blending of the two into schizo-affective disorders when the parents are one of each (Gottesman and Bertelsen, 1991). Clearly, it is as important to avoid exaggeration in the denial of genetic differentiation between psychoses as it is to avoid denial of common features (Crow, 1987).

What would happen if we relied exclusively on the statistical study (factor analysis, multiple discriminant analysis) of reliably rated symptoms in a very variegated psychotic population, including depressives and schizophrenics of all kinds? Lorr, Klett and McNair (1963) have carried out such a study, as well as including discussions of previous works employing similar methodologies. They isolated 10 syndromes: excitement; hostile belligerence; paranoid projections; grandiose expansiveness; perceptual distortions; anxious introspectiveness; retardation and apathy; disorientation; motor disturbance; and conceptual disorganization. Symptoms were carefully rated by specially trained interviewers, and these 10 syndromes emerged after oblique rotation of the factors emerging. The intercorrelations between the factors gave rise to three second-order factors: excitement vs. retardation, schizophrenic disorganization, and paranoid process. They again correlated positively together, to form a higher-order factor which the authors labelled 'schizophrenia', although 'psychosis' would have been a more appropriate name in view of the explicit inclusion of depressive patients (see their statement on p. 32). This analysis thus eventuates in a hierarchical model, very much as hypothesized here.

It is only fair to mention studies which suggest a lack of generality between different types of psychosis. Gottesman and Bertelsen (1991) report on in-patient psychotics who had children by other in-patients, thus making possible a variation of the diallele cross-method used in plant and animal research. In homotypic manic-depressive couples, the kind of risk (68%) in offspring for manic-depressive diagnosis approached the value expected for a dominant gene with complete dominance (78%); as the writers emphasize, 'the results do not support the view of a continuum of psychopathology between affective and schizophrenic psychoses because the risk of schizophrenia from such dual-mating manic-depressives is close to the base rate for the general population' (p. 95). The numbers involved are too small as yet to constitute a definite contradiction to the view here taken that we have *both* generality *and* specificity, but the results should act as a warning not to disregard genetic specificity.

It is often believed that the so-called 'new genetics' might, in principle at least, settle issues of dimensionality vs. categorical discrimination between hypothetical disease entities (Pato, Lander and Schulz, 1989). Roberts and Claridge (1991) have taken up this challenge and argue convincingly that the 'new genetics' is quite compatible with a dimensional view of schizophrenia, and need not favour the single gene hypothesis.

It is also important to realize that the inclusiveness of criteria used to define schizophrenia may determine in part the outcome of genetic studies. An interesting comparison of inclusiveness criteria is the comparison by Farmer *et al.* (1991) of the heritability (MZ/DZ concordance ratios) of different DSM-III categories. Heritability goes up from a simple 'schizophrenia' diagnosis to one adding schizotypal personality, with another increase by adding affective disorder with (mood-incongruent) psychosis, reaching its highest value with the addition of a typical psychosis, with a slight decline when schizophreniform disorder is included. The inclusion of any form of affective disorder produces a distinct lowering of the concordance ratio, which is again lowered by the inclusion of paranoid disorder, with a final lowering by the inclusion of any axis-I DSM-III category (Farmer *et al.*, 1987; Farmer *et al.*, 1991). The results suggest both a general psychoticism factor, and also a fair degree of specificity.

We have looked at eight different types of studies which might give evidence concerning the continuity vs. category conception of the major functional psychoses. It is possible in each case to evaluate the contribution made by each type of study in numerical form; of course these evaluations are inherently subjective, but being based on factual material rational discussion is possible. If we denote the continuity hypothesis G (for generality), and the Kraepelinian theory of complete distinction S (for specificity), we can assess the support

Table 6.1 *Empirical evidence favouring continuous (G for general) or categorical (S for specific) theories of psychosis (Eysenck, 1992a)*

	G	S
1. Distribution of symptoms	4	1
2. Symptom similarity	4	1
3. Outcome	3	2
4. Medication	1	4
5. Biological abnormality	2	3
6. Genetic research: markers	2	3
7. Genetic research: familial incidence	2	3
8. Diagnostic stability	2	3
Total	20	20

given by each of our eight types of study to G or S on a five-point scale, on which 5:0 would mean complete support for G; 4:1 strong support for G; 3:2 weak support for G; 2:3 weak support for S; 1:4 strong support for S, and 0:5 complete suport for S. Table 6.1 shows the outcome.

Clearly the outcome does not favour either side at the expense of the other. There is strong evidence for the existence of a continuum covering all the functional psychoses, and ordering them in relation to severity. But there is also strong evidence for the reality of differentiation, following Kraepelinian lines. It would not be reasonable to stress either line to the exclusion of the other; psychoticism is a reality, but so is the distinction between schizophrenia (and indeed different types of schizophrenia – Kendell, 1987) and manic-depression (and indeed unipolar or bipolar types of affective disorder – Kendell, 1987). For special purposes we may wish to emphasize one line of argument or the other, but clearly there is no victory in all these studies for either rigid Kraepelinian distinctions or for the ancient 'Einheitspsychose'.

Psychoticism and schizotypy

The theory of 'psychoticism' has two major components. The first of these, the existence of some degree of generality among the functional psychoses, has been dealt with in the preceding discussion. The second postulates an extension of this continuum to less serious disorders ('spectrum' disorders), and indeed to normal behaviour. Some material regarding 'spectrum' disorders has already been cited, but this section will deal specifically with the problem of extending the continuum from psychosis to normality.

Before returning to the psychiatric literature, I shall mention a study specially designed to answer the question of continuity from psychosis to normality, and introducing a method of analysis specially created to make such an answer possible. I have called the method 'criterion analysis' (Eysenck, 1950), because it uses a criterion (psychosis–normality) to furnish us with a model which can either show continuity or discontinuity.

The study was designed to test Kretschmer's (1946, 1948) theory of a schizothymia–cyclothymia continuum, as well as my own theory of a normality–psychosis continuum. Kretschmer was one of the earliest proponents of a continuum theory linking psychotic and normal behaviour. There is, he argued, a continuum from schizophrenia through schizoid behaviour to normal dystonic (introverted) behaviour; on the other side of the continuum we have syntonic (extraverted) behaviour, cycloid and finally manic-depressive disorder. He is eloquent in discussing how psychotic abnormality shades over into odd and eccentric behaviour and finally into quite normal typology. Yet, as I have pointed out (Eysenck, 1970a,b), the scheme is clearly incomplete. We cannot have a single dimension with 'psychosis' at both ends; we require at least a two dimensional scheme, with psychosis–normal as one axis, and schizophrenia-affective disorder as the other.

In order to test this hypothesis, I designed a method of 'criterion analysis' (Eysenck, 1950, 1952a,b), which explicitly tests the validity of continuum vs. categorical theories. Put briefly, we take two groups (e.g. normal vs. psychotic), and apply to both objective tests which significantly discriminate between the groups. We then intercorrelate the tests *within each group*, and factor analyse the resulting matrices. If and only if the continuum hypothesis is correct will it be found that the factor loadings in both matrices will be similar or identical, and that these loading will be proportional to the degree to which the various tests discriminate between the two criterion groups.

An experiment has been reported, using this method. Using 100 normal controls, 50 schizophrenics and 50 manic-depressives, 20 objective tests which had been found previously to correlate with psychosis were applied to all the subjects (Eysenck, 1952b). The results clearly bore out the continuum hypothesis. The two sets of factor loadings correlated .87, and both were proportional to the differentiating power of the tests r = .90 and .95, respectively). These figures would seem to establish the continuum hypotheses quite firmly; the results of the experiment are not compatible with a categorical type of theory.

Another study investigated 153 psychotic patients prior to the application of therapy (Verma and Eysenck, 1973). These patients were interviewed and rated on the In-patient Multidimensional Psychiatric Scale (IMPS), published by Lorr *et al.* (1963), and were administered the PEN Inventory (an early form of the Eysenck Personality Questionnaire – Eysenck and Eysenck, 1975), as well as several other tests and questionnaires. A factor analysis was carried out on altogether 34 tests or interview scores, and two major factors emerged. The first was clearly a general psychosis factor, with its highest loading on the psychoticism scale, and high loadings also on ratings, objective tests and other questionnaires. Factor 2 discriminated the outgoing, extraverted, extrapunitive type of psychotic from the inward-looking, introverted, intropunitive type, this factor loaded on the extraversion–introversion side. The P factor did not discriminate between the types of psychosis, but the E factor did. These results suggest that within the psychotic field extraverted and introverted behaviour patterns may be distinguished with a considerable degree of clarity, and thus reinforce the similar findings of Armstrong *et al.* (1967) and Venables and Wing (1962). The former concluded, from an examination of the MPI scores of schizophrenic patients, that 'these results raise the possibility that a significant degree of what is included within the process-reactive frame of reference may be considered a function of extraversion–introversion' (p. 69).

Our results (Verma and Eysenck) suggest that some even broader grouping is possible, embracing not only schizophrenics, but also other types of psychotics. Thus depressive disorders had relatively high E scores, paranoids and schizophrenics low ones. (It is interesting to note that in the Kendell and Gourlay study the item most highly correlated with the affective end of the continuum was 'More outgoing and gregarious recently', a typical extraversion item.) The possibility certainly exists, and should be investigated, that the major difference between functional psychoses are related to the other major dimensions of personality, i.e. Extraversion and Neuroticism.

When we turn from psychotic states to types of personality lying close to them on the psychoticism continuum, we encounter specifically the so-called 'schizoid personalities', 'spectrum' and personality disorders. Manfred Bleuler (1911, translation 1978) first described the schizoid personality:

> He is taciturn or has little regard for the effect on others of what he says. Sometimes he appears tense and becomes irritated by senseless provocation. He appears as insincere and indirect in communication. His behaviour is aloof and devoid of human warmth; yet he does have a rich inner life. In this sense he is introverted . . . Ambivalent moods are more pronounced in the schizoid than in others, just as he distorts the meanings of, and introduces excessive doubts into his own concepts. But on the other hand, the schizoid is also capable of pursuing his own thoughts and of following his own interests and drives, without giving enough consideration to other people and to the actual realities of life. He is autistic. The better side of this autism reveals a sturdiness of character, and inflexibility of purpose, an independence, *and a predisposition to creativity*. The worse side of it becomes manifest in a lack of consideration for others, unsociability, a world-alien attitude, stubbornness, egocentricity, and occasionally even cruelty.

(The emphasis here is mine. The statement is important in linking psychoticism to creativity.)

Bleuler concluded that at least half of his patients had shown some degree of schizoid behaviour before their psychotic breakdown, and he noted that similar characteristics were also very noticeable in their siblings and their offspring. This seems a clear indication of an extension of the psychotic *Erbkreis* to non-psychotic individuals.

Early work has been well summed-up by Reich (1976) who concentrated on the 'schizophrenia spectrum'; thus the studies reviewed are also very relevant to the continuity vs. categorical discrimination (Gottesman, 1987). Reich defines the spectrum concept as a theory which maintains that their exists a cluster or spectrum of psychopathological states, some characterized by psychosis and others not, which share a genetic aetiology with schizophrenics, and which, therefore, constitute, together with schizophrenia itself, a 'spectrum of schizophrenic disorders'. The theory implies a diathesis-stress conception (Gottesman and Shields, 1972), with the genetic diathesis being a requisite for the development of a spectrum-related illness.

> Any particular schizophrenic spectrum disorder is, therefore, seen as representing not a discreet state which is unrelated to the other disorders, but a point on a genetic continuum, with differences among the points reflecting differences in intensity, or in some other clinically evident quality, which may be environmentally and/or genetically determined (Reich, 1976, p. 4).

(My own theory would extend this 'spectrum' beyond schizophrenic disorders to *all* functional psychoses (Eysenck, 1952a,b; Eysenck and Eysenck, 1976).)

Tests of the 'schizophrenic spectrum' theory may be carried out along several lines. The first consists of the use of 'markers', i.e. characteristics of schizophrenics which are also found in a significant amount in relatives of schizophrenics not themselves psychotic. Examples are deviant eye-tracking, which is frequently found in schizophrenics (Holzman, Proctor and Hughes, 1973), as well as in their relatives (Holzman *et al.*, 1974). Another example would be a reduced level of platelet monoamine oxidase, which is found not only in schizophrenics but also in their non-schizophrenic monozygotic twins (Wyatt *et al.*, 1974). The study of 'markers' will be found to be a favourite method of analysis to show that psychoticism scores do indeed map into the psychoticism continuum, as discussed in a later section.

The second method uses the concept of familial incidence of spectrum cases, based on the assumption that they represent 'thresholds' along a genetic continuum. This notion is shown in Fig. 6.1, where different spectrum disorders have differential 'thresholds' on the P continuum. In other words, the threshold for schizophrenia is higher than that for psychopathy, or schizoid disorders, the latter requiring less genetic predisposition and/or environmental stress to appear. Although the model has only recently been specified in testable form, the notion itself goes back a long time (Planasky, 1972). Thus Rudin (1916) had already looked for psychiatric pathology among the relatives of (strictly defined) schizophrenics; so had Kallman (1938), who searched the families of schizophrenics for many kinds of pathology, and found two frequently occurring types of relatives whom he called 'eccentric personalities' and sufferers from 'schizoid psychopathy'. Slater (1953) used spectrum notions in studying behavioural traits in discordant twins of schizophrenic probands, as did Gottesman and Shields (1972) in looking for 'schizophrenic equivalents'.

Reich (1976) also cites a number of adoption studies in which the adopted children of schizophrenic parents are found to have a variety of spectrum disorders. In particular, the Extended Family Study (Kety *et al.*, 1968) and the Adoptees Study (Rosenthal *et al.*, 1968) gave strong evidence for the existence of such spectrum disorders. The adoptees were personally examined and compared with adoptees with non-schizophrenic parents; significantly more diagnoses of schizophrenic spectrum disorders were made in index adoptees than control cases. (For later analyses, and a more detailed collection of studies, see McGuffin and Murray (1991).)

Let us now turn to a consideration of the psychological literature. Much research has gone into the measurement of schizotypy and schizotypal personality disorders in recent years (Claridge, 1985), and we know much more about it as a result.

There are many schizotypy scales which have proliferated in recent years, and which owe much to Bleuler's description, although they tend to go beyond schizophrenia into 'psychosis-proneness', and usually concentrate on a single concept, like perceptual aberration (Chapman, Chapman and Raulin, 1978), magical ideation (Eckbald and Chapman, 1983), impulsive non-comformity (Chapman *et al.*, 1984), intense ambivalence (Raulin, 1984), social fear (Raulin and Wee, 1984), cognitive slippage (Miers and Raulin, 1985), etc. More general are scales like the STA and STB scales of Claridge and Broks (1984). All these scales seem to measure much the same construct, as Kelley and Coursey (1994) have shown by intercorrelating and factor analysing 11 such scales. Only anhedonia (Chapman, Edell and Chapman, 1980) failed to correlate with the other scales; the Claridge and Broks scale (STA), being the most general, had the highest loading (0.86), followed by cognitive slippage (0.79), intense ambivalence (0.78) and magical ideation (0.71). Non-conformity (0.65) also had a sizeable loading and the MMPI Mini-Mult (1968), i.e. the sum of the Depression, Psychasthenia and Schizophrenia scales, had a loading of 0.74. Other factorial studies (e.g. Muntaner *et al.*, 1988, and Bentall, Claridge and Slade, 1989) gave similar results.

What is noticeable in the studies mentioned is that while the psychoticism scale is uncorrelated with neuroticism, schizotypy scales usually show high correlations with N; indeed, these are sometimes so high as to suggest that what is being measured is N, rather than P. Thus in the Mutaner *et al.* (1988) study, the first factor (schizotypy) had a high loading on N, but not on P. In the Bentall *et al.* (1989) study, their second factor had high loadings on N, the Claridge and Broks STA and STB, the schizoidia and the schizophrenism scales. In view of the genetic distinctiveness of neurotic and psychotic disorders this suggests that many schizotypy scales measure two dimensions of personality simultaneously, which is an undesirable state of affairs (Eysenck, 1993b).

The Bentall *et al.* (1989) study also shows that introversion, as well as N and P, is involved in typical schizotypy scales. They factor analysed 10 such scales, as well as P, E and N, and found three factors, each of which had high loadings for some but not all of the schizotypy scales. Factor 1 was labelled 'perceptual cognitive', with a high loading (0.72) on psychoticism. Factor 2 was labelled 'social anxiety', with its highest loading on neuroticism (0.85!). Finally, factor 3 was labelled 'introverted anhedonia', with a high loading on extraversion (−0.61). It is difficult to agree to factor 3 having much to do with schizophrenia or schizotypy. Similarly, factor 2 is simply a factor of neuroticism, with N having much the highest loading. This leaves factor 1 as roughly representing psychoticism, although 'hypomanic personality' has the highest loading – hardly suggestive of 'schizotypy'.

This point is well demonstrated in a factorial analysis of some data collected by Kendler and Hewitt (1992), who administered 10 separate self-report scales supposedly measuring schizotypy, as well as shortened measures of extraversion and neuroticism, and a depressive and an anxiety scale, making 14 scales in all. (A short form of the P scale was included among the 10 schizotypy scales.) The factor analysis produced three factors. The first had neuroticism, depression and anxiety, as well as hallucinations, perceptual aberration and magical ideation as its highest loaders, suggesting again that many of the schizotypy scales are really measuring neuroticism. The highest loading scale for the second factor was extraversion, and the highest loading scale for the third factor was psychoticism, clearly identifying the factors. The highest commonality for the whole set of factors (0.75) was shown by the psychoticism scale, demonstrating that whatever was in common to all the schizotypy scales was found in the psychoticism scale. Thus the factor space defined by the schizotypy scales is clearly marked out by P, E and N, with P marking out the major part of this space.

If these data have anything to tell us, it may be that differences between different *types* of psychosis may be mediated by differences in N and E, at least partially (Eysenck, 1972b). This is not the place to argue the case, but just as there are different kinds of psychopathy defined by differences in N and E (Eysenck, 1987a), so there may be similar associations between different types of psychosis and the major dimensions of personality (Eysenck, 1993b).

There are clear connections between the concept of schizotypy and that of personality disorder, as is clear from Bleuler's description of the schizoid personality. Kallman (1938) was one of the first to recognize a relationship between psychopathy and schizophrenia in his description of what he calls 'schizoid psychopaths', and since then retrospective and longitudinal high risk studies of schizophrenics, together with studies of delinquents and criminals, have confirmed the association between psychopathy and schizophrenia (Eysenck and Gudjonsson, 1988).

Of particular importance in this connection has been the work of Heston (1966, 1970) who studied the children of schizophrenic mothers who were adopted away within three days of birth; there were 58 subjects who comprised the first experimental group, with another 58 matched controls whose mothers were not schizophrenic. Blind evaluation of subjects followed after they had reached maturity. Many differences were found. Controls had much lower scores on the Menninger Mental Health–Sickness Rating Scale; had no member diagnosed as schizophrenic, compared to five for the experimentals; had two as opposed to nine members diagnosed as sociopathic personality ($p < 0.017$); had two as opposed to 11 members spending more than one year in an institution (jail or psychiatric); and had two as opposed to seven labelled 'felon' ($p < 0.054$). Heston describes in detail personalities more frequently found in the experimental group who, he states, fit the older diagnostic category of 'schizoid psychopath' (Kallman, 1938). Rosenthal *et al.* (1968) and Kety *et al.* (1968) have also found evidence for such 'schizophrenic spectrum' disorders in adopted children of schizophrenic mothers. These data would seem decisive in extending the 'spectrum' from psychotic to non-psychotic disorders of a schizoid, psychopathic or criminal nature.

Do psychopaths have elevated levels of psychoticism? Hare (1985) showed that criminals with some features of psychopathy showed some signs of paranoid schizophrenia, schizotypal personality, and a relatively high incidence of neuropsychological and neurophysiological abnormalities. Raine (1994) studied 36 prisoners rated on the Hare Psychopathy Check List (Hare, 1980) and divided them into high, medium and low scorers. They were also given four schizotypal personality scales, which were summed to give an overall index. Prisoners were also assessed according to DSM-III criteria for borderline and schizotypal personality by interviewers blind to the questionnaire scores. The results showed significant relations between psychopathy and borderline disorder, schizotypal disorder, and schizotypal personality. There was also support for Heston's (1966) finding that an unstable, impulsive lifestyle lacking in commitments and long-term plans, represents that element in psychopathy most related to schizotypal personality, and also to psychoticism (Eysenck and Gudjonsson, 1989).

The evidence supporting the view that psychoticism is strongly related to psychopathic, antisocial and criminal behaviour is reviewed in detail by Eysenck and Gudjonsson (1988); it appears at all ages (childhood, youth, maturity) and results in sizeable correlations. The concept of 'personality disorder' or psychopathy is of course rather fuzzy; Eysenck and Eysenck (1978) have suggested that it is no more than a rather arbitrary combination of high P, high N and high E, a suggestion along lines of dimensional description which agrees quite well with DSM-III description, which isolates three separate clusters to characterize personality disorders, corresponding closely to P, E and N personality description (Eysenck, 1987a).

We may conclude that there is good evidence to suggest an extension of the psychotic continuum into non-psychotic types of behaviour variously described as psychopathic, schizoid, criminal, alcoholic, etc., but always genetically linked with psychosis through close relatives of one kind or another. It is curious that advocates of the concept of schizotypy, from Meehl (1962, 1989) onwards have linked this extension to non-psychotic diseases with schizophrenia alone, not with the whole psychotic continuum. Thus the subject matter of this and the preceding section, although obviously closely related, has been looked at in isolation; it is only the concept of psychoticism (Eysenck, 1952a, b; Eysenck and Eysenck, 1976) which has brought them together. This, plus the failure to separate psychoticism from neuroticism, suggests that much of the schizotypy literature is only partially relevant to the classification of the concept of psychoticism (Eysenck and Barrett, 1993a).

To end this discussion, it may be useful to consider a criticism of the P scale that has been made quite frequently from its earliest days of inception (e.g. Davis, 1974; McPherson *et al.*, 1974). The criticism is based on the hypothesis of a linear relation between P score and the position of various groups on the psychoticism continuum; it is suggested that schizophrenics should have the highest P scores being furthest to the right on the continuum (Fig. 6.1), and the fact that other groups (e.g. criminals, psychopaths) tend to have higher scores on P (Eysenck and Eysenck, 1976; Claridge, 1981, 1983) is taken to disprove the identification of P with psychoticism. Perhaps, it is sometimes argued, P should be renamed psychopathy, because of the high scores of psychopaths on this scale.

Psychoticism: the proportionality criterion

We will go on to a discussion of some of the reasons why P is indeed a *psychoticism* factor; here I only wish to state some reasons why P scores of psychotics are often lower than those of other groups. The first reason, of course, is simply that we attempted to construct a scale of psychoticism, not of psychosis; hence we deliberately left out all the typical symptoms which go to make up such diagnostic scales as the MMPI schizophrenia scale. Our aim was to construct a scale which would measure psychoticism *in normal groups* (i.e. in non-psychotic groups); the scale was not intended as a diagnostic clinical device. The second reason is that psychosis may easily reduce the patient's insight (a well-known schizophrenic symptom), thus making it more difficult for him to fill in the questionnaire truthfully. The third reason is that patients often have a high Lie score (e.g. McPherson *et al.* (1974) found an L score of 13 for psychotic patients, six for normal controls); this would automatically rule out any meaningful comparison. The fourth reason is the simple fact that psychotics are nearly always under drug treatment, thus altering their mental state in ways that are likely to interfere with accurate answers to P-related questions. The fifth reason is the fact that most of the psychotics tested were held in mental institutions, and institutionalization is likely to affect the responses of inmates to such questions in unpredictable ways. These and many other reasons caused me originally (Eysenck, 1952a) to abandon all thoughts of producing a questionnaire or inventory of psychoticism, and rely rather on experimental tests not open to such objection – or at least not to the same extent. I would thus argue that on *a priori* grounds we would have expected a linear increase in P score with change of group from left to right on the continuum in Fig. 6.1, but would not expect a continuation of this increase into the psychotic range. A reasonably high P score, correlated with severity of illness, might be found (Verma and Eysenck, 1973), but not a score higher than anything found in other groups, such as psychopaths, criminals, etc.

This leaves me with the most crucial question of all – how can one *prove* the identification of a statistical factor (P) with a concept like psychoticism? Let us turn to a description of the methodology used, and a brief account of the results achieved.

The most crucial property of a scale designed to measure psychological traits is of course *validity*, but this is difficult to establish. Construct validity is difficult to prove, in the absence of agreement on theoretical constructs. Concurrent validity assumes the validity of some already existing measure, which is usually hard to find. Predictive validity assumes the existence of external criteria, but this poses difficult problems, as we have seen. I wish here to introduce a rather novel, and certainly unusual measure of validity which is based on the nature of the nomological network surrounding the concept (psychoticism) under investigation. This method (proportionality analysis) is related to criterion analysis (Eysenck, 1950, 1952b), and has the added advantage that it allows the objective determination of the psychological nature of the factors that emerge from factor analysis.

It is well known that the psychological identification of factors is a very difficult task, with few agreements and many doubts. The Wechsler Test subscales on analysis tend to divide, after extraction of *g*, into verbal and non-verbal tests. But is this the psychological principle involved? Verbal scales may measure crystallized ability, non-verbal scales fluid ability. Or the distinction may be between timed and untimed tests. Clearly interpretation is not obvious, and may require elaborate experimental follow-up. Examination of the *content* of the scale is certainly not sufficient. Analysing the Guilford Scale of Social Shyness, which had been declared by him to be a measure of a single entity, I found two uncorrelated factors correlating respectively with introversion and neuroticism, suggesting two quite separate types of social shyness (Eysenck, 1956). I have discussed in detail the inadequacy of purely psychometric analyses in discerning and identifying personality traits and dimensions (Eysenck, 1991b); what are the alternatives?

Scientific research should begin with a theory; the one here to be tested is incorporated in our Fig. 6.1, postulating a continuum from empathic, altruistic, socialized behaviour through average, schizoid and psychopathic behaviour to psychotic illness. We can deduce certain consequences which follow from the theory and are testable in relation to the claim that P is a valid measure of this continuum. Such deductions take a number of steps. (1) Select a theoretical concept which postulates a marked difference between schizophrenics and normals. (2) Construct a proper test of the concept in question. (3) Demonstrate that the test is valid, i.e. discriminates well between schizophrenics and normals. Let us call this test T. If the hypothesis of a continuum is correct, and if T and P are good measures of this continuum, then (4) P and T should correlate significantly *within the normal group*, and possibly also within the psychotic group (the latter prediction is subject to the problem outlined in the preceding section, i.e. the possible effects of the actual psychotic illness). Ideally, therefore, the proportionality of schizophrenia vs. normal T scores can be translated to within-group comparisons of high P scoring normals vs. low-P scoring normals, and high-P scoring psychotics vs. low-P scoring psychotics. If P does not measure the continuum in question, then none of these consequences follow. The prediction is that on test T schizophrenics:normals = P+ : P− in both psychotic and normal groups. We would thus have a powerful method of testing the theory in question, and providing evidence for the validity of the P concept. Alternative theories, e.g. that P is a measure of antisocial personality disorder, and 'that the label *psychoticism* isn't appropriate for the P dimension' (Zuckerman, 1991, p. 375), can thus also be subjected to factual scrutiny.

As an example, consider HLA B27, a subsystem of the human leucocyte antigen system, which is found more frequently in schizophrenics than in normal, non-psychotic subjects (McGuffin, 1979; Gattaz, Ewald and Beckman, 1980). Gattaz (1981) has shown that in a comparison of schizophrenic patients with and without HLA B27, those with the antigen had significantly higher P scores ($p < 0.02$, $n = 11{:}29$). In another study Gattaz, Seitz and Beckman (1985), 17 B27 positive and 16 B27 negative non-psychotic subjects showed a difference on P scores in the expected direction ($p < 0.01$). This example shows the expected effects of an association between P and T in both a normal and a psychotic group, and may serve to illustrate the method.

Another study concentrated on the prevalence of hallucinations. Slade (1976) contrasted three groups – normals, psychotics without and psychotics with auditory hallucinations; on the P test normals had the lowest scores (2.80), hallucinated psychotics the highest (7.25), with non-hallucinatory psychotics in between (4.80). In a later study, Launay and Slade (1981) correlated scores on a 12-item questionnaire testing hallucinatory predispositions with the P scale. The correlations were 0.21 for 100 male (non-psychotic) prisoners and 0.46 for 100 female (non-psychotic) prisoners, making a combined $r = 0.35$. Thus here again there is an association between T (hallucinations) and P in both psychotic and non-psychotic groups.

Eye-tracking is another variable that has been related to P and the general psychoticism continuum. Lipton *et al.* (1983) have shown that not only schizophrenics, but often also their relatives, show faulty lack of smoothness in the pursuit or tracking eye movements when required, say, to follow a swinging pendulum. Similarly, twins discordant on schizophrenia may nevertheless be concordant for this test. Iacono *et al.* (1982) found this symptom in patients with unipolar and bipolar affective disorder in remission. Bosch (1984) and Iacono and Lykken (1979) have reported positive correlations with psychoticism questionnaires in schizophrenic and normal subjects. There are some contradictions in the data, and a large-scale replication would seem suggested (see also Simons and Katkin (1989) and Siever *et al.* (1982)).

A rather different approach was used by Jutai (1988), who examined specifically the lateralized cerebral dysfunction in schizophrenia and affective disorder postulated by Flor-Henry and Gruzelier (1983), and based on a model suggested by Venables (1980). Jutai concluded that the results of his study supported Venables's notion that, in the development of schizophrenic disorders, there may be an initial disturbance of right-hemisphere mechanisms of attentional control. Psychosis-prone young adults diagnosed on the Chapman tests tended to use visual search strategies similar to those of right-brain damaged patients. He adds the usual disclaimer that at present it is not certain that they do so for similar reasons.

A different technique for investigating hemisphere differences is the dichotic shadowing technique (Rawlings and Borge, 1987). The theory that schizophrenics are characterized by left-hemisphere overactivation (Flor-Henry and Gruzelier, 1983) has been tested by Rawlings and Claridge (1984) that subjects scoring high on a measure of schizotypal personality responded more quickly to verbal material presented tachistoscopically to their left visual field than to their right, while subjects with low scores, showed the usual superiority for material presented to the right visual field. Brooks (1984) found a similar difference. Rawlings and Borge (1987) have reported two experiments using the dichotic shadowing technique. Both gave positive results, showing differential responding in the two ears, with high P scorers failing to show the right-ear superiority shown by the low P scorers; the second experiment gave similar results for the male subjects but gave little evidence for the females. Overall the studies give mild support for the theory; but leave many questions unanswered. It is worth mentioning that Hare and McPherson (1984) found that a group of criminal psychopaths showed a significantly smaller right-ear advantage on a dichotic listening task than did groups of criminals who were not clearly psychopathic; this adds to the evidence that psychopaths belong to the schizophrenic *Erbkreis*.

Related to these studies of attention deficit in schizophrenics is an experiment carried out by Hinton and Craske (1976), who argued that 'attentional effort' is positively correlated with the magnitude of action potentials in those muscles which are not involved in the tasks being undertaken (Eason and White, 1961), and that 'degreee of effort' in attentional tasks would lead schizophrenics to show higher action potentials in a simple attention task (Goldstein, 1965; Malmo, Shagass and Davis, 1951). He predicted and found positive correlations between his EMG index and the P score for both males (r = .56) and females (r = .44), concluding that P score related directly to increase in generalized muscle action potentials on attending to simple perceptual discrimination tasks.

Attentional processes may also be involved in an interesting experiment reported by Badcock, Smith and Rawlings (1988). The topic selected was the effect of a masking stimulus (backward masking) on exposure of a target stimulus, with a specified inter-stimulus interval intervening. Masking deficits, i.e. increased susceptibility to a mask, had been found prior to (Braff, 1981), during (Braff and Saccuzzo, 1981) and following (Miller, Saccuzzo and Braff, 1979) a schizophrenic episode. Saccuzzo and Schubert (1981) had used the presence of a masking deficit within various subgroups of schizophrenics and schizotypals to verify the existence of spectrum disorder. Badcock *et al.* successfully extended this research to include high P scorers, who showed significantly more deficit than low P scorers. They also argued that the results of such experiments might be simply the effects of (1) differential no-masking thresholds and (2) differential susceptibility to increasing task difficulty. Both possible determinants were shown to be active, with high-P scorers requiring longer target durations, at a particular level of accuracy, than low P scorers, and with high-P scorers showing greater effects for more difficult stimuli. These results make interpretation of the masking data more difficult, but agree with reduced original sensitivity levels in schizophrenics (Mannuzza *et al.*, 1980; Braff and Saccuzzo, 1981; Neuchterlein and Dawson, 1984) and in schizotypics (Merritt and Balogh, 1984). Whatever the correct interpretation, the results support a proportionality approach.

Word association tests show a similar result. It is well known that schizophrenics show unusual and rare responses to standard lists of words (Kent and Rosanoff, 1910; Tendler, 1945; Pavy, 1968). A similar effect has been observed in the biological relatives of schizophrenics (Ciarlo, Lidz and Ricci, 1967; Zahn, 1968; Mednick and Schulsinger, 1968; Griffiths *et al.*, 1980). Separate studies have extended this relationship to normal groups of students, finding P positively and significantly correlated with unusual and rare word associations (Upmanyu and Kaur, 1986; Ward, McConaghy and Catts, 1991; Merten, 1992, 1993). Here again we find agreement with the proportionality criterion.

Low platelet monoamine oxidase (MAO) has been found in psychotic patients, and also in their relatives and inpatients who have recovered, suggesting that low MAO activity may be a marker for 'vulnerability' (Buchsbaum, Coursey and Murphy, 1976; Schalling, Edman and Aesberg, 1987). In a recent study of 61 healthy high school volunteers, Klinteberg *et al.* (1987) found correlations of −.30 in female and −.27 in males between MAO and psychoticism. It may also be noted that low MAO activity was found related to extraversion, impulsiveness, and sensation-seeking, as well as monotony avoidance, and that Lidberg *et al.* (1985) found it related to psychopathy, again suggesting a relationship between psychopathy and schizophrenia. (See also Checkley (1980) for a review of MAO in relation to depressive illness.)

These results may be related to serotonin levels which seem to have similar behavioural correlates as MAO, and hence may be predicted to correlate inversely with P (Zuckerman, 1991). Schalling, Aesberg and Edman (1984) have in fact found that CSF 5-HIAA levels were inversely related to P scores; similarly, CSF levels of 5-HIAA were found to be positively related to a measure of inhibition of aggression, suggesting that in humans, P is related inversely to the functioning of the serotonergic system, as is much psychopathology.

The next measure is based on the Venables (1963–4) and Claridge (1972) theory that psychotic patients differ from normals not so much in their absolute levels on the range they cover on given psychophysical measures, but rather in the way in which different measures covary. Thus in psychosis, whether natural or LSD induced, there occurs a peculiar inversion of the covariation between autonomic and perceptual function. The most widely used measures were the two-flash threshold and the electrodermal response. Claridge and Chappa (1973) have extended this model to normal subjects, and have shown that high-P scorers do indeed behave, when compared with low-P scorers, as schizophrenics do when compared with normals. They conclude: 'The results provide evidence for psychoticism as a normal personality dimension having, as its biological basis, a particular kind of nervous typological organization seen, in its extreme form, in the psychotic disorders' (p. 175). Later studies have extended this peculiar inversion of perceptual and autonomic functioning to relatives of psychotic patients, i.e. to members of the psychotic *Erbkreis* (Claridge, Robinson and Birchall, 1985). A more detailed discussion of the whole theory and its relevance to the concept of psychoticism is given in Claridge's (1985) book on the *Origins of Mental Illness*. (It should be noted that a replication study of the Claridge and Chappa study was only partially successful, for reasons which are not immediately obvious.)

Another interesting variable frequently used in this context is the alleged inefficacy of 'filtering' mechanisms in schizophrenia (Hemsley, 1975, 1976). In this connection it may be useful finally to list two psychological systems which have received much theoretical attention, and which may have a causal influence on schizophrenia, as well as being related to P. The first of these is *negative priming* (Beech and Claridge, 1987), a concept widely used to explain the schizophrenic's failure to use inhibitory factors early in the information processing system, thus allowing material in the preconscious to gain conscious representation (Frith, 1979). The general nature of the effect is as follows. First, a distractor is used in a priming display; when next used as a target stimulus, response latency to the latter is increased compared with trials where no such relation is present (Tipper, 1985; Tipper and Cranston, 1985). This effect may be used as a measure of individual differences (Tipper and Baylis, 1987), and if it is true that schizophrenia is associated with a weakening of the inhibitory mechanisms, then we would predict a *negative* correlation between measures of negative priming and interference, and a positive correlation between psychoticism or schizotypy and interference. In other words, negative priming is a precondition of effective inhibition of interfering stimuli, and inefficient negative priming, as in schizophrenics and high P scorers, would lead to interference and hence poor performance.

Several studies, reviewed in the next chapter, have shown that schizophrenics and high-P scorers, as well as high schizotypy subjects, do indeed show a distinct lack of negative priming, and may indeed show positive priming (Peters, Pickering and Hemsley, 1994).

Another concept to be discussed here is that of 'latent inhibition', a close relation of negative priming (Weiner, 1990). Passive pre-exposure of a stimulus reduces its ability to enter into new associations when that opportunity is offered in the same context as the initial pre-exposure (Macintosh, 1985; Pearce and Hall, 1980). This phenomenon, originally studied in animals, has now also been widely investigated in human subjects, both adults and children (Lubow, 1989). Lack of latent inhibition would promote attentional deficits, such as occur in schizophrenics, and it has been shown that schizophrenics not under medication, or at an early stage of medication, do indeed show less latent inhibition than controls (Baruch, Hemsley and Gray, 1988). It was found that medication, as expected, reversed this trend.

When the same procedure was tried on normal subjects, using the Claridge schizotypy scale and the Eysenck psychoticism measure as psychosis-prone scales, these were negatively correlated with latent inhibition, supporting the hypothesis. Lubow *et al.* (1992) have replicated the Baruch *et al.* (1988) study, showing that latent inhibition was weaker in high-P than in low-P subjects. Here also predictions of proportionality are successfully verified.

Another important finding is that the P300 wave on the averaged evoked potential, which has frequently been found to be of lower amplitude in schizophrenic subjects (Pritchard, 1986), is also lower in high-P subjects (Stelmack, Houlihan and McGarry-Roberts, 1993). As the latter authors state, 'the negative association of P300 amplitude here might provide some support for the view that the psychoticism dimension is a dispositional trait for the development of psychosis' (p. 407), although they add correctly that 'further investigation is required for clear understanding of the smaller P300 amplitude that is associated with the higher psychoticism scores' (p. 407). Tradition favours an interpretation that smaller P300 amplitude is indicative of *less sensitivity to task demands*. This would fit in well with the behaviour of both schizophrenics and high-P scorers.

A final test of the hypothesis that psychoticism is related to psychosis, particularly schizophrenia as the most virulent manifestation of that illness, relates to gender differences. Males have scores on P twice as high as females; is there any evidence that schizophrenia is more frequent, and develops earlier, in males than in females? Murry (1991) has written a review paper on the topic which goes into more detail than is possible here. In brief, he concludes that (1) there is an excess of males in strictly defined schizophrenic cohorts, the excess in some studies amounting to a 2:1 ratio; (2) male schizophrenics tend to manifest a more severe form of the disease; (3) males show an earlier onset; (4) males show more and poorer pre-morbid characteristics. Whatever the reasons (and there are many different theories) there appears to be a roughly similar disproportion of the two sexes in schizophrenia and in high-P scores.

Efforts to investigate the proportionality criterion have not always been successful. Thus the Kamin blocking effect (impaired learning of an association between a conditional stimulus (CS2) and an unconditional stimulus (UCS) if CS2 is presented simultaneously with a different CS (CS1) already associated with the UCS) is absent in acute schizophrenia (Jones, Gray and Hemsley, 1992a), but there was no systematic relation between the blocking effect and any of four measures of psychoticism (Jones, Gray and Hemsley, 1990). The reasons for this failure are not known, but the theoretical model is a promising one that deserves to be studied more intensely. (A more recent reanalysis of the data gave much more promising results – Jones, Gray and Hemsley (1992b).)

Much more could of course be said about the theories involved in these studies, the experimental difficulties of taking into account drug administration in chronic schizophrenics, or indeed the theoretical prediction of changes in experimental behaviour to be expected when acute psychosis becomes chronic (Gray *et al.*, 1991). Many of the questions find at least a tentative answer in the Gray *et al.* paper, which attempts the construction of a neuropsychological model of schizophrenia (or perhaps psychosis?). This model includes animal studies, amphetamine effects on psychotic-like behaviour, and several other topics indirectly relevant to our purpose, but not sufficiently so to deserve detailed comment here; it will be referred to again in the next chapter.

We may summarize the findings of our discussion by stating that the methodology of proportional effect has been surprisingly successful in showing that schizophrenic-normal differences are reproduced when comparing high- and low-P scorers, both in normal and (less frequently) in psychotic groups. While not universally successful, the great majority of comparisons have shown the expected effects, and it would seem difficult to account for these findings on grounds other than the admission of a *continuum* ranging from the normal to the psychotic, with gradings both within the normal and the psychotic portion. Many details remain to be sorted out, and many other hypotheses remain to be tested, but the outline is becoming clear.

Conclusions

We thus arrive at certain conclusions. In this chapter I have tried to determine the degree to which the empirical evidence supports three major portions of the dimensional or continuity hypothesis which I originally advanced some 40 years ago (Eysenck, 1952a, b). The three components state:

(1) Psychotic symptoms and illnesses do not form completely separate diagnostic entities, unrelated to each other, but are genetically related and form a general cluster with severity of illness the major distinguishing marker. It is not part of the theory to deny specificity of genetic origin also existing and contributing to the total variance; it is merely asserted that in addition to specificity there also exists a certain amount of generality, suggesting that the term 'psychosis' contains a meaningful generalization.
(2) Psychotic disorder, of whatever kind, is not a separate diagnostic entity which is categorically separated from normality; it is merely an extreme along a continuum of abnormality shading into schizoid personality, 'spectrum' disorders, psychopathy and personality disorders, criminality and alcoholism, and average types of behaviour right to the other extreme of empathy, altruism and selflessness.
(3) This continuum is collinear with the concept of psychoticism, embodied (however imperfectly) in the P scale of the EPQ, and also in a number of 'schizotypy' constructs and scales. Proof for this proposition makes use of criterion analysis and its derivative, the proportionality criterion. All the elements of this theory are empirically testable, and have been so tested on numerous occasions.

As regards the generality of 'psychosis', it seems clear that there are definite genetic links between different diagnostic categories (schizophrenia, manic-depressive disorder, schizo-affective disorder, unipolar disorder) which make it impossible to regard them as entirely separate disease entities. Some specificity there undoubtedly is, but there is also a generality of disorder which links all these disorders and their sub-classifications and diagnoses together to form one end of the psychoticism continuum, with a severity gradient placing schizophrenia at the extreme end, followed by schizo-affective disorder, manic-depressive disorder and finally unipolar illness.

It equally appears clear that there is no absolute barrier between this concept of 'psychosis' and borderline disorders linking these psychoses with more normal behaviours. Many different names have been given to these transitional states (schizoid personality, 'spectrum' disorders, personality disorders, psychopathy) which in turn connect intimately with alcoholism, criminality, eccentricity and anti-social behaviour generally. Again there is probably some degree of specificity connected with all these types of behaviour but there is also the continuum which links them together, and with psychotic states (Rieder, 1979).

Is this continuum adequately measured by the EPQ–P scale? Because of the novelty of the concept, and because of the short time during which it has been investigated, the scale clearly has many faults, but nevertheless when tested it has proved surprisingly successful in marking the continuity from schizophrenia to normal behaviour. Undoubtedly, it is likely to benefit from continuous improvement (as in the EPQ–R scale – Eysenck, Eysenck and Barrett (1985)) but it seems already to have many of the attributes required by an instrument designed to investigate the properties of the psychotic continuum. Various schizotypy scales appear to fulfill a similar function, but they suffer from correlating so highly with neuroticism that we must seriously doubt their adequacy as measures of psychoticism; as Kendler, Gruenberg and Strauss (1981) have shown in their analysis of the Copenhagen adoption study, there were no genetic or familial environmental links between anxiety and schizophrenia. Schizotypy scales seem to measure aspects of both psychoticism and neuroticism, with an emphasis on the latter; this does not suggest that they would be well equipped to measure the former. However, only continued work with all kinds of 'psychosis-prone' measures will ultimately determine which is closest to collinearity with (true) 'psychoticism'.

In the list of variables used to illustrate the proportionality criterion, I have on purpose included several different types of measures. One class deals with biological variables (HLA B27, MAO; serotonin) of different kinds. A second deals with laboratory behaviours (eye-tracking; dichotic shadowing; sensitivity levels). A third is concerned with learning-conditioning variables (latent inhibition; negative priming). Yet a fourth is concerned with psychological variables (creativity, hallucinatory activity; word association). Physiological variables (EMG, autonomic-perceptual inversion) constitute yet a fifth set of variables. It is the variety of variables which makes the results impressive, together with the theoretical congruence; to obtain successful results over such a wide array of variables suggests that the underlying hypothesis may be along the right lines.

What are the advantages of the perspective here suggested over the traditional categorical viewpoint of psychiatric diagnosis? In the first place, it is more in line with reality, as the experiments and investigations listed in the text suggest. In the second place, it suggests experimental investigation which the orthodox model would fail to generate, or regard as important. In the third place, it obviates certain difficulties in experimentation, such as institutionalization and drug-treatment of patients, which have made proper experimental study of psychotic patients very difficult – it is always problematic whether observed differences between patients and controls are due to some disease process, or to drug and/or institutionalization (iatrogenic) effects. If the theory here offered is anywhere near correct, we can test our hypotheses by investigating high-P vs. low-P normals, or even animals (Gray *et al.*, 1991). This greatly expands the horizon of our theory-testing paradigms, and may hopefully lead to a better understanding of psychotic disorders, and their treatment.

The advantages of joining the psychiatric and psychological research efforts devoted to schizophrenia and psychosis generally, and normal behaviour and the rest of the psychoticism continuum, are likely to go both ways. Thus the question of the biological basis of psychoticism may find a solution based on biological model building in the schizophrenic compartment (Gray *et al.*, 1991; Schmajuk, 1987; Swerdlow *et al.*, 1988; Weinberger, 1989; Frith and Done, 1988; McKenna, 1987; Joseph, Frith and Waddington, 1979). Work already referred to concerning MAO and serotonin fits in well with at least some of these models, and it may not be too long before an agreed theory of biological causation for P arises to take its place with the biological theories giving a causal basis for N and E (Eysenck, 1967, 1981). Zuckerman (1991) has attempted to give a local habitation and a name to the entities involved in his book on the *Psychobiology of Personality*, written from the same point of view underlying the planning of this chapter, and the reader looking for further enlightenment is referred to his summary of the evidence. By accepting the continuity hypothesis, and by working towards a proper theory from both ends (psychosis and normality), we are more likely to arrive at the desired end.

One final comment may be in order. Factor analysis has often been criticized because of its lack of objectivity; the number of factors extracted, the mode of rotation adopted, and the naming of the resulting factors is to some extent at least subjective. The methods and theories described in this paper attempt to avoid such subjectivity; the factor isolated and named 'psychoticism' is firmly based on empirical and experimental studies which have tested a large number of deductions from the original theory. The resulting factor has not been *named* psychoticism *post hoc* and by simple inspection of its contents; it was conceived on theoretical grounds, and on the basis of a large body of empirical evidence, and constantly revised to accord with new evidence. The method of criterion analysis, in its varied forms, tests the fundamental correctness of the assumptions underlying the factor. It is suggested that the objections often made to factorial investigations do not affect in any way the concept of psychoticism, because of the efforts made to avoid precisely those criticisms. It is also suggested that the method used may be of much wider applicability in the personality sphere.

The discussion of the nature of psychoticism has been much fuller and longer than many readers might have preferred, but there are three reasons for this prolixity. The first is that in view of the prominence of psychopathology in the personalities of geniuses and highly creative individuals, clarification of the *nature* of this psychopathology is essential. In the past there has been a failure to distinguish neuroses, functinal psychoses, effects of toxic substances like lead or mercury, syphilitic infection, and a host of others. Furthermore, intepretations were usually in terms of psychiatric nosology having little empirical support, little reliability, and less than satisfactory validity. Not of such things is the Kingdom of Heaven made! What is required are concepts that fit the available data, and psychoticism seems to approach such a concept more closely than others. If we can link P with creativity, as we shall try to do in the next chapter, we would have some claim to have clarified this particular nebulous area.

There is, however, another and even more important consideration. Most authors have drawn attention to the observed correlation between genius or creativity, and psychopathology, without attempting to show just what the causal link consisted of. An attempt to do so will be made in the next chapter, based on the material discussed in this chapter.

A final reason for considering the clarification of the P construct as important and relevant is that it may lead us to a link between creativity and a large body of experimental literature which has not been drawn into the discussion previously. Concepts like latent inhibition have theoretical potential for clarifying the nature of creativity which may lead the discussion on to a much more factual plateau, as well as suggesting experimental developments not previously considered. One of the criticisms of much of the empirical work on creativity has been its purely correlational character, its lack of any link with the experimental psychological literature, and its failure to generate theoretical concepts to tie together the many dissociated findings. Psychoticism may be the bridge between genius and creativity, on the one hand, and such theoretical and experimental developments, on the other. The next chapter will deal with this extension.

7 The roots of creative genius

> *In the discovery of secret things and in the investigation of hidden causes, stronger reasons are obtained from sure experiments and demonstrated arguments than from probable conjectures and the opinions of philosophical speculators of the common sort.*
>
> William Gilbert

Creativity and psychoticism

In this chapter I will review the evidence that P is indeed associated with creativity, both conceived as a *trait*, and conceived of as achievement. This demonstration is central to the causal theory developed later in the chapter, trying to link creativity with *biological mechanisms* and the *genetic basis* of creativity with creative behaviour.

Any theory linking creativity and psychoticism must of course be validated by empirical research, along a number of different lines. One line of research has consisted of using *psychosis-prone subjects* as high psychoticism probands. As already mentioned, Heston (1966) studied offspring of schizophrenic mothers raised by foster-parents, and found that although about half showed psychosocial disability, the remaining half were notably successful adults, possessing artistic talents and demonstrating imaginative adaptations to life to a degree not found in the control group. Karlsson (1968, 1970) in Iceland found that among relatives of schizophrenics there was a high incidence of individuals of great creative achievement. McNeil (1971) studied the occurrence of mental illness in highly creative adopted children and their biological parents, discovering that the mental illness rates in the adoptees and in their biological parents were positively and significantly related to the creativity level of the adoptees.

These findings clearly support our main theory, as well as prior hypotheses such as those of Hammer and Zubin (1968) and Jarvik and Chadwick (1973) to the effect that there is a common genetic basis for great potential in creativity and for psychopathological deviation that are inimical to creativity and achievement; it appears to be psychoticism in the absence of psychosis that is the vital element in translating the *trait* of creativity (originality) from potential into actual achievement.

A second line of investigation would suggest a significant correlation between P and creativity as measured by current creativity (trait) tests, such as the Wallach and Kogan (1965) and Torrance (1974) tests. Several such studies have been reported. These studies also make it clear that actual *psychosis* works in ways hostile to creative achievement. The first was carried out by Kidner (1978), in a study using 37 male and 31 female subjects, mostly students, nurses and teachers. They were administered the EPQ (Eysenck Personality Questionnaire, a measure that includes a P scale), as well as the following tests: (1) Acceptance of Culture scale – an 18-item scale intended to measure the degree to which a person accepted the prevailing British culture; (2) 'creativity' or 'originality' tests – this set contained three of the Wallach and Kogan (1965) tests (unusual uses, similarities, and pattern meanings), scored for both associational fluency and originality, to give a total of six scores; (3) two IQ tests, viz. the advanced Matrices and the Mill Hill Vocabulary scale; and (4) an education scale, relating to the amount of education experienced, and the enjoyment and acceptance of the experience. These items were correlated and factor analysed, and in addition an index of 'creativity relative to intelligence' was computed, using standard scores on the Wallach and Kogan tests, and subtracting from their sum the standardized and summed intelligence test scores. This 'creativity index' was not included in the factor analysis, as it was based on scores already included.

The raw correlations indicate that the index of creativity was correlated significantly with P ($r=.31$); the correlation with L (Lie scale, a measure of conformity) was negative ($r=-.23$). The correlation with E was .21, that with N .09. P also correlated with the creativity tests, values of r ranging from .16 to .33, all positive, and with acceptance of culture, $r=-.52$! L showed negative correlations with the creativity tests, values ranging from $-.18$ to $-.30$; the correlation with acceptance of culture was .44. These correlations suggest that high P scorers are original and unsocialized, while high L scorers are unoriginal and socialized. (L, as always, correlates negatively with P.) The results of the factor analysis bear out this provisional conclusion; two factors were extracted which were clearly recognizable as *originality* (loaded on all the creativity tests, and also, much less highly, the IQ tests), and *socialization* (loaded on acceptance of culture, education, and vocabulary).

P and L are situated just where theoretical considerations suggest they ought to be; P combines originality with lack of socialization, while L combines socialization with lack of originality. L and acceptance of culture have almost identical loadings on the two factors, while P has loadings identical in size, but opposite in sign, to those of L and acceptance of culture. N is not related to either factor; E is characterized by lack of socialization, which again agrees with prediction. Kidner concludes his discussion by saying that 'the analysis supports the viewpoint that the element common to "creativity" and psychoticism concerns the associational mode of thought, whilst a crucial difference concerns the success or otherwise of socialization processes'.

In another experiment, Kidner was concerned with performance on a tachistoscopic Reaction Time task in which the subject had to decide whether or not the word shown was a member of a particular category. There were 62 subjects, and they were administered a series of scales, including most of those used in the preceding experiment. Among tests not used previously was an 'adjective relevance' test; this was intended to measure the subject's ability and willingness to think metaphorically. It was in the form of an Adjective Checklist with six nouns and eight adjectives in grid fashion. The subject simply had to tick each grid space where he thought that the adjective was relevant to the noun. High scores on the test might perhaps be considered evidence of 'over-inclusiveness', a psychotic symptom studied in detail by Payne and Hewlett (1960). P was found to correlate with acceptance of culture ($r = -.53$), with the Wallach and Kogan originality tests (values all positive, but rather lower than before), authoritarianism ($-.57$), adjective relevance (.34), and average reaction time (.33). These results to some extent confirm those of the previous experiment, and add over-inclusiveness and slowness in categorization. But of course the main finding, corroborating several studies already mentioned, is that high P scorers have unusual associative processes, resembling in this typical psychotic (particularly schizophrenic) patients.

Another early study was carried out by Farmer (1974). Using 40 students as his sample, Farmer also administered a battery of divergent-thinking tests, and factor analysed the resulting matrix of intercorrelations; also included in the matrix were scores of P, E, N, and L and Little's Person-Orientation index, which on *a priori* grounds was thought to be opposed to P. The divergent-thinking tests fell into two main groups, 'fluency' and 'originality'. On the 'fluency' factor, P had only a small loading of .24. On the 'originality' factor, P had a loading of .74, and Personal Orientation a loading of $-.66$. Thus, there is here some evidence that originality (as defined by the tests used) is indeed correlated with P. The number of cases involved is too small to allow us to consider this experiment more than suggestive, but it will be seen to align well with other evidence to be considered presently.

Next we may concentrate on what is perhaps the most impressive study so far done, namely the work of Woody and Claridge (1977). The subjects of their study were 100 university students at Oxford, both undergraduate and graduate. The students constituted a wide sampling of the various fields of specialization at the university. The writers chose students as their subjects because of (possibly doubtful) evidence that creativity is significantly related to IQ up to about IQ 120, but that it becomes independent of IQ above this level (Canter, 1973; Heansley and Reynolds, 1989). The tests used by them were the EPQ (Eysenck and Eysenck, 1975) and the Wallach–Kogan Creativity Tests, somewhat modified and making up five different tasks (instances, pattern meanings, uses, similarities, and line meanings). Each task was evaluated in terms of two related variables: the number of unique responses produced by the subject (originality), and the total number of responses produced by the subject (fluency).

The Pearson product moment correlation coefficients between psychoticism and creativity scores for the five tests are as follows: P with the 'number of responses' scores: Instances = 0.32; Pattern Meanings = 0.37; Uses = 0.45; Similarities = 0.36; Line Meanings = 0.38. P with 'uniqueness' scores: 0.61, 0.64, 0.66, 0.68, 0.65. It will be seen that all the correlations are positive and significant, and those with the uniqueness scores (which are of course the more relevant of the two) are all between .6 and .7. These values are quite exceptionally high for correlations between what is supposed to be a cognitive measure, and a test of a personality trait, particularly when general intelligence has effectively been partialled out from the correlations through the selection of subjects. There were effectively no significant correlations between E and N, on the one hand, and creativity on the other. It is interesting to note, however, that the L score of the personality questionnaire, which up to a point is a measure of social conformity, showed throughout *negative* correlations with creativity scores, seven out of ten being statistically significant. L is known to correlate negatively with P (Eysenck and Eysenck, 1976).

A partial replication of the Woody and Claridge study was carried out by Stayte (1977) who used the Wallach and Kogan tests and the EPQ score as a measure of psychoticism. 'All the correlations are positive, and a fair proportion are significant or near significant' (p. 49). She also used two other measures of psychoticism, which however correlated poorly with P. Only the 'total uniqueness' global score on the creativity tests correlated positively with all three psychoticism tests. Rawlings (1985) also provided some replication of the Woody and Claridge finding, correlations between P and creativity centring around .20.

In a recent study of the relationship between P and creativity, Furnham and Yazdanpanaki (1994) looked at the effects of *brainstorming* in nominal and real groups. (Nominal groups are comprised of individuals who are tested individually, but whose scores are added to those of other individuals in the 'group'. Nominal groups, as usual, out-performed real groups, i.e. people actually interacting with each other.) P scorers generally produced high creative quality answers. This effect was found in several of the measures used, such as superiority of responses, percentage of superior responses, and mean creativity rate. 'Those that were classed as having high P scores . . . tended to produce a higher level of creative achievement. Their ideas were constantly of a high quality and there was a larger percentage of them'. There was also a tendency for high-P scorers to do better when working alone, than when working in groups, as compared with low-P scorers

Similar results to those of Woody and Claridge (1977) were found by Stavridou and Furnham (1994), who correlated P and scores on the Wallach–Kogan Creativity test, for a sample of 37 students. Uniqueness scores on the five subtests of the Wallach–Kogan test correlated 0.49, 0.49, 0.51, 0.44 and 0.33 with P, averaging 0.51. Correlations wth E were all positive, but only averaged 0.17; those with the Lie scale, measuring conformity, were negative averaging −0.16. This study is a good replication of the Woody and Claridge study, with very similar results.

These studies demonstrate that there is a sizeable relation between P and tests of divergent thinking. We must next consider the literature concerning Word Association. There are quite a few studies in which significant correlations have been found between P (or measures of schizotypy) and unusual responses on the W-A (Word Association) test (Routh, 1971; Miller and Chapman, 1983; Hundal and Upmanyu, 1981; Upmanyu and Kaur, 1986; and Upmanyu and Upmanyu, 1988). These are direct tests of the hypothesis; in addition there are studies linking unusual responses with parents of schizophrenics (Ciarlo *et al.*, 1967; Zahn, 1968), schizophrenia high-risk children (Griffiths *et al.*, 1980), and relatives of schizophrenics (Callahan and Saccuzzo, 1986).

Of particular interest are two more recent studies by Merten (1992, 1993) carried out in Germany; they show a much more rigorous methodology than most previous studies, many of which are subject to criticisms such as those voiced by Schwartz (1978a,b; 1982). Merten used several different methods of testing word association parameters, and it is important to look at the differential results for these different methods in order to be able to integrate the results within a theoretical framework. Beginning with Mannhaupt's (1983) norms for verbal reactions, Merten constructed six word lists of 25 words each, carefully equated for the categories of words used (e.g. tools, insects, musical instruments). Norms were established of the usual free association mode, and associations described as frequent, medium, and rare.

Subjects were tested along several different lines. (1) Free association – respondent answers with the first word that comes into his mind. (2) Individual response condition – respondent is asked to give responses which are unusual. (3) Usual responses – respondent is asked to give responses which most people would give. There is a fair literature concerning these different types of response requirements (e.g. Jenkins, 1959; Lisman and Cohen, 1972; Rothenberg, 1967; Routh and Schneider, 1970). A fourth, original method was (4) response choice – respondents are offered a usual, a middling frequent and an unusual response, and are required to indicate which is which.

In addition to these different tests, questionnaires and IQ tests were administered, including Sullwold and Huber's (1986) Thinking and Speaking scales, the Brief Psychiatric Rating Scale (Overall and Gorham, 1962), and the Eysenck Personality Questionnaire (Eysenck and Eysenck, 1975).

Subjects were 46 healthy persons, 43 schizophrenics, and 15 manic-depressive patients. The main results are as follows. There are no differences between acute and chronic schizophrenics. Manic-depressive patients are not differentiated from schizophrenics, except for the individual response condition, where they are close to the normal group. Most imporant, normals with *high* P scores give more unusual answers in the free and usual association conditions, very much like the psychotics; in the individual association condition they give more original and hence better answers. This is in good agreement with the theory that high-P normals are more original but can also judge appropriateness well. High-L scorers did poorly on the individual association condition, tending to give unoriginal answers. This finding is in good agreement with Horton, Marlowe and Crowne (1963), who find high-L-scale scorers giving less unusual responses, and Routh (1971) who found high 'schizoid' subjects to give more unusual responses in a free test, but also to give even less usual responses on instruction.

Merten's (1993) second paper took up the search for the relation between unusual responses and personality, particularly P and L. Using 46 normal subjects, he found *negative* correlations between P and response commonality in free, common and individual response conditions. 'That means that they present the "psychosis-like" associative disturbance in the free and common response condition, and yet fulfil the individual response condition better since it is precisely in that condition that idiosyncrative responses are really demanded.' Again, therefore, we find original response creation joined with control of relevance, with the former linking the high-P response activity with that of schizophrenics, and the latter forming a crucial difference. Similarly, high-P scorers did not fail to react appropriately to questions about the commonness/uncommonness of their own associations; they clearly are aware, as schizophrenics are not, of responding more individually on the Word Association (W-A) Test than does the majority.

L scorers show a high positive correlation with the individual association conditions; in other words even when asked to give unusual responses they are loath to do so, and generally fail to respond appropriately, confirming the usual interpretation of high L scores as indicative of conformity. L correlates, as usual, negatively with P ($r = -.30$). These two studies are in good agreement with our theory.

Merten (1992, 1993) has constructed a model to incorporate the hypothesized active ingredients of his theory, as shown in Fig. 7.1. The stimulus activates the associative memory which generates a strongly activated response. This is sent to the comparator which checks the response against some self-chosen criterion of acceptability (relevance, decency). If the response is acceptable, it is sent on and constitutes the output; the associated word is said aloud. If it is rejected, we go back to the second stage, the associative memory, and another strongly activated response is chosen and sent on to the third stage. Schizophrenics and high-P scorers would have a pathologically weak comparator, allowing associations to pass which are not generally acceptable (Cohen, 1978a,b; Lisman and Cohen, 1972).

Word Association measures of creativity seem to be as much in accord with our theory as divergent thinking tests; we must next turn to the complexity

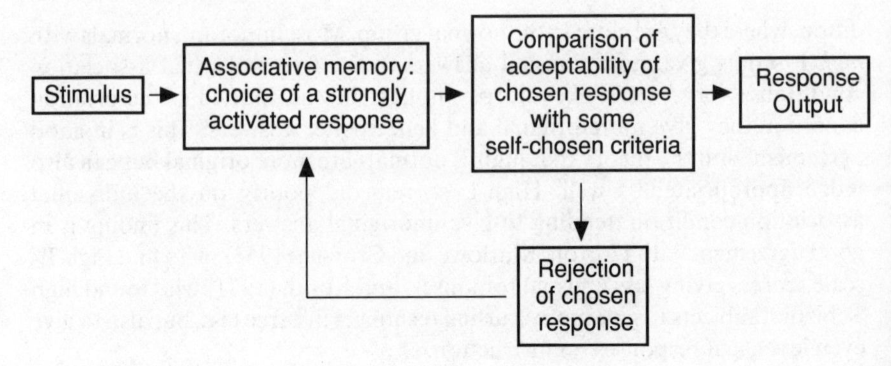

Fig. 7.1 Model of Word Association Responses (Merter, 1992).

score in the Barron–Welsh Art Scale (Eysenck and Furnham, 1993). Positive but low correlations were found for a student sample. A better support was found in another study using both the Word Association test and the Barron–Welsh Art Scale (Eysenck, 1993b). The population consisted of 100 non-academic adults who took an intelligence test giving verbal, non-verbal and total scores, a word association test and the Barron–Welsh Art scale; in addition, P, E, N and L were measured, as well as impulsiveness, ventursomeness, and empathy. The Word Association test was scored for common responses, rare and unique responses. Scores were submitted to a multidimensional scaling (smallest space) analysis. Results are shown in Fig. 7.2.

It can be seen that as hypothesized P, impulsiveness, venturesomeness, extraversion, the Barron–Welsh Art Scale complexity score, and the Word Association test rare and unique responses are placed together on the right side of the diagram. Conformity (as measured by the L scale), common Word Association test responses, empathy and neuroticism go together on the left side of the diagram. The IQ scores constitute a separate group; they clearly do not go with the creativity scores. It will also be seen that youth and male sex go with the creativity side, although not too much should be made of this agreement with what is known of genius. These results are certainly in agreement with theory, and we may say that measures of origence, W-A unusual responses, and divergent thinking all correlate to varying degree with high P scores. As far as these results go, they seem to support our theory.

Do schizotypal tests give similarly positive results? (They should to some extent, seeing that P is a central ingredient in such tests (Eysenck, 1993b).) Of interest here is a report by Schuldberg *et al.* (1988), demonstrating an overlap in creative and schizotypal traits. Schizotypy was measured on the Chapman Perceptual Aberration and Magical Ideation scales. Subjects high on these tests differed significantly from control subjects on the five creativity tests used, and the Barron–Welsh Art Scale. 'Measures of "impulsive non-conformity" were also high in these schizotypal positive symptoms and creativity scores' (p. 655).

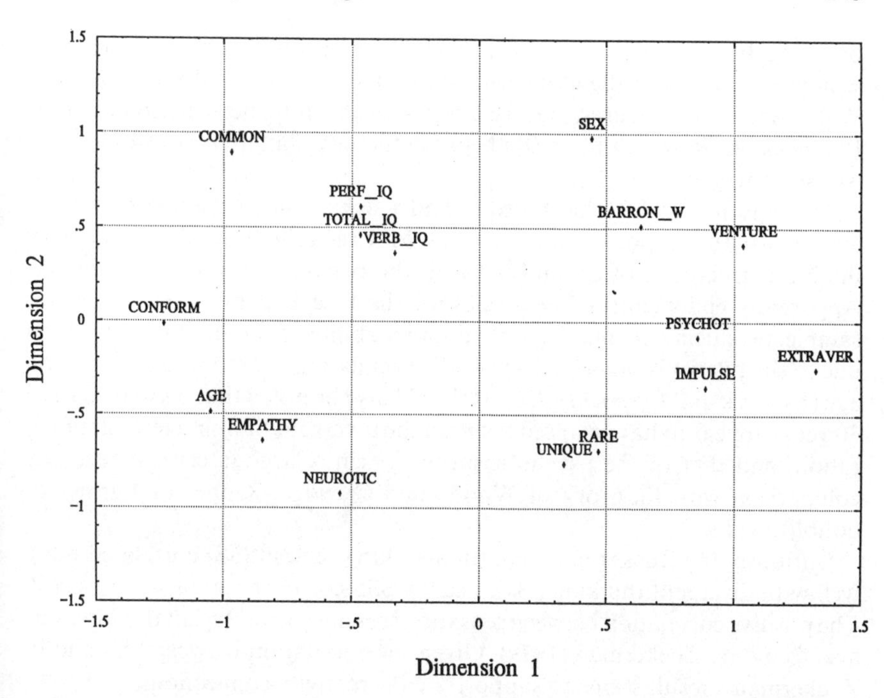

Fig. 7.2 Multidimensional scaling analysis of personality and creativity variables (Eysenck, 1993a).

In a later paper, Schuldberg (1990) enlarged his findings by using larger groups, and again found that 'unusual perceptual experiences and beliefs, Hypomanic traits and Impulsive Nonconfirmity are associated with creative attitudes and activities . . . Subclinical "positive" symptoms are associated with higher levels of creativity' (p. 218). In a final paper, Schuldberg (1993) has demonstrated the relative independence of ego-strength from creative aspects of cognition; presumably ego-strength acts mainly so as to counteract the debilitating aspects of psychoticism.

Most of the experimental studies of creativity have used verbal or pictorial tests; an exception is the work of Rawlings, Sherr and Dempsey (in 1994), which examined in four different studies the relationship between P and various aspects of musical preference. Using recorded musical excerpts, study 1 found that P is related to liking for heavy metal music and dislike for easy listening music; study 2 used a music preference scale and found the same. Study 3 examined preferences for recorded triads. Low-P subjects preferred *consonant* triads (major, minor), while relatively speaking high-P scorers preferred dissonant triads (augmented, diminished, atonal). Study 4 showed that chord preferences are related to music preferences. Explanation of these correlations may be in terms of the arousal properties of dissonant chords and heavy metal music, or such stimuli may come to represent for certain groups a breaking from convention and an aggressive revolutionary attitude. This would fit in well with the liking for complexity in visual art that the Barron–Welsh Art Scale measures, and the aggressive and anti-social attributes of the P dimension. Research to decide between these two alternatives would seem well worthwhile.

The only study which has failed to find a significant correlation between P and creativity is reported in a paper by McCrae (1987) which succeeded in finding correlations of around .40 with the personality trait of 'openness'. Apparently endorsement frequencies for the P scale items were low in this sample, producing an unacceptably poor reliability of .47; this may have been due to the relatively advanced age of the population – P scores go down with age (Eysenck and Eysenck, 1976). It should also be noted that between 20 and 30 years appear to have elapsed between the divergent thinking test administration, and that of the psychoticism test; such a large interval makes any comparison with the work of Woody and Claridge, Kidner or Farmer of doubtful value.

Curiously the Zuckerman Sensation-seeking scales (SSS) correlated quite well with divergent thinking (.31); usually SSS correlate significantly with P. The positive correlation between sensation seeking and divergent thinking was also found by Zuckerman (1979). Given the correlation between SSS and P, Zuckerman's result seems to support the P–creativity connection.

We must now turn to the third line of research suggested by our theory, namely the correlation of P with creative achievement of a high order. The only study of what most lay people would consider genuine creativity has been reported by Götz and Götz (1979a,b). Their work significantly extends that of other investigators who tried to link creativity in the arts with personality (e.g. Csikszentmihalyi and Getzels, 1973; Barron, 1972; Eysenck, 1972a; Eysenck and Castle, 1970; and Drevdahl, 1956). Some of these studies are difficult to interpret, but we may note that Eysenck (1972b) and Eysenck and Castle (1970) found that art students were significantly more introverted and neurotic than non-art students. Götz and Götz (1973) pointed out in criticism that art students in general may not be particularly creative, but when a group of highly gifted art students were compared with less gifted and ungifted subjects, they found that the highly gifted students also had low scores on extraversion and high scores on neuroticism. It may be noted that neuroticism would seem to be related *positively* to creative work in the arts, *negatively* to creative work in science; the reason may be the emotional involvement in art, and the explicit rejection of emotion in science. The point would be that neuroticism is not related to creativity *per se*, but to the *direction* of creativity. Introversion is characteristic of high creative achievement in art and science, but probably by way of concentrating the mind on the tasks in hand, and preventing the dissipation of energy on social and sexual matters unrelated to work (Glover *et al.*, 1989).

In the study under review Götz and Götz (1979a,b) administered the EPQ to

337 leading professional artists living in West Germany, of whom 147 male and 110 female artists returned to the questionnaire; their mean age was 47 years. One outstanding result of this work was that male artists were significantly more introverted and significantly more neurotic than non-artists, while for females there was no difference on either of these dimensions. As the authors suggest, it is perhaps true that in our Western World it is mainly women with average or higher scores on extraversion who have the courage to become artists, while the more introverted and possibly more artistically gifted women do not dare to enter the precarious career of the artist.

We must now turn to scores on psychoticism. Here the results are very clear; male artists have much higher P scores than male non-artists, and female artists have much higher P scores than female non-artists. As Götz and Götz point out, 'these results suggest that certainly many artists may be more tough-minded than non-artists. Some traits mentioned by Eysenck and Eysenck may also be typical for artists, as for instance they are often solitary, troublesome and aggressive, and they like odd and unusual things' (p. 332).

The work of Götz and Götz (1979a,b) thus offers important support for the results of Woody and Claridge, and the other authors cited above, in that their work uses actual artistic achievement as a criterion for the measurement of creativity and originality. In doing so they give credibility to the validity of divergent thinking tests as measures of creativity and originality, and the fact that both in the artistic and in the non-artistic population studied by other investigators significant correlations are found between psychoticism and creativity and originality very much strengthens the hypothetical link between the personality trait and the behavioural pattern. We may thus be justified in concluding that originality and creativity are the outcome of certain traits, rather than (or as well as) aspects of cognitive ability.

Given, then, that P is genuinely a predictor and correlate of creativity, it follows that provisionally at least we may in our search for *causal* factors in creativity use P as a link to suggest the nature of such factors. When Glover *et al.* (1989) in their *Handbook of Creativity* wrote that the field of study of creativity 'had come to be a large-scale example of a "degenerating" research program' (p. XI), they were probably referring to the mainly psychometric nature of the enterprise, and to the extremely 'bitsy' nature of research and findings. There were absent any signs of an overarching theory linking together the many different aspects studied, and facts discussed, and clearly the absence of such a theory leaves the whole field unstructured and confused.

Overinclusion and the associative gradient

So far, the contents of this book are very likely to give a similar impression of 'bitsyness'. The two major findings have been (a) that creativity is closely related to psychoticism, and (b) that creativity is based on a flat associative gradient, or a wide associative horizon. What has been lacking is a theory binding two apparently quite separate variables together. Why should psychotics, and non-psychotic people with high-P scores, have a wide associative horizon? Are there any theories in clinical and experimental psychology to help us construct a link between (a) and (b)? Are there underlying physiological or hormonal factors which can explain the observed relationships? It is to these questions that the present chapter addresses itself. Clearly it is more theoretical than the preceding chapters, which have in the main been highly factual, relying on psychometric or experimental studies. In looking for a theory bringing together isolated pieces of factual knowledge, I must obviously rely on the reader's indulgence; while the theory is eminently testable, and indeed directly suggests appropriate empirical tests, such tests are mainly for the future. Thus it is with some hesitation that I am putting forward the theory, and attempting to furnish it with some empirical underpinnings wherever possible. But the major test of the theory will come if and when experimentalists take it seriously enough to test some of its consequences.

The answer to our question may lie in considering the nature of psychotic (mainly schizophrenic) thinking. If the theory is correct, or at least along the right lines, then there should be some connection between what characterizes such thinking, and original and creative cognition. It may be useful to start with a well-established theory, namely Cameron's notion of 'overinclusion' (Cameron, 1947; Cameron and Magaret, 1950, 1951). Cameron believes that schizophrenics' concepts are over-generalized. Schizophrenics are unable to maintain the normal conceptual boundaries, and incorporate into their concepts elements, some of them personal, which are merely associated with the concept, but are not an essential part of it. Cameron used the term 'overinclusion' to describe this abnornmality, and reported that in working on the Vigotsky test, and a sentence completion test, schizophrenics were unable to preserve the 'conceptual boundaries' of the task. In solving a problem, the schizophrenics '. . . included such a variety of categories at one time, that the specific problems became too extensive and too complex for a solution to be reached' (Cameron, 1939).

A fair number of experiments have been carried out to investigate this theory. These have been reviewed elsewhere (Payne, 1960; Payne and Hewlett, 1960; Payne, Matusek and George, 1959). The results obtained have consistently supported the theory, e.g. Moran (1953) and Epstein (1953); see also Chapman (1956, 1958) and Chapman and Taylor (1957).

Payne *et al.* (1959) have suggested that it is possible to reformulate Cameron's theory of overinclusion in a slightly more general way so that a number of predictions follow from it. Concept formation can be regarded as largely the result of discrimination learning. When a child first hears a word in a certain context, the word is associated with the entire situation (stimulus compound). As the word is heard again and again, only certain aspects of the stimulus compound are reinforced. Gradually the extraneous elements cease to evoke the response (the word), having become 'inhibited' through lack of 'reinforcement'. The 'inhibition' is in some sense an active process, as it suppresses a response which was formerly evoked by the stimulus. 'Overinclusive thinking' may be the result of a disorder (failure) of the process whereby 'inhibition' is built up to circumscribe and define the learned response (the word or 'concept'). In short, it could be an extreme degree of 'stimulus generalization'.

The same theory can be expressed in different terms. All purposeful behaviour depends for its success on the fact that some stimuli are 'attended to' and some other stimuli are ignored. It is a well-known fact that when concentrating on one task, normal people are quite unaware of most stimuli irrelevant to the task. It is as if some 'filter mechanism' cuts out or inhibits the stimuli, both internal and external, which are irrelevant to the task in hand, to allow the most efficient 'processing' of incoming information. Overinclusive thinking might be only one aspect of a general breakdown of this 'filter' mechanism.

The notion of overinclusion as being characteristic of normal as well as schizophrenic thinking ultimately derives from Rapaport's (1945) suggestion that at least two quite different types of formal thought disorder contributed to the disturbances of thinking found in schizophrenics, neither of which was in fact specific to schizophrenia. One of these defects, demonstrated clearly in object-scoring tests, consisted in a tendency to function more at a concrete than an abstract level (Vigotsky, 1934). The other consisted of a 'loosening of the concept span', in that schizophrenics included objects in the various groups of the test to which they did not strictly belong. This 'looseness of thinking' is what others have called overinclusive or allusive thinking, and it occurs in normal people as well as in schizophrenics. Looseness of thinking, as measured by sorting tests, correlates well with clinical assessments of that behaviour (Lovibond, 1954). 'Looseness' may be suggested to be a normal type of thinking related to psychoticism, and fundamental to creativity; concrete thinking is characteristic rather of psychosis, and has no link with creativity, but rather precludes it. (Sorting tests present the subject with objects, or words, or pictures, which he is asked to sort into meaningful groups of his own devising. Such groups may be more or less sensible, or 'correct'.)

A similar concept to overinclusion is that of 'allusive' thinking, characteristic of many schizophrenics on object sorting tests. McConaghy and Clancy (1968) demonstrated that this type of thinking existed widely in less exaggerated forms in the normal population, showed similar familial transmission in schizophrenics and non-schizophrenics, and was akin to creative thinking. Dykes and McGhie (1976) actually demonstrated that highly creative normals scored as highly 'allusive' on the Lovibond object sorting tests as do schizophrenics. The low creative normals tended to produce conventional, unoriginal sortings, while the highly creative normals and the schizophrenics tended to give an equal proportion of unusual sortings. 'This supports strongly that a common thinking style may lead to a controlled usefulness in normals and an uncontrollable impairment in schizophrenics' (Woody and Claridge, 1977).

An interesting study which demonstrates the *dependence* of creativity (as shown by fluency and unusualness of word associations) on degree of psychosis and also the relevance of bipolar disorders, is the work of Shaw, Mann and Stokes (1986). They found that lithium treatment *decreases* both the number of productions and the idiosyncrasy of production. Lithium, of course, also serves to reduce the severity of mania and depression: Merten (1992) found schizophrenics and manic-depressives differing to the same extent from normals. Thus the link with creativity need not be via schizophrenia, but may also be via psychotic depression.

Whatever may be the most appropriate name for the thinking characteristics which link schizophrenics and highly creative normals (overinclusiveness, allusive, looseness, etc.), there clearly is a marked similarly between these concepts. Furthermore, this view supports the notion of schizophrenia as a genetic morphism (Huxley *et al.*, 1964), whose frequency results from a balance between selectively favourable and unfavourable properties.

The term 'overinclusion' has long since been abandoned, and new theories and experiments developed to include what are essentially similar ideas and conceptions; I have discussed these in some detail elsewhere (Eysenck, 1993a). Our theory would thus demand that some good and appropriate measure of 'overinclusion' should (a) be commonly found in schizophrenics and/or in other psychotic patients; (b) should correlate with measures of psychoticism in normal people; and (c) should correlate with creativity. We have shown in some detail that tests such as the word association test do so correlate; the question is *why* they should do so.

I have suggested that creativity is closely related to psychoticism, that underlying both is a cognitive style which may be loosely identified as 'overinclusiveness', i.e. a tendency to have a flat associative gradient which allows the individual a wider interpretation of 'relevance' as far as responses to stimuli are concerned. This overinclusiveness may be due to a failure of inhibition, characteristic of psychotics, high-P scorers, creative people and geniuses. Clearly there must also be further characteristics of the cognitive apparatus which make the difference between a psychotic patient and a genius; presumably these include high intelligence (and the other variables associated with creative achievement listed in Fig. 1.4) and an ability to reject responses which are too far removed from the stimulus to make a genuine contribution to the solution and the problem under consideration.

Latent inhibition and psychoticism

I shall now discuss two candidates for the role of inhibitor of remote associations. Both fulfill this role to an extent indicated by a great deal of experimental work; both have been shown to be linked with schizophrenia (by their absence, or at least weakness); and both have been shown to be equally related to psychoticism. First, I shall deal with *negative priming*, and second with *latent inhibition*. (See Chapter 6 for definition and evidence of relation to P and schizophrenia.) Both discussions are theoretical in nature; there has not been any direct study of creative as opposed to non-creative people, as far as I know. However, the theory underlying negative priming and latent inhibition presents a possible answer to our problem, as well as an experimental paradigm which can be used to test the theory here presented; it is hoped that this may encourage readers to carry out the necessary experiments to disprove or support the theory.

The concept of *cognitive inhibition* is only one of the many experimental paradigms used to investigate the stages and selectivity of processing information which have recently been applied to the assessment of possible creative dysfunction in mental health (Power, 1991). Incoming information has to be narrowed down, and irrelevant information selectively excluded, a process which is postulated to occur through a balancing of facilitatory processing of task-relevant stimuli and the inhibition of task-irrelevant ones. Schizophrenia is postulated to be characterized by a *breakdown* of this balance, in the sense that the inhibitory part of the mechanism is not working properly, so that the failure of inhibitory processes produces over-inclusiveness (Frith, 1979; Bullen and Hemsley, 1986; Beech *et al.*, 1989a). This line of argument originated with a paper by Treisman (1964), who suggested that selecting certain specific stimuli for analysis might involve the exclusion or alteration of others. Keele and Neill (1978) produced a similar argument for the activation of memory traces; activated memory traces inappropriate to the task in hand have to be actively inhibited. An experimental paradigm for such cognitive inhibition is that of *negative priming*.

Negative priming may be defined in terms of the experimental paradigm in which a distractor object which had previously been ignored is subsequently re-presented as the target object to be named, classified, or otherwise dealt with. These processes take longer than if there had been no prior presentation; in other words, the subject has associated the prime with negative salience (disregard, omit, discount, ignore, overlook), so that it is more difficult (takes longer) to make it salient when required. As a typical defining experiment we may cite the Stroop colour naming task, in which a colour-word (e.g. RED) is presented, written in green ink; the task is to disregard the word and call out the colour of the ink. If now the next word is printed in red ink, the response of normal subjects is significantly slowed; in other words, the to-be-ignored word RED has acquired negative salience which inhibits cognition associated with it. Hence the term 'negative priming'; the irrelevant stimulus word acts as a prime for later recognition and meaning, but negatively so – it partly inhibits such reaction.

Frith (1979) suggested that schizophrenia is associated with a weakening of the inhibitory selection mechanisms that are active in the early phases of information processing, giving rise to some of the positive symptoms of schizophrenia, such as hallucinations, delusions and formal thought disorders, such as over-inclusiveness. Cognitive inhibition is vital for normal thought processes to occur; its absence (lack of negative priming) would therefore characterize, and be causally related to, the vagaries and excesses of schizophrenic thinking. Beech *et al.* (1989a) used a negative priming task to differentiate a group of schizophrenics from a matched group of mixed diagnosis psychiatric patients. As predicted, the schizophrenics shared *reduced inhibition.* We would expect that high-P, or schizotypal normal subjects would show a similar lack of cognitive inhibition, and this has been shown to be so (Beech and Claridge, 1987; Beech *et al.*, 1989b; Beech *et al.*, 1991). High schizotypes not only showed failure of negative priming, but even *positive* priming effects; in other words, the supposedly 'negative priming' had facilitatory rather than inhibitory effects for this group.

Curiously enough the failure of negative priming was less noticeable in the schizophrenic subjects studied in these experiments than in the (normal) high schizotypes. This may be explained in terms of the medication effects shown by the schizophrenics; as Beech *et al.* (1990) have shown, a small dose of chlorpromazine in normal subjects significantly *increased* the negative priming effect, as compared to placebo.

The nature and definition of the negative priming effect are fairly clear, but the actual processes involved are still a matter of debate. Neill (1977) has put forward the view that priming effects occur as a result of *active inhibition* of the irrelevant stimulus, making possible an efficient response to the target stimulus. The need to undo the inhibition produced a response cost on the subsequent trial which is measured in terms of increased reaction time. Tipper (1985) has suggested an alternative theory according to which what is inhibited is the access of the activated structure to the mechanisms required for an overt response, de-coupling the representations from the construct of action. For the purpose of this section, we need not prefer one theory over the other.

What, essentially, does the negative priming paradigm say in general terms that would make it applicable to our problem? It is based on the view that both facilitatory and inhibitory processes are involved in selectivity determining attention to relevant information input, relevance being decided by prior experience. There are marked individual differences in the degree of cognitive inhibition, measured by negative priming, with schizophrenics/schizotypes failing to show such inhibition, and consequently becoming overinclusive. In other words, the flat associative gradient characteristic of creative people may be the result of lack of cognitive inhibition, as measured by negative priming. At present this is clearly only an hypothesis, there being no direct evidence on the assumed relation between creativity and negative priming. However, the fact that high-P scorers have been shown to be creative, to have flat associative gradients, and also to have low negative priming scores, gives at least indirect evidence to firm up the general theory.

The theories of Tipper (1985) and Neill (1977) mentioned in the previous section as explaining negative priming effects are clearly *cognitive* theories, yet there is another obvious candidate which may explain these phenomena along the lines of classical conditioning theory. As far as I know this line of argument has not previously been followed, but clearly the theory and phenomena of *latent inhibition* bear a remarkable similarity to negative priming. (Lubow's (1989) book on latent inhibition has no mention of negative priming.)

Latent inhibition is defined by an experimental paradigm which requires, as a minimum, a two-stage procedure. The first stage involves stimulus *pre-exposure*, i.e. the to-be-CS (conditioned stimulus) is exhibited without being followed by any unconditioned stimulus (UCS); this leads theoretically to the CS acquiring a negative salience, i.e. it signals a *lack* of consequences, and thus acquires inhibitory properties. The second stage is one of *acquisition*, i.e. the CS is now followed by a UCS, and acquires the property of initiating the UC response (UCR). Latent inhibition (LI) is shown by increasing difficulties of acquiring this property, as compared with lack of pre-exposure. With humans, there is a masking task during pre-exposure to the CS. For instance, the masking task might be the presentation of syllable pairs orally, while the CS would be a white noise randomly super-imposed on the syllable reproduction. The LI group would be exposed to this combined recording, while the control group would be exposed only to the syllable pairs, without the white noise. In the test phase the white noise is reinforced, and subjects given scores according to how soon they discover the rule linking CS with reinforcement. LI would be indicated by the group having the pre-exposure of the white noise discovering the rule later than the control group.

There are more complex, three-stage procedures, but these complications are not crucial to our argument (Lubow, 1989). Is it possible to classify negative priming as a variant of latent inhibition? There are obviously close similarities. Conditioning or performance is preceded by the exhibition of the to-be-CS, or action-stimulus, under conditions which endow the to-be-CS with negative salience; the subject is either instructed to disregard it (negative priming), or learns independently that the to-be-CS does not signal anything specific; hence in both cases the link sign–significate is counterindicated, and the establishment of such a link is made more difficult. Applying the nomenclature of conditioning to the negative priming paradigm is permissible in the recent climate of SS theorizing and inclusion of cognitive elements in the conditioning process (Mackintosh, 1975, 1983, 1985; Gray, 1975). The cognitive elements in latent inhibition theory are emphasized by Lubow (1989) in terms of his conditioned attention theory. According to this theory, non-reinforced pre-exposure to a stimulus retards subsequent conditioning to that stimulus because during such pre-exposure the subject learns not to attend to it. The theory is based on the use of attention as a hypothetical construct, with the properties of a Pavlovian response, and on the specification of reinforcement conditions that modify attention. (For the latest theory of latent inhibition, taken in the context of a real-time neural network, see Scharajuk, Lam and Gray, in press.)

The same theory may be used to explain negative priming effects. Non-reinforced pre-exposure in this case is the not-to-be reacted to part of the red–green Stroop combination, which retards subsequent conditioning to that stimulus because during such pre-exposure the subject learns not to attend to it. The general view of the importance of changes in attention to stimuli, which underlies this theory, goes back to Lashley (1929) and Krechevsky (1932), and may be traced through Lawrence (1949) to Mackintosh (1975), Frey and Sears (1978), and Pearce and Hall (1980). Granted the similarities, we would expect (1) less latent inhibition in schizophrenics, and (2) less latent inhibition in high-P scorers. The evidence supports both deductions.

Baruch *et al.* (1988) found an abolition of latent inhibition in acute schizophrenics, but not in chronic schizophrenics or normals. Lubow *et al.* (1987) also failed to find such abolition in chronic cases, presumably due to the fact that such patients are on a dopaminergic antagonist, neuroleptic drug regime that would normalize attentional processes (e.g. Braff and Saccuzzo, 1982; Oltmanns, Ohayon and Neale, 1978). There is a large body of evidence to show that LI can be attenuated or abolished in rats by dopamine agonists such as amphetamine, and can also be increased with dopamine antagonists such as haloperidol and chlorpromazine; see discussion by Lubow *et al.* (1992). In this respect, too, latent inhibition closely resembles negative priming.

Regarding their high vs. low psychoticism group, Baruch *et al.* (1988) found the expected negative correlation between LI and P; the greater the proneness to psychosis (high P), the less latent inhibition. Similar results have been reported by Lubow *et al.* (1992), using two different experimental procedures; high-P subjects showed an attenuated latent inhibition effect compared with low-P scores. Both auditory and visual stimulus pre-exposure resulted in slower acquisition of new associations as compared with non-pre-exposure to the test stimulus, but to a much lesser extent in high-P than in normal and low-P subjects. Lubow *et al.* (1992) argue that

> the idea that schizophrenics fail to filter out irrelevant stimuli is congruent with the phenomenology of schizophrenics, and with a considerable variety of data on the differential effects of distractors on the behaviour of schizophrenics and normals. Frith (1979) cogently and succinctly described this type of result as reflecting an inability to limit the contents of consciousness (p. 570).

This, of course, is precisely what is characteristic of the mechanism needed to explain the overinclusiveness of schizophrenics and high-P scorers; the failure of negative priming and/or latent inhibition to limit associationist spreading (flat associationist gradient) would appear to account for the prominent symptoms of psychotic cognition and the major feature of creativity. Accordingly, this may be the missing link between psychopathology and genius. Of course, as already explained, creativity based on a flat associative gradient produced by an absence of cognitive inhibition is not enough *by itself* to produce creative achievement; other components, such as those listed in Fig. 1.4 are needed. Among these the ability to weed out unsuitable and unusable associations must be the distinguishing mark between the word salad of the schizophrenic and the utterances of the poet.

Latent inhibition and negative priming of course have a biological basis, and this seems firmly related to dopaminergic transmitters. As reported, dopamine agonists such as amphetamines attenuate or abolish LI, while dopamine antagonists (heloperidol, chlorpromazine) increase LI, just as they affect psychotic behaviour. As Lubow *et al.* (1992) say,

> these data are in accord with the premise that schizophrenia has a major attentional deficit component . . . and that the disorder is mediated by a dopamine system dysfunction. While other neurotransmitter involvements in schizophrenia have been proposed . . . the dopamine hypothesis remains a leading component in understanding schizophrenia (Gray, Feldon, Rawlins, Hemsley & Smith, 1991) (p. 503).

The suggested relevance of LI to creativity is of course similar to that suggested for negative priming. Cognitive inhibition characteristic of most people is lessened or removed in creative people, and hence the associationist gradient is flattened, criteria for 'relevance' are reduced, and 'overinclusiveness' appears. Again, it should be emphasized that there is no *direct* evidence in favour of the theory; it is based essentially on the strong association between creativity and psychoticism, the finding that psychoticism (like schizophrenia) is characterized by low degrees of LI and negative priming, and that LI and negative priming account for the lack of cognitive inhibition apparent in schizophrenics and high-P scorers. Direct evidence is of course needed before the theory can be accepted as a true, rather than a possible, account of the observed relation between personality and creativity.

There is an obvious question which arises from this discusison of latent inhibition and negative priming, namely: do they refer to one and the same phenomenon, measured in slightly different ways, or are they quite different phenonema? The answer would seem to be that they are probably different phenomena, although no studies have been reported in which both were measured. (An as yet unpublished study from our laboratories has done so, and suggests that there is no significant correlation between the two.) Lubow (personal communication) has argued as follows in favour of the conclusion that they are essentially *different* phenomena.

(1) Although the term 'inhibition' is used in both contexts, it is used *descriptively* in connection with LI, as the opposite of facilitation

(Lubow, 1989); on the other hand, the normal explanation of NP (negative priming) invokes inhibition as an underlying explanatory concept. While this is true of Lubow's concept of LI, it is possible to formulate alternative theories involving inhibition as a causal concept for LI; this argument clearly is not conclusive.

(2) An examination of the basic procedures and data from the two paradigms suggest the absence of a common underlying process. LI is a two-stage procedure, with a relatively large number of stimulus pre-exposures preceding the test phase. The dependent measure is one that reflects ease of association, such as percent correct or number of trials to reach a criterion. Furthermore, the effects of such pre-exposures are relatively long-lasting. NP, on the other hand, is basically a one-stage procedure, with single stimulus pre-exposure immediately preceding the test-probe. Here the response measure is reaction time. In LI experiments reaction time has never given positive results in discriminating between pre-exposed and not predisposed groups, even when direct associative strength measures do effectively discriminate.

(3) Finally, to obtain LI in adult humans, the pre-exposed stimulus has to be presented so that the subject does not attend to it, i.e. some masking task has to be used; without such a task, LI is not obtained; at least in instrumental learning tests. NP has no such requirement.

These are powerful arguments, but of course they cannot settle the issue in the absence of direct focussed experiments. Fortunately the issue is not crucial to our theory; if Lubow is right, then the correlation between P and either LI or NP should be significantly raised by correlating P with LI + NP; no such study has been reported. Experiments studying LI and NP jointly would seem likely to advance our understanding of both.

There is one study linking negative priming with P, and attempting to link it also with creativity. This study, carried out by Stavridou and Furnham (1994), has already been mentioned; it duplicated Woody and Claridge's (1977) classical study showing that P and creativity were closely related. They also used a test of negative priming, which correlated significantly (negatively) with P, as theory predicts. The predicted relationship between creativity and negative priming, although present, was not statistically significant. A rough calculation suggests insufficient statistical power for the test; it would take about twice the number of subjects actually used to ensure statistical significance in 95% of trials. This study should certainly be repeated, with a larger number of subjects.

One difficulty in theoretical discussion of these phenomena is that while work on LI with animals has been theory-driven and highly productive, work with humans has been largely demonstrative (showing that the phenomenon occurs in humans). This makes it difficult to make meaningful comparisons

(Siddle and Remington, 1987). Even in the animal field, there are different theoretical positions which have not been clarified (e.g. Wagner, 1978; Pearce and Hall, 1980; Mackintosh, 1985; Lubow, 1989). In what follows I will be more concerned with LI than with NP; mainly because there is a much broader experimental and theoretical background, but also because the phenomenon seems more directly relevant to the broadening of the associative horizon. Furthermore, there is an abundance of evidence on the neural substrates of latent inhibition (Weiner, 1990) which links the concept with genetic antecedents, and enables us to examine in some detail whether and to what extent the biological substrates of schizophrenia resemble those of LI. Before turning to this task, we must note one further phenomenon.

To be fair, a third effect, similar in some ways to both latent inhibition and negative priming, should be mentioned, if only because, unlike LI and NP, attempts to fit it into the theoretical system I have described have been unsuccessful so far. This is the Kamin Blocking Effect (Kamin, 1969). This type of experiment uses a classical conditioning procedure of some sort, and involves again a pre-exposure stage in which the experimental group learns an association between a conditioned stimulus (CS_1) and an unconditioned stimulus (UCS), while control subjects learn either no association or a different one. Both groups are then presented with a series of pairings between a compound of two conditioned stimuli, including the first, i.e. $CS_1 + CS_2$, and the same UCS as before. As the final stage, both groups are tested for what they have learned about the CS_2–UCS relationship. The group pre-exposed to the CS_1–UCS association demonstrates less learning about the CS_2–UCS relationship; this is the blocking effect, and like latent inhibition, it is abolished by amphetamine, a dopamine agonist (Crider, Solomon and McMahon, 1982). This suggests that as with LI, the effect should be observed in acute, but not chronic schizophrenics, and this has indeed been found (Jones *et al.*, 1992a).

I would also expect that, as in the case of LI, high-P scoring subjects should demonstrate that Kamin blocking effect, but two experiments failed to demonstrate any such correlation (Jones *et al.*, 1990). A later re-analysis showed that results assumed a more positive aspect if only certain aspects of psychosis-proneness were considered (Jones *et al.*, 1992b), but until matters are clarified further it would not be correct to adduce these studies as supporting the general theory outlined.

Genetics and creativity

I have already mentioned some of the phenomena which indicate the broad genetic background of genius and creativity, such as the fact that genius very frequently springs from poor soil; very few of the parents of great mathematicians showed any great ability themselves. Intelligence, a vital component of genius of course has a very strong genetic component (Eysenck, 1979; Brody,

1992). Training seems to have little effect on creativity (e.g. Mansfield, Busse and Krepelka, 1978; but see Stein, 1975, and Frederickson, 1984). Overall effects are slight and evanescent. Gender differences have been found to be of supreme importance, and are of course genetic. The personality factors related to creativity have been shown to be strongly determined by genetic factors (Eaves *et al.*, 1988). These and other findings are reviewed in detail by Vernon (1989), who points out the difficulties of research in this area, but also favours a view that recognizes the importance of genetic influences.

Most of the material reviewed by Vernon is only indirectly relevant to the question of whether or how genetic factors determine individual differences in creativity. The recent work of Waller *et al.* (1993) has added flesh to these bones, and allows us to advance some more firmly based conclusions. Waller *et al.* begin by reviewing familial studies of creative talent which did not seem to bear out Galton's view of familial aggregation (Bullough *et al.*, 1981). Bramwell (1948) explicitly re-examined Galton's study of hereditary genius and concluded that, of all the professions studied by Galton, only judges seemed to congregate within families. The Bullough study bore this out in their compilation of additional data; 'creative achievement was rarely carried on in the same family beyond one generation', and 'contrary to the assumptions of Galton . . . creative achievers did not usually have children who also achieved' (p. 109). Waller states that 'consequently, much of this literature concludes that hereditary factors play a minor role at best in the determination of creativity'.

Clearly this conclusion follows as little from the facts as did Galton's opposite conclusion from the facts as he perceived them. Familial aggregation is irrelevant to the genetic hypothesis, as I have already argued in an earlier chapter; if it occurs it can be due to *either* genetic or environmental factors, or any combination. Indeed, given that children will tend to regress to the mean for partially genetic traits, and that they acquire half their genes from the mother who presumably is less talented than the father who has been selected as highly creative, there is no reason on genetic grounds to expect familial aggregation to any marked extent. Environmental factors would be more likely to produce such aggregation, through provision of educational facilities, role models, and direct 'old school tie' assistance. Thus one might argue that the *absence* of familial aggregation would confirm environmental hypotheses, and thus strengthen genetic ones. Genius would be seen as a highly unlikely segregation of genes, occurring very rarely for a few individuals only.

The literature covering creativity tests certainly suggests some genetic influence. Twin studies (Barron, 1972; Reznikoff *et al.*, 1973) give generally positive heritabilities for tests of creative thinking and ideational fluency, and Nichols (1964) reported average MZ (monozygotic) and DZ (dizygotic) intraclass correlations of .61 and .50 for ten studies of divergent thinking. When corrected for error of measurement this would suggest that just above a quarter of the phenotypic variance is due to genetic determinants. Most studies are on a small scale, and results not too dependable. More impressive is the work of Waller *et al.* (1993).

They used the Creative Personality Scale (CPS), constructed by Gough (1979). This is a 30-item adjective check-list made up to measure trait creativity, devised through empirical keying using data from more than 1700 persons, many of whom were architects, mathematicians, outstanding scientists and graduate students, taking part in the IPAR studies already mentioned on several occasions. Waller *et al.* (1993) report the factor loadings of the items on their sample of 157 reared-apart twins; these are given in Table 7.1 below.

We find here again many of the personality traits which we have previously encountered as characteristic of creative people – intelligent, self-confident, original, egotistical, unconventional, insightful, resourceful, reflective, wide interests, humorous. (Adjectives with loadings below .40 may be disregarded.)

The test was administered as part of an extensive battery of psychological tests to the participants of the Minnesota Study of Twins Reared Apart (Bouchard *et al.*, 1990). Having partialled out the effects of age and sex, Waller *et al.* calculated the intraclass correlations for reared-apart identical (MZ) and fraternal (DZ) twins. For MZ twins the correlation was .54; in other words, the best estimate of additive genetic variance is 54% of phenotypical variance. If corrected for measurement error, this would almost certainly go above .60, demonstrating very clearly the strong effect of genetic factors on creativity. There is, however, an important proviso to this conclusion.

If we were dealing only with additive genetic variance (V_A), and possibly error variance (V_E), then we would expect the correlation for the DZ twins to be roughly half that for MZ twins, i.e. .27. In reality, however, the correlation is $-.06$, ruling out any possibility of accounting for the facts by postulating only V_A and V_E on the genetic side (as well as environmental factors, of course). Waller *et al.* in fact anticipated such a finding and would explain it in terms of the concept of 'emergenesis' (Lykken, 1982; Lykken *et al.*, 1992). This bears a close resemblance to the concept known as *epistasis* to geneticists, which is defined as non-additive genetic variance due to interaction between different gene loci. In other words, human traits are unlikely to be governed by single genes, they are much more likely to be governed by assemblies or combinations of genes whose interaction may not be *additive* but *synergistic*. It may be useful to spell all this out more in detail, because it may help us to interpret the mode of genetic determination of creativity differences more closely. I have followed the Lykken *et al.* (1992) account, which should be consulted for a much more extensive treatment.

We start with the *genome*, i.e. the entire collection of genes arranged in their 46 chromosomes. As Lykken *et al.* remark, they can be thought of as a blueprint or, better still, a very large book of instructions, each of whose 100 000 or so pages represent a different gene. Of these, about three-quarters are identical for all normal human individuals (*monomorphic* genes), determining that we have 2 legs, 2 arms, 1 nose, etc. Some of these we share with

Table 7.1. *Factor loadings on the Creative Personality Scale*

Factor	Nonlinear loadings
Resourceful	.82
Insightful	.76
Individualistic	.72
Reflective	.68
Intelligent	.67
Interests – wide	.66
Humorous	.64
Clever	.62
Inventive	.58
Self-confident	.57
Original	.55
Interests – narrow	−.54
Confident	.53
Egotistical	.47
Unconventional	.49
Sexy	.45
Mannerly	.40
Capable	.38
Dissatisfied	.33
Sincere	.31
Snobbish	.31
Conventional	−.28
Informal	.28
Submissive	.28
Honest	.22
Suspicious	.14
Conservative	.13
Affected	.09
Cautious	.09
Commonplace	−.06

Notes
From Waller *et al.* (1993), with permission.

other animals, many of them with the higher apes; only some are quite specifically human. Other genes are *polymorphic*, and responsible for genetically based individual differences within a given species. Genes have different *alleles*, i.e. different variations or forms of expression, varying from 2 to 20 or so.

Genetic effects tend to be *additive*. We inherit 'tall' genes from father or mother, on a random basis, and our height is determined by the *sum* of these genes (and some small environmental input via nutritional differences). Phenotypic height correlates about .95 for MZ twins, .50 for DZ twins who only inherit half the same genes as their siblings. Height aggregates within a family because parents share 50% genetic variance with their children. This is the model that is normally applied to intelligence, but it is not the only one, and it does not fit differences in creativity.

Many bodily features, like the eyes, are not constructed on this additive model, but rather on a *configural* model, in which all the component genes are essential, and the absence of, or change in any one can produce a large and possibly disastrous change in the result. Polymorphic genes, as well as monomorphic genes, can behave in this configural manner. Traits and abilities that depend on *configuration* and polymorphic genes segregating independently would be shared (to the extent of genetic contribution) by MZ twins, who share all their genes and hence all gene configurations, but would not be likely to be shared by DZ twins, siblings, or parents and offspring. 'Such traits, while genetic, would not tend to run in families' (Lykken *et al.*, 1992). Lykken *et al.* give as an example facial beauty. This depends on the *configuration* of many different components (nose, forehead, ears, chin, lips, hair, etc.); any of these is inherited independently of the others, and hence their *configuration* will be identical for MZ twins (who inherit all the components in an identical manner), but not for DZ twins. The heritability of *individual elements* will be half that of the MZ twins, but the inheritance of the *configuration* will be practically nil for the DZ twins.

This configural inheritance is called 'emergenesis' by Lykken (1982), who defines it as

> arising as a novel or emergent property resulting from the interaction of more elementary and partly genetic properties. An emergent trait might be determined by a configuration of independently segregating polygenes in the sense of epistasis, or, at a more molecular level, an emergent trait might be a property of a configuration of independent traits that are themselves in part genetically determined. In some instances situational factors may figure as configural components. The distinctive feature of emergenesis is the notion of configurality which implies that any change in any one component may result in a qualitative, or a large quantitative change in the emergenic trait. Emergenic traits may – but are likely not to – run in families. On the other hand, marked MZ similarity combined with weak similarity of DZ twins is strongly suggestive of an emergenic trait.

Lykken *et al.* (1992) apply this argument to the emergence of genius, very much as I have done in my previous discussion. It is not that genius or 'emergence' is a function of familial descent that makes it likely that there is a strong genetic component; it is rather the fact that familial aggregation is largely *lacking* that provides the clinching argument. Thus modern research stands Galton's argument on its head; familial aggregation, had it been a fact, would suggest environmental rather than genetic influences! For a more detailed discussion, Lykken *et al.* (1992) should be consulted.

Neurophysiology of creativity

If genetic influences are indeed important, then there must exist physiological–hormonal–enzymatic structures or substances mediating the effects of DNA on behaviour. Recent attempts to elaborate a psychophysiological theory have shown some congruence (Schmajuk, 1987; Swerdlow and Koob, 1987; Weinberger, 1989; Gray *et al.*, 1991), and it may be useful to examine briefly the latest of these to discover to what extent the major findings integrate with the neurophysiology of latent inhibition, and explain the biological nature of psychoticism. If any congruence can be found, this would suggest that here perhaps we may have the beginning of a neurophysiology of creativity. Inevitably much of what is said here is speculative, but there are enough hard facts to make the excursion profitable.

What concerns us here is *the nature of schizophrenic cognitive impairment*, our hypothesis being that something similar to that impairment may be responsible for the wider associative horizon characteristic of the creative person. What precisely is the major cognitive impairment in schizophrenia! Gray *et al.* (1991) give a table listing seven major current views; most of these can be brought together in a formulation stating that *it is a weakening of the influence of stored memories or regularities of previous input on current perception* that is basic to the schizophrenic condition (Hemsley, 1987, 1991). Similarly, Patterson (1987) has suggested that there is 'a failure in the automaticity with which prior experience may be recreated in parallel with current stimulus input in schizophrenics (with concomitant failures in future orientation or contextually generated expectancy).'

Gray *et al.* (1991) generalize the large literature on this point as follows: 'There is a weakening of the capacity to select for cognitive processing only those stimuli that, given past experience of similar contexts, are relevant' (p. 3). This might be regarded as a statement of schizophrenic deficit, but it also fits in perfectly with our view of creativity. If we select for cognitive processing only those stimuli that, given past experience of similar contexts, are relevant, then clearly we are not going to use those stimuli in a novel, original, creative manner! The formulation of 'normality' implicit in it would condemn us to forever do the customary, ordinary, accepted, familiar, habitual, regular, tradition; never the novel, adventurous, imaginative, inventive, visionary or stimulating sort of thing we call 'creative'. Perhaps here we have the esssential link between genius and madness, creativity and psychoticism, originality and cognitive impairment. The influence of past experience can be useful for everyday living, but a dead hand for the creation of novel experiences and difficult problem solutions. Perhaps plotting the importance of past experience against mental health gives us a *curvilinear* regression; too much leads to rigidity, fixed behaviour, lack of flexibility, intransigence, undeviating courses of action, while too little leads to gross abnormality, as in schizophrenia. Best is a middle position, making use of prior experience, but not being a slave to it. Gough's (1976) interesting finding may be relevant; the most creative among his subjects were not those with *unique* word associations (too little constraint by past experience), but those with slightly less unique associations – unusual but not outrageously so.

Gray *et al.* (1991) recognize the strength of this argument by continuing their definitions:

> Note, however, that this statement begs the question of what constitutes 'relevance'. The most obvious way to remedy this deficiency is to define relevance in terms of the subject's current 'plans' or 'motor programs' (these being defined broadly to cover all action plans, including those in the purely cognitive sphere, such as the selection of stimuli for attention or the programming of streams of speech or thought), established and guided by positive instrumental reinforcement (p. 3).

The cognitive defect underlying schizophrenia may also be defined in a slightly different manner; different but not contradictory. There are several models of (normal) cognition that suggest that awareness of redundant and irrelevant information is inhibited in order to reduce information processing demands on a limited capacity system (Broadbent, 1977; Neuchterlein and Dawson, 1984). A similar hypothesis states that we must distinguish automatic processes from conscious attention, the latter involving awareness. Such awareness is closely associated with a self-limiting 'general inhibitory process' (Schneider and Schiffrin, 1977; Posner, 1982). On this hypothesis, cognitive abnormalities in schizophrenics would then be considered as related to a weakening of inhibitory processes important in conscious attention, and leading to the intrusion into awareness of aspects of the environment (outer and inner) not normally considered. This phenomenon is called 'overattention' by Gray *et al.* (1991); cognitive performance is disrupted by the intrusion of material normally below awareness (automatic processing).

Here again the stress on the pathological (overattention) recalls our discussion of 'overinclusiveness', in both cases we may ask about the weaknesses implied in cognitive functioning of 'underattention' and 'underinclusiveness'. Clearly too great an amount of cognitive inhibition may also be deleterious, although perhaps not to the same extent as lack of cognitive inhibition; too narrow an associative horizon, or too steep a gradient, may spell the absence of novel, original, creative ideas. What is vital is a person's position on the curve from over-inhibition to under-inhibition. Hardy, in our example from mathematical genius, would tend in the direction of rigidity, while of course perfectly normal psychiatrically; Ramanujan would tend in the direction of creativity. What has usually been assumed, at least implicitly, namely that we are dealing with a linear regression where the opposite of schizophreniform thinking must be good, may in fact be erroneous; both extremes may be bad, and an intermediary position preferable.

Somewhat similar to the above is a theory of attentional strategies of schizophrenic and highly creative normal subjects offered by Dykes and McGhie (1976). They investigated the hypothesis that individuals in both their populations habitually employ common attentional strategies which cause them to sample an unusually wide range of available environmental stimuli. They used three tests designed to assess attentional and other cognitive styles, and employed three groups of subjects: (1) a group of highly creative adults; (2) a group of equally intelligent but low creative adults; (3) a group of acute non-paranoid schizophrenic adults.

> The results offer support to the view that both highly creative and schizophrenic individuals habitually sample a wider range of available environmental input than do less creative individuals. In the case of schizophrenics this involuntary widening of attention tends to have a deleterious effect on performance, while, in contrast, the highly creative individual is more able to successfully process the greater input without thus incurring a performance deficit (p. 50).

(See also Hasenfus and Magaro, 1976).

Whatever the origin of individual differences along this axis from overattention to underattention, from overinclusiveness to underinclusiveness, from overinhibition to underinhibition, it must be based on some neural function or dysfunction. Again, most writers simply consider 'dysfunction' by looking at the schizophreniform end of the dimension; possible causes (e.g. dopaminergic hyperactivity in schizophrenics), may be complemented by its opposite – dopaminergic hypoactivity in rigid, non-imaginative, inflexible people. Both may be dysfunctional. In fact, there is a good deal of evidence implicating dopamine in schizophreniform cognition (Carlsson, 1988; Swerdlow and Koob, 1987).

Gray *et al.* (1991) have well and succinctly summarized the evidence in favour of the dopamine–schizophrenia hypothesis. The major points are as follows: (1) There is a good correlation between the antipsychotic and dopamine-receptors blocking activities of neuroleptic drugs. (2) Indirect dopamine agonists, e.g. amphetamine and cocaine, give rise to or exacerbate psychotic behaviour. (3) There are reports of elevated numbers of dopamine D_2 receptors in schizophrenic brains, both *post-mortem* and *in vivo*. There are of course also problems with this identification, which are considered by Gray *et al.* (1991); they are not regarded as insurmountable.

These general formulations have led to the postulation of an *animal model* of schizophreniform behaviour; an animal model obviously can only provide indirect evidence for testing hypotheses derived from human subjects. In particular, cognitive processes are more difficult to study in animals than in humans, as already Pavlov emphasized with his insistence on the importance of the second signalling system (language) in the latter. It is here that the importance of such phenomena as latent inhibition, Kamin's blocking effect, negative priming, and the partial reinforcement extinction effect (Gray, 1975) come in. The former three have already been discussed, the PREE paradigm is one which shows that *partial reinforcement* leads to greater resistance to extinction than continuous reinforcement; the animal, having learned to ignore non-reward, is exhibiting cognitive inhibition. If we regard the *absence* of such cognitive inhibition in animals subjected to these manipulations as evidence for psychotomimetic behaviour, then according to the dopamine–schizophrenia theory dopamine agonists should lower or abolish such inhibitory behaviour, while antagonists should increase it. There is good evidence in favour of this deduction for LI, the Kamin effect, and PREE. These phenomena are *abolished* by amphetamine, a dopamine-reducing drug, while these effects are increased by administering neuroleptic dopamine-receptor blocking drugs. When neuroleptic drugs are administered to normal animals, the effects are the opposite to that of amphetamine. Thus haloperidol produces latent inhibition, and gives rise to larger PREE.

Gray *et al.* add one further consideration. In their theory of schizophrenia, the subiculo-accumbens projection plays a large part, and consequently the general hypothesis clearly requires that *disruption of the subiculo-accumbens should have the same consequences as amphetamine administration* in the tests considered. Large hippocampal lesions have been used to destroy the source of this projection, and as a result latent inhibition, Kamin's blocking effect and the PREE have all been disrupted by such damage to the hippocampal formation. (Note that other projections are also destroyed by hippocampectomy, so that the connection may be more complex.) Gray *et al.* (1991) would wish to interpret the effects of amphetamine on cognitive inhibition, as tested in these three paradigms, as reflecting the release of dopamine in the nucleus accumbens, and the effects of hippocampal damage in terms of the disruption of the projection to nucleus accumbens from the subiculum. It would take us too far afield to follow their extended argument.

If creativity is determined to a significant extent by heredity, and if dopamine plays an important part in producing differences in creativity, then one would expect to be able to find evidence of (1) genetic determinants of dopamine levels, and (2) a link between dopamine and psychoticism. As regards (1), there is the D_2 dopamine receptor (DRD2) gene, with two alleles – the less prevalent A1 allele and the more prevalent A2 allele. Noble (1992) has published evidence that this DNA probe showed polymorphism significantly associated with alcoholism, particularly for the Al allele. This is relevant to (2), because drug addiction has been found to be strongly related to psychoticism (Eysenck and Eysenck, 1976; Gossop and Eysenck, 1980). Noble (1993) lists nine additional studies, using in all 986 subjects, an association of the DRD2 A1 allele with alcoholism, and six independent case-control studies (501 subjects) of such an association with severe alcoholism, demonstrating that a strong relationship existed.

Evidence now exists to demonstrate that the connection with the DRD2 gene is not specific to alcohol abuse but is with drug addiction generally (e.g. Cummings, 1991; Smith, 1992; Noble, Blum and Khalsa, 1992). Uhl *et al.* (1993) conducted a meta-analysis on all the reported studies in the literature where the DRD2 alleles had been determined in samples of substance users and controls, with a sample size of 2189 American individuals. A significantly higher prevalence of the A1 allele ($p < .0001$) was found when any substance user was compared either to any controls (odds ratio = 2.09) or to assessed controls (i.e. controls purged of substance users), where the odds ratio was 3.48. Differences of similar magnitude were also found for the B1 allele. Uhl *et al.* conclude that the most severe substance abusers have up to three times the likelihood of displaying DRD2 markers compared to controls free of significant additive substance use.

The causes of this link between additional dopamine use are presumed to lie in the dopaminergic reward system of the mesocorticoachic pathway (Wise, 1987; Koob and Bloom, 1988). Drugs are nearly always reinforcing substances, producing pleasure or euphoria, and hence their use is facilitated by the greater reinforcement produced by the dopaminergic reward system. Another shared effect is the release of dopamine in the brain when substances like alcohol, cocaine, nicotine etc., are administered. Gray (1970, 1972, 1976) has suggested that extraverts are more responsive to rewards, introverts to punishment; it seems likely that high-P scorers may also belong to the reward-responsive type of person, and that the dopamineic system is responsible for this connection.

It may be a remote possibility, but a possibility nevertheless that the prospective genius is work-addicted, and receives positive reinforcement through his labours. The 'ninety-nine percent perspiration' characterization of the genius finds support in the concentration on professional activities recorded over and over again for geniuses in science and the arts, and this addiction to creative work may be facilitated or even produced by an excess of dopamine. The theory may not appear a very likely one, but it seems eminently testable.

In conclusion, it has to be stated that direct evidence is lacking for a link between the DRD2 gene and schizophrenia (Moises, 1991; Sarkar, 1991) or manic-depressive disorder (Byerley *et al.*, 1990; Nothen, 1952) or for such a link with P directly (i.e. other than through the connection of P with addiction). Here again it is to be hoped that future research will provide answers to these questions. The studies quoted may be faulted for using psychotic subjects receiving drug treatment which may have affected negatively the measures of dopamine receptors, as it is known to do in rats. Future research will have to be done with acute cases prior to treatment, a difficult task nowadays.

On the other hand, there is now evidence on a fairly close relation between dopamine and psychoticism (Gray, Pickering and Gray, 1994). As the authors point out, neurochemical functions have homeostatic properties: an increase in dopamine results in down-regulation of post-synaptic receptors, i.e. a decrease in post-synaptic receptor sensitivity, or number of receptors (Creese, Burt and Snyder, 1977). A reasonable prediction would then be that the assumed increased dopamine activity in high-P subjects would be indicated by a relatively *low* number of post-synaptic dopamine receptors. Such a result would be consistent with the hypothesis of increased dopaminergic activity in schizophrenia (Crow, 1980; Meltzer and Stahl, 1976) and with reports of dopaminergic dysfunction in clinical populations with psychotic, or psychotic-like symptoms (e.g. Siever *et al.*, 1993).

Dopamine D_2 binding in the striatum was measured by single positron emission tomography (SPET), subjects being scanned for at least 80–100 minutes. As predicted, there was found to be a significant (negative) correlation between the psychoticism scale and dopamine D_2 binding in the left ($r = -0.75$) and right ($r = -0.75$) basal ganglia. (Dopamine binding in the left and right basal ganglia correlates 0.82.) This dopamine assay did not correlate at all with any other personality dimension, and thus seems specific to the P dimension. The study of course requires replication, in view of the small numbers used, but results are statistically significant at the 0.02 level, and are replicated from left to right hemisphere. As far as they go, these data seem to support the general hypothesis linking P and schizophrenia through the effects of dopamine.

Excess of dopamine is not the only factor involved in the cognitive dysfunctions we are considering; a lack of serotonin may also be involved. Solomon, Kiney and Scott (1978) and Solomon *et al.* (1980) have shown that latent inhibition is abolished by systematic injection of parachlorophenylalanine, a drug that depletes central stores of serotonin. Quite generally, a variety of drugs that reduce serotonergic activity reduce or abolish the effects of pre-exposure. Gray *et al.* (1991) summarize a lengthy review of various studies to conclude that 'the exclusive disruption of serotonergic function in the hippocampal formation is sufficient to eliminate latent inhibition' (p. 9). It seems likely that dopaminergic over-activity and serotogenic under-activity jointly and severally constitute the basic causes of schizophreniform cognition. For a more detailed discussion of some of these issues, Weiner (1990) may be referred to.

The link between low serotonin levels and schizophrenia extend to P (Zuckerman, 1991); Pritchard (1993) has reviewed the literature and concludes that 'in humans, P is indeed related to functioning of the serotonin system'. He also links P and smoking, explaining the well-documented correlation between smoking and high P by postulating that nicotine increases serotonin levels, and that high-P scorers find it socially useful to thus increase serotonin levels in order to decrease overly impulsive actions which lead to unpleasant consequences.

This lengthy and complex, and inevitably highly technical, discussion is necessary to understand the relevance of the data presented a few pages earlier relating the phenomena of latent inhibition, in particular, to psychoticism. If cognitive inhibition is fundamentally linked (negatively) with high P, and acute schizophrenia, which in turn is linked with high dopamine–low serotonin, it is not unlikely that creativity, also linked to high P, is the effect of high dopamine–low serotonin. If true, that would give us a biological basis for a type of behaviour that has attracted a great deal of speculation, and has been declared variously to be divine and inexplicable. Our discussion suggests that we have a curvilinear connection between the dopamine–serotonin balance and cognitive behaviour. High dopamine–low serotonin levels lead to schizophreniform thinking and a complete lack of cognitive inhibition. Very low levels of dopamine, accompanied by high levels of serotonin lead to stereotyped, rigid forms of thinking. Both are disadvantageous. Low middling ratios are 'normal', not inhibiting thinking but also not leading to creative 'or original' types of work. It is the high middling ratios, high but not excessively so, that lead to high trait creativity, and in combination with the other qualities specified earlier to creative achievement.

There is little doubt that dopamine and serotonin serve to oppose each other in the control of many biological and behavioural variables (Depue and Spoont, 1986; Depue and Zald, 1993); thus for instance dopamine inhibits and serotonin activates prolactin release. The issue is discussed in detail by Depue (1993). There are of course considerable problems in actually carrying out research linking personality and biogenic amines; thus specific-transmitter interpretations of *resting* levels are unlikely to prove useful, because of the interaction between transmitters. The preferred method of substantially increasing interpretative power is the use of a *challenging protocol*, where an agonist of the transmitter in question is administered in order to assess the functional *reactivity* of the transmitter system (Oades, 1985). A good example of such use in relation to personality is a study by Depue *et al.* (1994), using bromocriptine as a direct D_2 receptor agonist. The study demonstrated the immense possibilities inherent in such an approach for testing specific hypotheses linking personality (in this case extraversion) and transmitter activity; it argues well for the possibility of testing the theory outlined above.

There are a few additional studies which support this general hypothesis. There is for instance Stein's (1978) theory that dopamine controls the readiness to explore and approach novel stimuli – although whether these stimuli can be cognitively novel is a moot point. Coccaro *et al.* (1989) reported strong correlations between inventory trait measures of irritability and impulse aggressive behaviour and the magnitude of prolactin secretion in response to a serotonin agonist; these are typical P components. Finally both

Cloninger (1987) and Zuckerman (1991) have proposed a role for dopamine in novelty seeking and sensation seeking behaviour. Similarly, Mason (1984) stresses the cognitive function of amplification of attention to all stimuli of potential interest in the environment – a direct link with overinclusiveness and latent inhibition.

Many of the steps involved in this argument are of course speculative, but many others have good empirical support. The theory is far from proven, but it is testable in all its parts, and hence may justify the sub-title of this book – the natural history of creativity. It would be surprising if personality, particularly P, regulated through the dopamine–serotonin balance, were not instrumental in determining a person's degree of cognitive inhibition, and through it his or her creativity. Details may require confirmation, change and alteration, but the very general statement of the theory may survive experimental testing. Critics have suggested that it may be 'overinclusive', too much lacking in cognitive inhibition; that may very well be so. Something very different from the usual seems required to breathe life into the 'degenerating research program' mentioned by Glover, Ronning and Reynolds (1989). If it has no other value, it may serve at least to solve the ancient paradox of genius and madness, by rejecting the either–or, disjunctive categorical approach, and favouring instead a *dimensional* approach. People are not 'sane' or 'mad'; they differ along a continuum of psychoticism (as well as other continua, of course!), and cognitively both extremes may be disadvantageous.

Cortical arousal and creativity

An additional psychological variable that has been connected both with creativity and personality is *cortical arousal*. Theoretically this link between creativity and (lack of) arousal dates back to Hullian theory (Hull, 1943), which postulated a 'behavioural law' according to which increases in drive (arousal) make the dominant response to a stimulus even more dominant, i.e. increase the steepness of the associative gradient. Anxiety, acting as a drive, has a similar effect (Eysenck, 1973a). Easterbrook (1959) put forward the hypothesis that arousal causes attentional narrowing, again suggesting an increase in the steepness of the associative gradient. Martindale (1981, 1989) has rephrased this general law, stating that in the information network more nodes will be activated and to a more equal degree in a state of low as compared with a state of high arousal.

Martindale (1981) has provided some empirical evidence that defocussed attention, flat associative hierarchies and 'primary process thought' are indeed associated with states of low cortical activation. This law would imply that anything that increases arousal impairs performance on tests of creativity. Positive evidence for such a deduction from the general principle has been found for stress (Dentler and Mackler, 1964), the simple presence of other people (Lindgren and Lindgren, 1965), noise (Martindale and Greenough, 1973), extremes of temperature (Lombroso, 1901), and even reward (extrinsic motivation) (Amabile, 1983a).

There is here an apparent contradiction. It would not be true to say that generally creative people are in a state of low arousal. Maddi and Andrews (1966) found that creative people are more anxious than uncreative people; they also tend to show slightly higher levels of resting (basal) arousal on physiological measures. Similarly, creative people like scientists tend to be introverted (Eysenck, 1973b), very much like artists (Götz and Götz, 1979a,b). Introversion of course is linked with *high* levels of arousal (Eysenck and Eysenck, 1985; Strelau and Eysenck, 1987). Götz and Götz (1979a,b) also found successful artists to be high on neuroticism. Clearly there is a paradox here.

Martindale has suggested the solution of this paradox. As he points out, as compared with less creative people, those who are more creative do show low levels of cortical arousal while performing creative tasks (Martindale and Hines, 1975). Martindale and Hasenfus (1978) found that low levels of arousal were found precisely where they were expected to occur – during creative inspiration rather than during the elaboration stage. He goes on to suggest that creative people may be more *variable* in their level of arousal, and show more extreme fluctuations. This is a psychophysiological restatement of Kris's (1952) contention that creative people are more variable on the primary process–secondary process continuum. While there is no direct evidence for this hypothesis, Goodwin and Jamison (1990) have shown that highly creative people tend to fluctuate between states of excessive energy and excessive apathy, abulia and depression – perhaps the effect of high P (manic-depressive abnormality).

Perhaps historical reports of actual discoveries may help; all the evidence seems to exist in anecdotal accounts of famous scientists. One of the best known stories is that told by Poincaré (this, and many others, have been assiduously collected by Koestler (1964)). (The Poincaré account has already been reproduced in an earlier chapter.)

Most famous perhaps is Kekule's dream-like reverie which led him to the discovery of the 'benzene ring'. He relates:

> I turned my chair to the fire and dozed . . . Again the atoms were gambolling before my eyes. This time the smaller groups kept modestly to the background. My mental eye, rendered more acute by repeated visions of this kind, could now distinguish larger structures, of manifold conformation; long rows, sometimes more closely fitted together; all twining and twisting in snakelike motion. But look! What was that? One of the snakes had seized hold of its own tail, and the form whirled mockingly before my eyes. As if by a flash of lightning I awoke . . . Let us learn to dream, gentlemen.

Those, and many other introspective accounts and memories (Ghiselin, 1954) do appear to give some support to the view that *conditions of low arousal give rise to creative ideas*, provided a firm basis of knowledge has been laid containing all the relevant elements. Yet an experimentalist instinctively distrusts such accounts, knowing the vagaries of memory, the inaccuracies of introspection, and the motivated distortion of past events. Is it possible to argue that the repetitive quality of the accounts indicates their validity? Perhaps it is the opposite instead. Perhaps it is because these events were *unusual* that they were remembered; the possibly much more casual discovery of new truths in the course of deliberate cogitation might not be thought worthy of being written up, being only what everybody expects to happen! The eagerness with which many writers on the subject have accepted these tales as evidence may merely argue for a preference for magic over everyday simplicity. I think there is probably a kernel of truth in these stories, but belief is not evidence, and cannot be relied upon. What is needed is a whole body of experimental evidence, of the kind collected by Martindale, to set our minds at rest.

Quite generally, people tend to influence their level of arousal by choosing activities which raise or lower arousal to approach optimum levels. Hence introverts seek solitude, extraverts company (Wilson, 1990). While anecdotal, the evidence of supreme acts of intuition/creativity on the part of scientific geniuses suggests that very frequently they occurred in states of low arousal – dreamy pre-sleep, sitting in train or bus, on holiday. High arousal accompanied the elaboration stage, when creative people attempt to prove their intuitive insights, search the literature, argue with sceptics, etc.

While the evidence is less strong than in the case of introversion, it seems clear that P is related to low arousal–arousability (Zuckerman, 1991; see also Robinson and Zahn, 1985). The physiological mediators are again dopamine and monoamine oxidase (MAO). There has been less study of the low arousal–P connection than of the low arousal–E connection, and in particular the possibility of rapid change from high to low arousal, suggested by Martindale, has not been investigated in relation to personality, although Pavlov's notion of excitation–inhibition equilibrium may be relevant. Here, as in so many other aspects of this theory, future research must come to the aid of Martindale's view and support it – or not, as the case may be. That there is a connection between arousal and creativity is very likely; whether this connection is similar to that suggested by Martindale remains to be seen. He has certainly made an important beginning in the direction of testing it.

We may here add one further point. The concept of arousal has many similarities with the concept of 'drive' in the Hullian sense, and an attempt has been made to see if schizophrenics are characterized by low drive, as has often been suggested in explanation of their frequently poor performance on various tasks. I have reported an experiment to test this hypothesis (Eysenck, 1961), using as a measure of drive the amount of reminiscence on a pursuit-rotor task. (Eysenck and Frith (1977) have summarized the evidence that reminiscence can be used as a good measure of drive.) The results showed no evidence of low drive in schizophrenics. Another, more appealing explanation of the observed retarded appearance of reminiscence in schizophrenics and manic-depressives may be a *slow process of consolidation of the memory trace* (Eysenck and Frith, 1977). This might lead to a comparative failure in psychotics and possibly high-P subjects to form firm memory structures which might help the construction of steep association gradients. Unfortunately little work has been done in this field, so this possibility must remain a suggestion.

Hemispheric laterality

One additional field has given cause for considerable research in this area, and that is the problem of hemispheric laterality, already mentioned in an earlier chapter. Ever since Sperry (1974) originated research in this area, there has been a growing tide of contributions to demonstrate that the two isolated hemispheres of 'split brain' patients are strikingly different in higher cognitive functions such as language and spatial reasoning. Since then, research has concentrated on hemispheric specialization in normal or psychiatric (not 'split brain') subjects. Much of this work is of dubious value (Efron, 1990), although there can be no doubt about a certain amount of hemispheric specialization, with rational speech and logical sequencing associated to some degree with the left hemisphere, spatial and imaginative activity with the right hemisphere (Kitterle, 1991). There has been a good deal of speculation linking right-hemisphere functioning with creativity and giftedness generally (accompanied by many attempts to discover special brain functions linked with giftedness and creativity). Methods of study have involved EEG, evoked potentials, positron emission tomography, regional cerebral blood flow, single photon emission computed tomography, computed axial tomography, etc. The field has been surveyed in detail by Eysenck and Barrett (1993b), and it would take us too far afield to duplicate that survey. However, some of our conclusions may serve as an opener for our discussion.

Lateralization of cerebral hemisphere functioning, and general cerebral dominance, have often been suggested as being closely related to giftedness. While the evidence does not rule out such a possibility, support is meagre, and the theory has not been stated in an acceptable form. Resting EEG recordings and evoked potentials have been found to correlate quite highly with intelligence, but anomalies are still too frequent to give us a convenient and acceptable EEG measure of creativity, although the time may not be far off when such a measure may become feasible. Information processing investigations using nuclear magnetic imaging, positron emission tomography and magnetic resonance spectroscopy all hold out great promise of precise location and measurement of mental activity and individual differences, but here too it is too early to be certain of replicability. Much of the work reported in all these fields is of inferior quality, done on too small and poorly selected samples, and poorly carried out and reported. No such studies should be carried out and reported unless the number of subjects was at least 50, groups clearly defined along measurable continua, and procedures and statistical analyses outlined in detail. Personality correlates of giftedness in the biological field have a good theoretical basis, and have received considerable empirical support. They form an important aspect of giftedness, particularly in the sense of creativity, and have a good theoretical basis in psychophysiology.

In addition to the more strictly biological, neurophysiological methods surveyed by Eysenck and Barrett (1993b), many other, less direct methods of assessing laterality have been employed (Katz, 1994). Among these are handedness (left-hand superiority being associated with right-brain functioning because of the cross-over of neurons); dichotic listening (different speech presented to the two ears; left-ear advantage signals right-hemisphere superiority; Kimura, 1966); conjugate lateral eye movements (CLMs), where movements to the right indicate left-hemisphere dominance (Kinsbourne, 1974; Zenhausern and Kraemer, 1991); the AIP ratio (Bogen *et al.*, 1972), i.e. the ratio of performance on right-hemisphere (oppositional) to left-hemisphere (proportional) tasks; and one or two other, less frequently used methods.

Results are not always easy to interpret. No test is a perfect measure of 'hemisphericity'; indeed, the concept is probably many-faceted and lacking in uniqueness (Hellige, 1993). A given test may only measure one aspect of this complex concept, and give very different results from another (Coren, 1990). Intelligence is not always partialled out, and may confuse the picture. Thus O'Boyle and Benbow (1990) compared 'gifted' children and average-ability controls on a verbal dichotic listening task; the gifted children showed a tendency to show a right-hemisphere advantage – but this may not involve creativity. Results are often complex and contradictory. Katz (1986) found hemisphericity to be related to objective indices of creativity (achievement), but not to tests. Furthermore, the direction of the relationship between hemisphericity and creativity was incompatible with the view that right hemisphere dominance would be directly related to creativity; rather, it was related to right-hemisphere dominance for architects and left-hemisphere dominance for mathematicians and scientists.

Katz (1994) summarizes the rather undistinguished literature on people with creative achievements, as compared with controls, by stating that the pattern is one in which cognitive processes for which the right hemisphere is most specialized may play some privileged role in creativity. However, this role is integrated and co-ordinated with cognitive activities subserved by the left hemisphere, and the emergence of either a left- or right-hemispheric dominance may depend on the type of creative activity being performed (Hines, 1991; Gardner, 1982).

More frequent than studies of achieving creative people have been studies of high test creatives. Here the work of Martindale has been the most relevant and detailed. Using the EEG, he and his co-workers found greater right-hemisphere activity for the more highly creative test scorers, an effect often observed while the subjects were actually engaged on doing the creativity tasks (Martindale and Hasenfus, 1978; Martindale and Hines, 1975; the work is reviewed by Martindale, 1977). In a later study, Martindale *et al.* (1984) examined the question of whether the hemispheric asymmetry would be observed in both the *inspirational* and *elaborative* phases of creative performance, and whether the hemispheric asymmetry was specific to creativity, and would not emerge with non-creative work. High levels of right-hemisphere activity were observed during creative production, but not during basal recordings or during a reading task. However, there were no differences between inspirational and elaborative phases of creative work, leading Martindale to suggest that creative production involves an alternation back and forth between the two phases.

Correlations of creativity as a trait and handedness have shown little congruence (e.g. Katz, 1980; Hattie and Fitzgerald, 1983). Dichotic listening tasks have not produced much positive information (Katz, 1994), but conjugate eye movement studies have been more successful, high creatives habitually using right-hemisphere supported processes in thinking (Katz, 1983; see also Harnard, 1972, and Falcone and Loder, 1984). The AIP ratio also gave positive results (Harpaz, 1990), sometimes quite impressive. Altogether, the results either favour a right hemisphere–creativity relation, or they show no relation; hardly ever is there a suggestion of an adverse relation.

Thus on the whole, the hypothesis of a right hemisphere–creativity relationship remains tenable, and indeed receives mild support. It is usually observed when non-verbal tests are used, and is less likely to emerge with verbal tests of creativity. When batteries of creativity tests are given, hemispheric relationships are usually only found with a few of these tests. Clearly what is needed is more specifically directed research attempting to formulate general theories as to what are the most promising relationships in this field, and why they are more successful. We are coming to the end of the first, preliminary phase of research in this field, where progress is along arbitrarily chosen paths; the second phase must adopt a more clearly theory-driven approach, and benefit from suggestive results of the first phase. We must certainly abandon the notion that creativity is located in the right hemisphere (e.g. Edwards, 1979; Hendron, 1989); this is a meaningless proposition. The two hemispheres always interact; at most there may be a slight superiority of one or the other for certain tasks, under certain conditions. Research suggests that some such preferences exist, but little certainty attaches to the conditions under which such preferences are realized.

An interesting study suggests the existence of correlations between schizotypy, right-hemisphere dominance and belief in ESP (which is certainly unusual and may be considered 'creative', or at least 'unorthodox'!). Possibly this may be a line worth following up, although of course direct measures of creativity would be preferable to very indirect ones like belief in ESP (Brugger *et al.*, 1993)!

The whole area is in a state of growth, and promises to give us many answers to causal questions raised by the problems associated with giftedness and creativity. The technical problems are considerable, but not insurmountable, and it seems likely that the next few years will see a marked increase and improvement in the number and quality of papers devoted to this topic (Eysenck and Barrett, 1993a,b). There may be a tentative argument for hemispheric dominance being related to creativity, but it is too early to come to any firm conclusions.

If the hemispheric differences favouring the right hemisphere could be linked with similar differences characterizing schizophrenia, we might have a powerful support of our general theory, and an additional link between psychoticism and creativity. In looking at the literature on hemispheric differences in schizophrenia, we may well remember Plum's (1973) warning concerning schizophrenia, which he called 'a graveyard of neuropathologists'. Not being a neuropathologist, I shall tread lightly on this ground which is covered with mines. It is possible, however, to note that there are many indications that there may be some substance in the early adumbrations implicating the left hemisphere in support of theories of lateralized tempero-limbic abnormalities in schizophrenia (Flor-Henry, 1969, 1974; Gruzelier, 1973; Gruzelier and Venables, 1974; Jacob and Beckmann, 1986). There is a growing body of evidence associating imbalance of hemispheric functioning with active and withdrawn syndromes in schizophrenia (Gruzelier, 1984; Gruzelier and Liddiard, 1989; Gruzelier *et al.*, 1988; Andrews *et al.*, 1986, 1987).

The *active* syndrome (grandiose delusions, religious and sexual preoccupations, excitement, situational anxiety, cognitive acceleration and creativity) appear to be associated with *integrity of function* of the left hemisphere and losses of function by the right hemisphere. Schizophrenic patients with the passive-withdrawn syndrome, consisting of conventional features such as blunted affect, poverty of speech and social withdrawal, show the opposite functional imbalance.

Other studies also give support to functional hemispheric differentiation (Wexler, 1979; Drane, 1986; Jeste and Lohr, 1989; Kerwin *et al.*, 1988), with Sherman and Galaburda (1985) adding the suggestion that even in rats some such association between behaviour and laterality might be found. All these studies showed pathological features more prominent in the left hippocampus. Probably the majority of patients here were suffering from passive, withdrawn features; the contrast between different *types* of schizophrenics make any generalized conclusion difficult if not impossible.

There is more clear-cut evidence of right-hemispheric (or non-dominant) impairment in patients suffering from manic-depressive disorder. A recent review of 25 such studies (Goodwin and Jamison, 1990) disclosed a quite consistent pattern of such impairment.

Are these data related to psychoticism? Kidd and Powell (1993), using schizotypy tests, have shown that subjects with high schizotypy scores tended towards left hemisphericity and exhibited considerably less bilateral alpha power in the EEG than those with lowered schizotypy scores. 'This is interpreted as offering tentative support for a left hemisphere overactivation hypothesis of psychotic experience.'

It will be evident that no clear-cut conclusion emerges from all these studies. Laterality may be connected with psychoticism and creativity, but it would need much more targeted research to make such an hypothesis tenable. It has certainly been quite appealing to many people, but a critical spirit suggests that too many pieces are missing to allow us to accept it in its present non-specific form. Particularly worrying is the apparent distinction in functioning between different categories of schizophrenics. What is needed is direct evidence on P+ and P− subjects, with strong hemispheric differences favouring the right hemisphere. Work using schizotypal scales will not do, as I have argued in a previous chapter; these scales correlate vey highly with neuroticism, so that the results of Kidd and Powell above might indicate a relationship of left-hemispheric overactivation with N, rather than with P. The whole field is ripe for a determined attack along much more sophisticated lines than have been used in the past. Measures of coherence would be of particular interest.

A somewhat different approach to the vexed topic of hemisphericity has been taken in connection with the Myers–Briggs Type Indicator (MBT2), mentioned earlier, using the Human Information Processing Survey (Taggart and Torrance, 1984) scales designed to measure brain hemispheric dominance. This survey stresses the differences in information processing strategies associated with dominance of each of the hemispheres (allegedly!). The authors argue that left-dominant tactics include an approach that is structured, verbal, fact-oriented, sequential, logical, while right-dominant tactics are open-ended, spatial and intuitive. As we shall see, these dichotomies are grossly exaggerated by many writers, but they seem to agree with the distinction between the sensing and intuitive types. Possibly the thinking–feeling dimension is also involved, with thinking left-dominant and feeling right-dominant.

Shiflett (1989) calculated the relevant correlations on a very small set of 32 male and 35 undergraduate students. 'Intuitive mode on the MBTI was positively related to left hemisphericity ($p = .40$) and negatively related to right dominance ($r = -.54$)' (p. 743). These correlations are claimed to be 'in the expected direction' (p. 741), but it is difficult to see how; intuitive mode should correlate positively, not negatively, with right dominance! Taggart, Kroeck and Escoffier (1991) repeated the study on a much larger sample of 284 females and 270 males, with much more credible results. Intuition correlated −.48

with left and .42 with right dominance. Sensing showed the opposite direction of correlation, as expected (.50 and −.42). Judgment showed a positive correlation with left, and a negative correlation with right dominance (.45 and −.44). All of these are in the expected direction; the only point of doubt is the questionable validity of the Torrance, Taggart and Taggart (1984) scales. They are based on questionnaire items, not psychophysiological examination, and hence of doubtful value. We may only be correlating one set of behaviours and experiences with another, rather than behaviour and experience with cortical activity! This point will be discussed in detail when we examine the psychophysiological evidence.

An alternative approach, not necessarily discordant or contradictory, to the problem of biological intermediary between DNA and creativity has been presented by Nyborg (1991). He bases his analysis on his General Trait Covariance–Androgen/Estrogen (GTC) model of development (Nyborg, 1983, 1987, 1988, 1993), and begins his study by considering the personalities and behaviour of high-ranking scientists, referring to the sources I have also quoted in previous chapters. Particularly important for him are the facts that such persons are not very interested in girls, marry late, have few children, and live stable solitary lives. (Social scientists apparently mature faster, are more popular, begin dating earlier, have more children, and become more often divorced – in other words they are more extraverted than 'hard' scientists.) With respect to sexual identity, highly creative people have generally been found to be androgenous (e.g. MacKinnon, 1961, 1962a,b, 1965, 1978; Hassler, Birbaumer and Nieschlag, 1992; Kemp, 1985).

Nyborg's GTC model is based on the theory that gonadal hormones explain much of covariant trait development, and that effects of gonadal hormones on the body, the brain and behaviour can be formalized in fairly simple ways (Nyborg, 1983). The fundamental hypothesis is that the development of the originally sexually neutral fetus is guided by genes, gonadal hormones and experience. Hormones mediate their organizational and activational effects on body and brain tissues by modulating available genes, by affecting neurotransmitter systems, and by changing cell membrane characteristics. Gonadal hormones, of course, are effective only in hormophilic tissues occupied by special receptors. This arrangement, Nyborg points out, makes hormones uniquely suited to selectively co-ordinate and pace brain, body and behavioural development and explains why a sexually neutral fetus transforms into a male, a female, or a 'something-in-between' pattern of phenotypic traits.

Males and females are classified into 'hormotypes' in accordance with their particular androgen/estrogen balance. A male with high plasma testosterone (t) is said to be hormotype A5, and a male with low t is hormotype A1. A female with high plasma estrogen (E_2:17-estradiol) is hormotype E5, and a female with low E_2 is hormotype E1. Given the hormotype, the GTC model generates rather precise predictions about body, ability and personality development, most of which fairly closely fit with the available evidence.

Nyborg's speculations are interesting and worth following up, but they also raise a very difficult question. We seem to be faced with a paradox. I have argued that there is an essential connection between psychoticism and creativity, and there is clearly much evidence to support this claim. Psychoticism is much higher in males than in females, and there is some evidence linking high P with high testosterone levels. Yet there is also evidence, as Nyborg points out, of creativity being linked with androgyny, or, as he puts it, creatives being low t (A2) males. It seemed only fair to mention this apparent contradiction because it is by no means clear how it can be resolved. Perhaps the answer here is, as in the case of the apparent contradiction between psychopathology and ego-strength, negatively correlated in the general population, but found closely associated in a small sub-group of creative people. High testosterone levels and 'cognitive androgyny' may be negatively correlated in the general population, but is closely associated in a small sub-group of creatives. Whether this is the solution is impossible to say at the moment; clearly only further research can answer the question raised by this paradox.

Evidence in favour of some such argument comes from the work of Bem (1974) who argued against the usual bipolar, unifactorial model of male–femaleness (Constantinople, 1973; Spence, 1984, 1993). According to this model, all of the traits and behaviours that distinguish between men and women in a given society correlate to determine a single Masculinity/Femininity factor on which every man and woman can be assigned a position. Bem (1981) assigned items on her BSRI (Bem Sex–Role Inventory) to *separate* and uncorrelated Masculinity and Femininity scales. Spence (1993) has opposed the Bem scheme, suggesting that gender phenomena are multifactorial; her theory is embedded in the PAQ (Personal Attributes Questionnaire – Spence and Helmreich, 1978).

It is not my intention to enter this battlefield, except to note that the unifactorial model has lost favour among all combatants. Bem's notion of four major types has some appeal, but is almost certainly too simplistic. According to her, we have a masculine and a feminine type, having high scores on the M and F scales respectively, but low scores on the alternate scale. *Androgynous* individuals have high scores on both scales, and *undifferentiated* individuals low scores on both scales. (In addition of course we have individuals with unequal scores in the counterstereotypic direction, i.e. cross sex-typed.) Spence's theory would not rule out the possibility of such types, but would presumably insist on a larger sampling of factors in this field.

There is thus some evidence in favour of our hypothesis that the apparent contradictions between different theories of M–F differences in respect to creativity can be reconciled by postulating some independence between the levels of male and female hormones, and that a creative person may have high levels of both. One of the many research problems thrown up by our theory is that of finding direct evidence linking high levels of male and female sex hormones with creativity.

It will be clear that of the three avenues of approach to the problem of DNA–creativity nexus surveyed, I would much prefer that involving cognitive inhibition, as indicated by latent inhibition, and the dopamine–serotonin ratios. The cortical arousal hypothesis also looks interesting, and would certainly deserve much further study. Least appealing at the moment is the hemispheric laterality hypothesis, but a more highly focussed approach, using greatly improved methodology, might advance its claims to consideration. It is perhaps worth noting that these three theories might not stand in contradiction to each other. It is perfectly possible that cortical arousal is not independent of the dopamine–serotonin ratio, and both may be influenced diffentially by laterality differences. So little is known about the functioning of these systems in normal people, and in relation to creativity and/or psychoticism, that the main value of this brief survey must lie in suggesting areas of research worth pursuing, and prompting workers in the field to try and search for the missing links between DNA and creativity. The task is difficult, but not impossible and the research in terms of better understanding should make it well worth while.

There is one final point which deserves at least some brief discussion, namely that of *treatment*. If psychoticism is *causally* linked with creativity, then might not psychiatric treatment lower the creativity of the afflicted person? It is only in recent years that effective drug treatments have been found, so that the empirical study of any effects on creativity is fairly recent, too. Most workers have concentrated on lithium treatment for depression, testing the hypothesis that such treatment might ameliorate the depression, but also reduce the manic component which is associated with many positive features. It is difficult to make any predictions; as we have seen, the connection between P and creativity is curvilinear; too much (psychosis) or too little (rigidity, conformity) are equally liable to lower creativity. Thus for those actually suffering from clinical psychosis, treatment with lithium, valproate, carbamazepine, the neuroleptics, and the antidepressants may *increase* creativity (through liberation from debilitating psychosis), while for high-P subjects not actually psychotic these drugs might lower creativity. It is not surprising that Jamison (1993) summarizes her discussion by stating that 'the short- and long-term effects on artistic creativity of the major drugs . . . remain unclear' (p. 241). Quite generally, low blood level lithium in normals has minimum effects, but at higher levels there was decreased responsiveness to the environment, increased indifference and malaise, and greater passivity, e.g. lowering of *motivation* (Schou, 1968; Judd *et al.*, 1977; Krapf and Muller-Derlinghaussen, 1979). Decreased social involvement, and concentration, as well as mood-lowering were other consequences. Jamison (1993) quotes evidence to show that creative individuals who do their best work in the course of a hypomanic period often complain that lithium acts as a 'brake'; they are willing to accept the terror of depression to obtain the 'high' of the manic state.

Two studies of the actual effects of lithium on the artistic productivity of artists and writers give a rather different picture (Schou, 1979; Marshall, Neumann and Robinson, 1979). Fifty-seven percent indicated their productivity while on lithium had increased; for twenty percent it had remained the same. Only twenty-three percent reported a decrease in productivity, while seventeen percent had stopped taking the drug due to its adverse effects. Clearly there are great individual differences, and generalizations are difficult to justify.

Effects in cognitive studies, summarized by Jamison (1993) give similar inconclusive and contradictory results. Some, like Judd *et al.* (1977) found no effects of short-term lithium use on creativity in normal individuals, while others found substantial detrimental effects of lithium on associational processing in patients with manic depressive illness. Clearly this whole area is of great potential importance for the study of creativity, but suffers from lack of clear-cut hypotheses. In due course, it may throw much light on the nature of the psychophysiological intermediaries between DNA and creativity.

8 From DNA to creativity and genius

Felix qui potuit rerum cognoscere causas.
(Happy he who can discover the causes of things)

Virgil

We are now in a position to consider *as a whole* the model of genius and creativity that I have been at pains to construct from a variety of writers, psychologists and interested scientists. There are some novel aspects, but essentially the novelty lies in my attempt to make personality differences central to the argument. Previously personality traits were indeed studied, but were never given the central position I believe they deserve. The early indication that mental ability (IQ) is only loosely correlated with creativity has not usually been interpreted to suggest that creativity is not an *ability*, but a *cognitive style* closely related with psychoticism. It may be useful to set out the general theory in diagrammatic form (Fig. 8.1).

We start inevitably with heredity, embodied in a person's DNA (deoxyribonucleic acid). Practically all the variables we have found associated with creativity and genius have a genetic component. Creativity is associated physiologically with the hippocampal formation, and with the level of activity of dopamine and serotonin, the former heightening, the latter lowering creativity (regarded as a trait). It is suggested that their influence is directly on *cognitive inhibition*, i.e. cognitive factors like latent inhibition and/or negative priming, which reduce the tendency towards over-inclusiveness indicative of psychoticism, and, when lacking, in extreme form produce functional psychosis (schizophrenia; manic-depressive illness), and in lesser extent creativity. Thus, through the widening associative horizon associated with lack of latent inhibition and/or negative priming, psychoticism is closely linked with the trait of creativity.

Possessing this trait, however, does not guarantee *creative achievement*. Trait creativity may be a necessary component of such achievement, but many other conditions must be fulfilled, many other traits added (e.g. ego-strength), many abilities and behaviours added (e.g. IQ, persistence), and many socio-cultural variables present, before high creative achievement becomes probable. Genius is characterized by a very rare combination of gifts, and these gifts function *synergistically*, i.e. they multiply rather than add their effects.

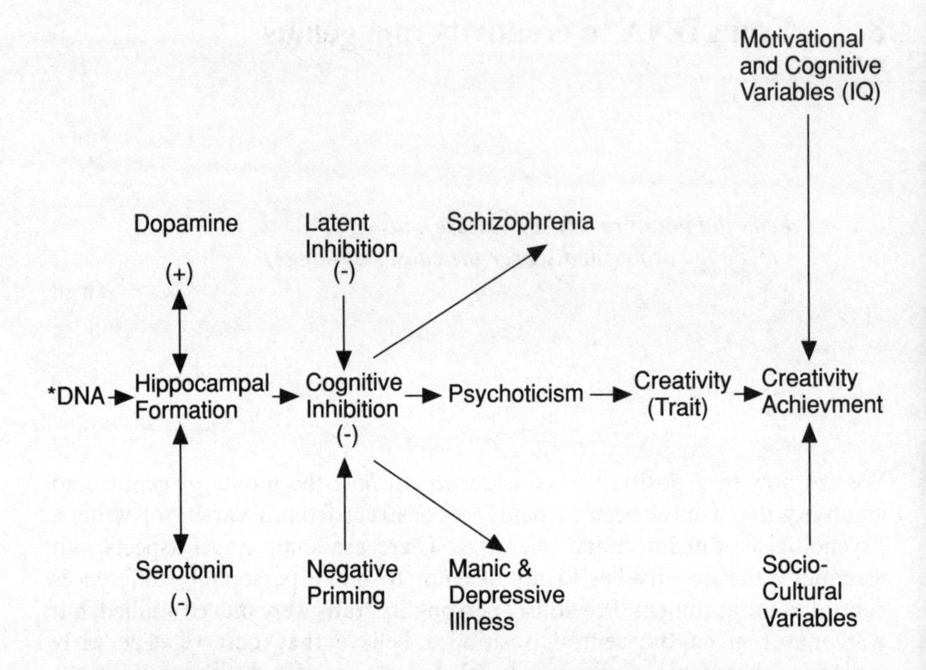

Fig. 8.1 A model of the causal theory of creativity.

Hence the mostly normally distributed conditions for supreme achievement interact in such a manner as to produce a J-shaped distribution, with huge numbers of non- or poor achievers, a small number of high achievers, and the isolated genius at the top.

This, in very rough outline, is the theory here put forward. As discussed, there is some evidence in favour of the theory, and very little against it. Can we safely say that the theory possesses some scientific credentials, and may be said to be to some extent a valid account of reality? There are obvious weaknesses. Genius is extremely rare, and no genius has so far been directly studied with such a theory in mind. My own tests have been done to study deductions from the theory, and these have usually been confirmatory. Is that enough, and how far does it get us?

Terms like 'theory', of course, are often abused. Thus Koestler (1964) attempts to explain creativity in terms of his theory of 'bisociation' according to which the creative act 'always operates on more than one plane' (p. 36). This is not a theory, but a description; it cannot be tested, but acts as a definition. Within those limits, it is acceptable as a non-contingent proposition (Smetslund, 1984), i.e. necessarily true and not subject to empirical proof. A creative idea *must*, by definition, bring together two or more previously unrelated concepts. As an example, consider US Patent 5,163,447, the 'force-sensitive, sound-playing condom', i.e. an assembly of a piezo-electric sound transducer, microchip, power supply and miniature circuitry in the rim of a condom, so that when pressure is applied, it emits 'a predetermined melody or a voice message'. Here is bisociation in its purest form, bringing together mankind's two most pressing needs, safe sex and eternal entertainment. But there is no proper theory here; nothing is said that could be disproved by experiment. Theory implies a lot more than simple description.

The philosophy of science has thrown up several criteria for judging the success of a theory in science. All are agreed that it must be testable, but there are two alternative ways of judging the outcome of such tests. Tradition (including the Vienna school) insists on the importance of *confirmation*; the theory is in good shape as long as results of testing deductions are positive (Suppe, 1974). Popper (1959, 1979), on the other hand, uses *falsification* as his criterion, pointing out that theories can never be proved to be correct, because we cannot ever test all the deductions that can possibly be made. More recent writers like Lakatos (1970, 1978; Lakatos and Musgrave, 1970) have directed their attention rather at a whole *research programme*, which can be either advancing or degenerating. An advancing research programme records a number of successful predictions which suggest further theoretical advances; a degenerating research programme seeks to excuse its failures by appealing to previously unconsidered boundary conditions. On those terms we are surely dealing with an advancing programme shift; building on research already done, many new avenues are opening up for supporting or disproving the theories making up our model.

It has always seemed to me that the Viennese School, and Popper, too, were wrong in disregarding the *evolutionary* aspect of scientific theories. Methods appropriate for dealing with theories having a long history of development might not be optimal in dealing with theories in newly developing fields, lacking the firm sub-structure of the older kind. Newton, as already mentioned, succeeded in physics, where much sound knowledge existed in the background, as well as good theories; he failed in chemistry/alchemy where they did not. Perhaps it may be useful to put forward my faltering steps in this very complex area situated between science and philosophy (Eysenck, 1960, 1985b).

It is agreed that theories can never be proved right, and equally that they are dependent on a variety of facts, hunches and assumptions outside the theory itself; these are essential for making the theory testable. Cohen and Nagel (1936) put the matter very clearly, and take as their example Foucault's famous experiment in which he showed that light travels faster in air than in water. This was considered a crucial experiment to decide between two hypotheses: H_1, the hypothesis that light consists of very small particles travelling with enormous speeds, and H_2, the hypothesis that light is a form of wave motion. H_1 implies the proposition P_1 that the velocity of light in water is *greater* than in air, while H_2 implies the proposition P_2 that the velocity of light in water is *less* than in air. According to the doctrine of crucial experiments, the corpuscular hypothesis of light should have been banished to limbo once and for all. However, as is well known, contemporary physics has revived the corpuscular theory in order to explain certain optical effects which cannot be explained by the wave theory. What went wrong?

As Cohen and Nagel point out, in order to deduce the proposition P_1 from H_1, and in order that we may be able to perform the experiment of Foucault, many *other* assumptions, K, must be made about the nature of light and the instruments we employ in measuring its velocity. Consequently, it is not the hypothesis H_1 alone which is being put to the test by the experiment – it is H_1 and K. The logic of the crucial experiment may therefore be put in this fashion. If H_1 and K, then P_1; if now experiment shows P_1 to be false, then either H_1 is false *or* K (in part or complete) is false (or of course *both* may be false!). If we have good grounds for believing that K is not false, H_1 is refuted by the experiment. Nevertheless the experiment really tests both H_1 *and* K. If in the interest of the coherence of our knowledge it is found necessary to revise the assumptions contained in K, the crucial experiment must be reinterpreted, and it need not then decide against H_1.

What I am suggesting is that when we are using H + K to deduce P, the ratio of H to K will vary according to the state of development of a given science. At an early stage, K will be relatively little known, and negative outcomes of testing H + K will quite possibly be due to faulty assumptions concerning K. Such theories I have called 'weak', as opposed to 'strong' theories where much is known about K, so that negative outcomes of testing H + K are much more likely to be due to errors in H (Eysenck, 1960, 1985b).

We may now indicate the relevance of this discussion to our distinction between weak and strong theories. Strong theories are elaborated on the basis of a large, well founded and experimentally based set of assumptions, K, so that the results of new experiments are interpreted almost exclusively in terms of the light they throw on $H_1, H_2 \ldots H_n$. Weak theories lack such a basis, and negative results of new experiments may be interpreted with almost equal ease as disproving H or disproving K. The relative importance of K can of course vary continuously, giving rise to a continuum; the use of the terms 'strong' and 'weak' is merely intended to refer to the extremes of this continuum, not to suggest the existence of two quite separate types of theories. In psychology, K is infinitely less strong than it is in physics, and consequently theories in psychology inevitably lie towards the weaker pole.

Weak theories in science, then, generate research the main function of which is to investigate certain problems which, but for the theory in question, would not have arisen in that particular form; their main purpose is not to generate predictions the chief use of which is the direct verification or confirmation of the theory. This is not to say that such theories are not weakened if the majority of predictions made are infirmed; obviously there comes a point when investigators turn to more promising theories after consistent failure with a given hypothesis, however interesting it may be. My intention is merely to draw attention to the fact – which will surely be obvious to most scientifically trained people – that both proof and failure of deductions from a scientific hypothesis are more complex than may appear at first sight, and that the simple-minded application of precepts derived from strong theories to a field like psychology may be extremely misleading. Ultimately, as Conant has emphasized, scientific theories of any kind are not discarded because of failures of predictions, but only because a better theory has been advanced.

There is a further characteristic of weak theories, as contrasted with strong, which deserves mention. In strong theories the different postulates are *interdependent*; it is not possible to change one without changing the rest, and indeed throwing overboard the whole theory. In weak theories, such interdependence is much less marked, and changes in one part of the theory are quite permissible without the necessity of altering other parts as well. Thus Hull's (1943) theory of inhibition is peripheral and 'work'-oriented; I have preferred (for various experimental reasons) to work with a central theory of inhibition rather more akin to Pavlov's. This substitution, as well as many others, can be made without extending the framework of Hull's theory unduly; nothing of this kind would have been possible with Newton's theory of gravitation. Weak theories are very flexible; that is why they are such good guides for research; strong theories have an air of 'take it or leave it' which makes them superior as guides to action, but also less likely to lead to important new discoveries.

This discussion is relevant to the arguments between adherents of the verification and falsification theories. Clearly for weak theories *verification* is much more relevant and important; falsification is almost meaningless because of the weak development of K. For strong theories, however, *falsification* becomes more crucial; K is well known, and if predictions don't work, H is much more likely to be implicated. The argument is also relevant to the perennial disputes about the roles of *induction* and *deduction*; the former is more appropriate for weak theories, the latter for strong theories.

Fig. 8.2 will illustrate the argument. When we know very little about a given discipline, we may have hunches and intuitions, based on unsystematic observations and based on induction. When we learn a little more, we may develop hypotheses which (hopefully!) lead to verification. If our hypotheses are generally supported, they lead to theories, and we can begin to talk about the importance of falsification. Finally, well-supported theories lead to scientific laws, like Newton's law of gravitation. If they are falsified, we have what Kuhn calls a paradigm-change, or a revolution (Krige, 1980; Cohen, 1985). This is a historical and evolutionary process in which psychology is inevitably located towards the lower end; we have outgrown the 'hunch' stage in some of our endeavours, and may be advancing towards the 'theory' stage, but our theories are still very weak, in the sense outlined above, and our

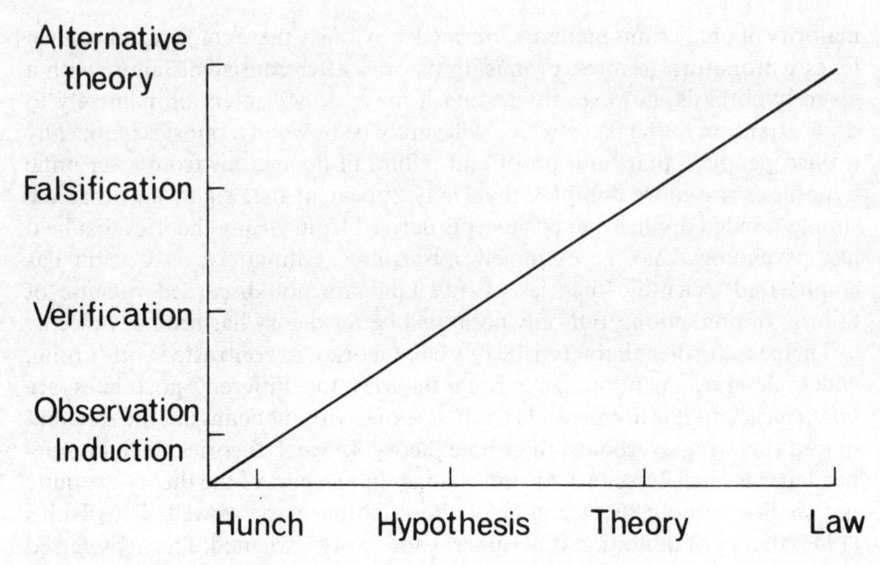

Fig. 8.2 Evolutionary sequences of theories and methods used for their testing (Eysenck, 1985b).

concern should be with verification, rather than falsification. The account given in this book contains hunches, hypotheses and theories; there is an encouragingly large amount of verification, but a lot more work needs to be done before we can hope to articulate a unified theory making possible serious efforts at falsification.

Does all the work cited in this book, and the theories developed on the basis of the facts disclosed, have any practical implications? It is often stated that we owe much to originality, creativity and imaginative innovation, and value it highly; I suggest that only the first part of the sentence is true. We *owe* much to creativity, but we seem to *value* it only in hindsight. There are many horrifying stories in the history of science and art about the fate of the innovator; I have mentioned some of them already, like that of Semmelweis. But that is all history, readers may say; nowadays we worship the great minds who create new theories, new works of art, new concepts. This has not been my experience. The major grant-giving bodies tend to support routine work that can safely be predicted to have positive outcomes; they shy away from true novelty. Novelty emerges from an *individual* mind; when it is judged by a *committee*, orthodoxy will usually prevail.

It may be argued that surely there are exceptions, and that is quite true. One learns to play the system. If you have a truly novel and revolutionary idea, you can disguise and hide it behind a more acceptable, ordinary appearance; when you have got your grant, you skilfully combine the two and satisfy the grant-givers and your own desires. But this is an awful waste of time, it does not always work, and not everybody has the needed skill at camouflage! You only have to look through the research approved by the Medical Research Council in Great Britain, or by the National Institutes of Health in the USA, to realize the truth of what I have been saying.

Of course the excuses are only too ready. Truly innovative ideas may fail; how can we risk public money on research that may not work out? But of course that is precisely the point; creative ideas are high-risk ideas. Furthermore, they inevitably go counter to orthodoxy – if they didn't, would they be original? And inevitably orthodoxy occupies all the seats of power, and would rather die than support the rebel who dares to question the status quo. The establishment hunting down the maverick is not a pretty sight; it can destroy careers, damage reputations, and squash originality flat.

The way research is often treated is exemplified by two examples. I once obtained a research grant from ONR (the Office of Naval Research, USA) for reasons unintelligible to me – the work had no conceivable connection with the Navy, or anything it could possibly be concerned with. However, they sent me a form to fill in, a task I found extremely difficult, because it was an all-purpose contract used *inter-alia* to purchase potatoes for the Navy! They bought research as they bought vegetables – a wonderful insight into official thinking about science.

The other example is the recent decision by the British Government to reduce or eliminate the research monies administered by the Universities, and have them go instead to the official research councils, which in turn are supposed more and more to give preference to areas indicated as useful by the government! In this way officialdom and orthodoxy reduce the already slender chances of original thinkers, particularly in the social sciences, to obtain the necessary funds for carrying out their research.

These facts and ruminations may be relevant to an issue often raised by educationists and others. Can we increase children's creativity through educational means! Cropley (1992) suggests an affirmative answer; he believes that every child can be trained to become more creative. Meeker (1985) asks directly: 'Can creativity be trained?', and answers very decisively: 'The answer is yes' (p. 792). She goes on to claim that 'there are many studies to substantiate this, based on validated, field-tested materials.' She further says, however: 'Whether society deems these people as creative, however, is another question' (p. 792). The problem is really a very difficult one, involving the ever-present danger of 'teaching the test.' As already explained, it is easy to obtain much higher scores on creativity tests by simple instruction; this can hardly be said to raise the subject's creativity, although it raises their creativity test *scores*! The same error is frequently committed by researchers who try to raise the IQ of deprived children by special educational involvement. What the children are taught is in part reflected in the test items, so they are in effect taught the test. Such improvement usually vanishes once the special education ceases; there is no long-term effect on IQ. Teaching 'creativity' has the same effect as instructions to respond with unusual responses; the effect is easy to produce but trivial.

To demonstrate a true improvement in creativity one would have to follow up a control group and a specially trained group and demonstrate differential creativity in activities *not* specially taught, such as being creative in real-life situations. This is hard to do, and has never been done. It seems doubtful to me whether much could be done to make people more creative (in the trait sense), but there are two important ways of improving creative achievement. To begin with, consider the effects of creativity-based teaching. What has become gospel in many teaching circles has been the notion that children are naturally creative, and should not be prevented from expressing their originality by forcing them to learn facts, principles and rules which might lower their creativity. There is no evidence that this procedure in fact has any effect on their creativity, but considerable evidence that their store of factual knowledge, their ability to acquire needed knowledge, their ability to communicate, and quite generally their scholarship decline disastrously under such a regime. But as many studies have shown, creativity (as achievement) must be based on a vast amount of knowledge and practice in order to produce original works of art and science; failing to give children the chance to acquire this basic knowledge condemns them to a life of complete failure as far as genuine, socially valued and creative activity is concerned.

Those who advocate allowing children to 'express themselves' (as if they had anything valuable to express!) and failing to teach them the basic skills required for such expression, and the needed knowledge to have worthwhile creative ideas, are not helping the child to become creative in any meaningful way; they are making certain that he would never have anything worthwhile to contribute to society. Needless to say, they have never done the obvious experiment of following up children allowed to 'express' themselves and children of equal IQ taught along traditional lines, to see which group would come up with more original ideas. Perhaps they fear what T. H. Huxley called 'the great tragedy of science – the slaying of a beautiful hypothesis by an ugly fact.' The tragedy is that for 30 years or more children in the UK and the USA have been prevented from learning all about the necessary *infrastructure* on which any creative endeavour must be based. This, surely, is not the way to creative achievement.

Is there nothing society can do to produce greater creativity in its citizens? There would seem to be two things, the first of which seems at first sight to be very much opposed to creativity, namely the creation of a first-class educational system geared entirely to the production of a firm knowledge based on science, mathematics, English, history, geography, foreign languages, music and the arts, a system developing each pupil's achievements to the limits of his abilities. This would produce the necessary basis on which the child could then proceed to exert any creative powers he might have. Without such a basis creativity has nothing to work on; with it the sky would be the limit.

The other element needed is much more difficult to supply; it is in fact not a positive but a negative one. At present, creativity is suppressed at every level by orthodoxy, by bureaucracy, by authority – all dislike change, innovation, revolution. Creativity, solemnly praised, is in fact anathema; it threatens the structure, and cannot be tolerated. The creative person is willy-nilly turned into a rebel, an outcast, a maverick; I have given some examples elsewhere (Eysenck, 1990a,b). Many no doubt give in and are absorbed into orthodoxy, with the accompanying loss to society of their original talents. Many others battle and are overcome by the forces of reaction. There is a famous poem by Matthew Arnold which pays tribute to those who lost the fight, as Semmelweis did.

> Creep into thy narrow bed,
> Creep, and let no more be said!
> Vain thy onset! all stands fast;
> Thou thyself must break at last.
>
> Let the long contention cease!
> Geese are swans, and swans are geese.
> Let them have it how they will!
> Thou are tired; best be still!
>
> They out-talk'd thee, hiss'd thee, tore thee;
> Better men fared thus before thee;
> Fired their ringing shot and pass'd,
> Hotly charged – and broke at last.
>
> Charge once more, then, and be dumb!
> Let the victors, when they come,
> When the forts of folly fall,
> Find thy body by the wall.

If society could remove this oppressive harness on creative thought, much might be gained. There are, of course, considerable difficulties in the way. The most obvious one is simply the fact that side-by-side with creative thoughts of great social value there is a much larger body of silly, absurd and indeed mad ideas which compete with sound ones for support. But how to separate the chaff from the wheat? This of course is what peer review in science is attempting to do, but with little success (Eysenck, 1992a,b,c). Agreement among expert judges is so small (around .2) that resulting group judgments almost inevitably favour the safe, the predictable, rather than the daring, the novel, the innovative. It would be inevitable that if committees became more adventurous, many of the new, creative ideas would fail, even if a few would emerge triumphantly from the fray. And the triumph might be delayed for a long time; if a research council had financed Mendel, it would have had to wait for 30 years before its investment would have been recognized as a wise one!

The difficulties and problems are widely recognized, and I have no immediate answer. But what I would like to see would be an abatement of the raw hostility which greets new ideas, the determination to stamp the creative suggestion into the ground, the anger and aggression produced by trying out novel and untrodden paths. A good example is Linus Pauling's fight against the cancer mafia (Richards, 1991), showing how even a double Nobel Laureate can suddenly become a 'non-'person' when he takes on orthodoxy. Few people who have not been at the receiving end of such militancy can imagine the many ways in which such hostility manifests itself. Journals suddenly close their pages to the offender, but print attacks, however scurrilous, to which no reply is permitted. Research funds are suddenly cut off, even though promised. Irrelevant and untrue rumours are spread to impugn the offender. He may lose his job, or at least fail to be promoted. He may be barred from library and other facilities; privileges of all kinds may be withdrawn. In extreme cases, he may be suffering bodily attacks, his family may be threatened, bombs may be planted under his car, he may be burnt at the stake – it is difficult to list all the sanctions orthodoxy can muster to assert its right to be regarded as the guardian of truth. These things do not happen only under the aegis of the Inquisition, in Stalin's Russia, or Hitler's Germany. I have seen them happen in democratic countries happily boasting of the right to free speech. No legislative change would alter this general rule, but perhaps education might attempt to teach children the wisdom contained in Cromwell's letter to the General Assembly of the Church of Scotland, 3rd August 1650; 'I beseech you, in the bowels of Christ, think it possible you may be mistaken!'

That, of course, is the religion of the open mind, so often asserted to be the hallmark of science, but so often disavowed by individual scientists. To the extent that we preserve and cherish an open mind, to that extent will creativity flourish. The best service we can do to creativity is to let it bloom unhindered, to remove all impediments, and cherish it whenever and wherever we encounter it. We probably cannot train it, but we can prevent it from being suffocated by rules, regulations and envious mediocrity.

References

Adler, A. (1927) *Practice and Theory of Individual Psychology*. New York: Harcourt Brace and Oswald.

Albert, R. S. (1971) Cognitive development and parental loss among the gifted and the creative. *Psychological Reports*, **29**, 14–26.

Albert, R. S. (1975) Towards a behavioural definition of genius. *American Psychologist*, **30**, 140–51.

Albert, R. S. (1978) Observations and suggestions regarding giftedness, familial influence and the attainment of eminence. *Gifted Child Quarterly*, **22**, 201–11.

Albert, R. S. (1980) Family position and the attainment of eminence: a study of special family position and special family experiences. *Gifted Child Quarterly*, **24**, 87–95.

Alexander, W. P. (1935) Intelligence, concrete and abstract. *British Journal of Psychology*, Monograph Supplement No. 19.

Alissa, I. (1972) Stimulus generalisation and over-inclusion in normal and schizophrenic subjects. *Journal of Clinical Psychology*, **34**, 182–6.

Allison, P. D., Price, D., Griffith, B. C., Moraoscik, M. J. & Stewart, J. I. (1976) Lotka's law: A problem in its interpretation and application. *Social Studies of Science*, **6**, 269–76.

Allport, F. H. (1934) The J-curve hypothesis of conforming behaviour. *Journal of Social Psychology*, **5**, 141–83.

Allport, G. W. (1961) *Pattern and Growth in Personality*. New York: Holt, Rhinehart and Winston.

Altus, W. (1966) Birth order and its sequel. *Science*, **151**, 44–8.

Amabile, T. M. (1983a) *The Social Psychology of Creativity*. New York: Springer Verlag.

Amabile, T. M. (1983b) The social psychology of creativity: A componential conceptualization. *Journal of Personality and Social Psychology*, **45**, 357–76.

Amabile, T. M. (1985) Motivation and creativity: Effects of motivational orientation on creative workers. *Journal of Personality and Social Psychology*, **48**, 393–9.

Amelang, M., Herboth, G. & Oefner, I. (1991) A prototype strategy for the construction of a creativity scale. *European Journal of Personality*, **5**, 201–16.

Anastasi, A. & Lenee, R. F. (1960) Intellectual defect and musical talent: a case report. *American Journal of Mental Deficiency*, **64**, 659–705.

Anderson, C. C. & Cropley, A. J. (1966) Some correlates of originality. *Australian Journal of Psychology*, **18**, 218–27.

Anderson, J. R. & Bower, G. H. (1973) *Human Associative Memory*. Washington, D.C.: Winston.

Andreasen, N. C. (1987) Creativity of mental illness: Prevalence rates in writers and their first-degree relatives. *American Journal of Psychiatry*, **144**, 1288–92.

Andrews, H. B., Cooper, J. E., Barber, C. & Raine, A. (1987) Early somatosensory evoked potentials in schizophrenics: Symptom patterns, clinical outcome and interhemispheric functioning. In: R. Takahaski, P., Flor-Henry, J. & I. S. Niora (Eds.), *Cerebral Dynamics, Laterality and Psychopathology*, pp. 175–86. Amsterdam: Elsevier.

Andrews, H. B., Haine, D. A., Cooper, J. E. & Barber, C. (1986) The symptom pattern in schizophrenic patients with stable abnormalities of laterality of somatosensory evoked responses. *British Journal of Psychiatry*, **146**, 46–50.

Angst, J., Felder, W. & Lohmeyer, B. (1979) Schizo-affective disorders. I. Results of genetic investigation. *Journal of Affective Disorders*, **1**, 139–53.

Angst, J. & Scharfetter, C. (1990) Schizoaffektive Psychosen. Ein nosologisches. Aergerniss. In: E. Lungerhansen, W. P. Kaschken & R. J. Witkowski (Eds.). *Affektive Psychosen*. Stuttgart: Schattener Verlag.

Ankney, C. D. (1992) Sex differences in relative brain size: The mismeasure of woman, too? *Intelligence*, **16**, 329–36.

Arasteh, A. & Arasteh, J. D. (1976) *Creativity in Human Development*. New York: Schenkman.

Arians, S. & Siller, J. (1982) Effects of subliminal oneness stimuli in Hebrew on academic performance of Israeli high school students: Further evidence on the adaptation-enhancing effects of symbolic fantasies in another culture using another language. *Journal of Abnormal Psychology*, **91**, 343–9.

Arieti, S. (1976) *Creativity: The Magic Synthesis*. New York: Basic Books.

Arlin, P. K. (1975) Cognitive development in adulthood: A fifth stage? *Developmental Psychology*, **11**, 602–6.

Arlin, P. K. (1977) Piagetian operations in problem finding. *Developmental Psychology*, **13**, 297–8.

Armstrong, H. E., Hotusson, M. H., Dies, H. A. & Holmes, D. S. (1967) Extraversion–introversion and process-reactive schizophrenia. *British Journal of Social and Clinical Psychology*, **6**, 69–74.

Arp, H. (1987) *Quasars, Redshifts and Controversies*. Berkeley: Interstellar Media. (Available from the Sourcebook Project, P.O. Box 107, Glen Arns, MD 21097.)

Asch, S. E. (1956) Studies of independence and conformity: I. A minority of one against a unanimous majority. *Psychological Monograph*, **70** (Whole No. 546).

Ashenfelter, O. & Krueger, A. (1992) Estimates of the economic return to schooling from a new sample of twins. *Princeton University: Working Paper 304, Industrial Relations Section.*

Ask, P. (1949) The reliability of psychiatric diagnosis. *Journal of Abnormal and Social Psychology*, **44**, 271–6.

Atkinson, R. C., Bower, G. H. & Crothers, E. J. (1965) *An Introduction to Mathematical Learning Theory*. New York: Wiley.

Babigian, H. M., Gardner, E. A., Miles, H. C. & Romano, J. (1965) Diagnostic consistency and change in a follow-up study of 1,215 patients. *American Journal of Psychiatry*, **121**, 895–901.

Bachtold, L. M. & Werner, E. E. (1970) Personality-profiles of gifted woman: Psychologists. *American Psychologist*, **25**, 234–43.

Bachtold, L. M. & Werner, E. E. (1973) Personality profiles of creative women. *Perceptual and Motor Skills*, **36**, 311–19.

Badcock, J. C., Smith, G. A. & Rawlings, D. (1988) Temporal processing and psychosis-proneness. *Personality and Individual Differences*, **9**, 709–19.

Baddeley, A. D. (1983) Working memory. *Philosophical Transactions of the Royal Society of London*, B**302**, 311–24.

Baeyens, F., Eelen, P. & Bergh, O. (1990) Contingency awareness in evaluative conditioning: A case for unaware affective-evaluative learning. *Cognition and Emotion*, **4**, 3–18.

Balay, J. & Shevrin, H. (1988) The subliminal psychodynamic activation method: A critical review. *American Psychologist*, **43**, 161–74.

Baltes, P. (1990) Entwicklungspsychologie der Lebeusspanne: Theoretische Leitsatze. *Psychologische Rundschan*, **41**, 1–24.

Baltes, P. B. & Kliegl, R. (1986) *On the dynamics between growth and decline in the aging of intelligence and memory*. In: K. Poeck (Ed.), Proceedings of the 13th World Congress of Neurology. Heidelberg: Springer.

Baltes, P. B. & Willis, S. L. (1982) Plasticity and enhancement of intellectual functioning in old age. In: F. Craik and S. E. Trebub (Eds.), *Aging and Cognitive Processes*, pp. 353–89. New York: Plenum Press.

Barker, B. (1952) *Science and the Social Order*. Glencoe, Ill.: Free Press.

Barker, B. (1961) Resistance by scientists to scientific discovery. *Science*, **134**, 596–602.
Baron, M., Gruen, R., Asnis, L. & Kane, J. (1982) Schizo-affective illness, schizophrenia and affective disorders. *Acta Psychiatrica Scandinavia*, **65**, 253–62.
Barrett, G. V. & Depinet, R. L. (1991) A reconsideration of testing for competence rather than for intelligence. *American Psychologist*, **46**, 1012–24.
Barrett, P. & Eysenck, H. J. (1992) Brain electrical potentials and intelligence. In: A. Gale & M. W. Eysenck (Eds.), *Handbook of Individual Differences: Biological Perspectives*. London: John Wiley.
Barrett, P. & Eysenck, H. J. (1993) Sensory nerve conduction and intelligence: A replication. *Personality and Individual Differences*, **15**, 249–60.
Barrett, P. T. & Eysenck, H. J. (1994) The relationship between evoked potential component amplitude latency, contour length, variability, zero-crossings, and psychometric intelligence. *Personality and Individual Differences*, **16**, 3–32.
Barrett, P. & Eysenck, S. B. G. (1984) The assessment of personality factors across 25 countries. *Personality and Individual Differences*, **5**, 615–32.
Barrett, P., Daum, I. & Eysenck, H. J. (1990) Sensory nerve conduction and intelligence: A methodological study. *Journal of Psychophysiology*, **4**, 1–13.
Barron, F. (1953) Complexity-simplicity as a personality dimension. *Journal of Abnormal and Social Psychology*, **48**, 103–72.
Barron, F. (1963a) The disposition toward originality. In: C. W. Taylor and F. Barron (Eds.), *Scientific Creativity: Its Recognition and Development*, pp. 139–52.
Barron, F. (1963b) *Creativity and Psychological Health*. New York; van Nostrand.
Barron, F. (1965) The psychology of creativity. In: *New Directions in Psychology*, pp. 1–134. New York: Holt, Rhinehart and Winston.
Barron, F. (1968) *Creativity and Personal Freedom*. Princeton: van Nostrand.
Barron, F. (1969) *Creative Person and Creative Process*. New York: Holt, Rhinehart & Winston.
Barron, F. (1972) *Artists in the Making*. New York: Seminar Press.
Barron, F. (1988) Putting creativity to work. In: R. J. Sternberg (Ed.), *The Nature of Creativity: Contemporary Psychological Perspectives*, pp. 76–98. Cambridge: Cambridge University Press.
Barron, F. & Harrington, D. M. (1981) Creativity, intelligence and personality. *Annual Reviews of Psychology*, **32**, 439–76.
Barron, F. & Welsh, G. S. (1952) Artistic perception as a possible factor in personality style: its measurement by a figure preference test. *Journal of Psychology*, **33**, 199–203.
Barry, J. V. (1958) *Alexander Maconochie of Norfolk Island*. Melbourne: Oxford University Press.
Barsalla, G. (1988) *The Evolution of Technology*. New York: Cambridge University Press.
Barsalon, L. W. (1982) Context-independent and context-dependent information in concepts. *Memory and Cognition*, **10**, 82–93.
Barsalon, L. W. (1983) Ad hoc categories. *Memory and Cognition*, **11**, 211–27.
Bartlett, F. (1958) *Thinking*. London: George Allen & Unwin.
Baruch, I., Hemsley, D. R. & Gray, J. (1988) Differential performance of acute and chronic schizophrenics in a latent inhibition task. *Journal of Nervous and Mental Disease*, **176**, 598–606.
Beard, G. (1874) *Legal Responsibility in Old Age*. New York: Russell.
Beck, A. T. (1962) Reliability of psychiatric diagnosis: I. A critique of systematic studies. *American Journal of Psychiatry*, **119**, 210–16.
Becker, G. (1978) *The Mad Genius Controversy: A Study in the Sociology of Deviance*. Beverly Hills: Sage.
Beech, A. R., Baylis, G. C., Smithson, P & Claridge, G. (1989b) Individual differences in schizotypy as reflected in measures of cognitive inhibition. *British Journal of Clinical Psychology*, **28**, 117–29.

Beech, A. R. & Claridge, G. (1987) Individual differences in negative priming. *British Journal of Psychology*, **78**, 349–56.
Beech, A. R., McManus, D., Baylis, G. C., Tipper, S. P. & Agar, K. (1991) Individual differences in cognitive processes: Towards an explanation of schizophrenic symptomatology. *British Journal of Psychology*, **82**, 417–26.
Beech, A. R., Powell, T., McWilliam, J. & Claridge, G. (1989a) Evidence of reduced cognitive inhibition in schizophrenia. *British Journal of Clinical Psychology*, **28**, 109–16.
Beech, A. R., Powell, T., McWilliam, J. & Claridge, G. (1990) The effect of a small dose of chlorpromazine on a measure of cognitive inhibition. *Personality and Individual Differences*, **11**, 1141–5.
Behr-Pinnow, von C. (1933) Die mathematische Begabung in der Familie Bernouilli. *Archiv fur Rassen-mid Gesellschaftsbiologie*, **27**, 395–412.
Bell, E. T. (1939) *Men of Mathematics*. London: Camelot Press.
Bell, E. T. (1965) *Men of Mathematics*. London: The Scientific Book Club.
Bell, J. E. (1948) *Projective Techniques*. New York: Longmans, Green & Co.
Bem, S. L. (1974) The measurement of psychological androgyny. *Journal of Consulting and Clinical Psychology*, **42**, 155–62.
Bem, S. L. (1981) Gender scheme theory: A cognitive account of sex typing. *Psychological Review*, **88**, 369–71.
Benassy, M. & Chauffard, C. (1947) Le Test F de Cattell est-il un test objectif de temperament? *L'annee psychologique*, **43**, 200–30.
Benbow, C. & Lubinski, D. (1992) Gender differences in abilities and preferences among the gifted: Implications for the maths-science pipeline. *Current Directions in Psychological Science*, **1**, 61–6.
Benbow, C. P. & Lubinski, D. (1993) Gender differences in mathematical reasoning: Some biological linkages. In: M. Hang, R. Whalen, C. Aron & K. Olsen (Eds.), *The Development of Sex Differences and Similarities in Behaviour*. London: NATO Academic Publishers.
Bentall, R. P., Claridge, G. S. & Slade, P. D. (1989) The multidimensional nature of schizotypal traits: A factor analytic study with normal subjects. *British Journal of Clinical Psychology*, **28**, 363–75.
Berlyne, D. E. (1960) *Conflict, Arousal and Curiosity*. New York: McGraw-Hill.
Berlyne, D. E. (1965) *Structure and Direction in Thinking*. New York: Wiley.
Berlyne, D. R. (1971) *Aesthetics and Psychobiology*. New York: Appleton-Century-Crofts.
Berrrington, H. (1983) Prime ministers and the search for love. In: R. S. Albert (Ed.), *Genius and Eminence*, pp. 358–73. Oxford: Pergamon Press.
Berrios, G. E. (1987) Historical aspects of psychoses: 19th century issues. *British Medical Bulletin*, **43**, 484–98.
Berry, P. (1981) The Nobel scientists and the origins of scientific achievement. *British Journal of Sociology*, **32**, 381–91.
Biederman, J., Lerna, Y. & Belmaker, R. H. (1974) Combination of lithium carbonate and haliperidol in schizo-affective disorder. *Archives of General Psychiatry*, **36**, 327–33.
Bingham, M. T. (1953) Beyond psychology. In: *Homo sapiens audubonians: A tribute to Walter Van Dyke Bingham*, pp. 5–29. New York: National Audubon Society.
Bleuler, M. (1911, translation 1978) *The Schizophrenic Disorders*. New Haven: Yale University Press.
Blom, E. (Ed.) (1966) *Grave's Dictionary of Music and Musicians* (10 vols., 5th edition). New York: St. Martin's Press.
Bloom, B. S. (1963) Report on creativity research by the examiner's office of the University of Chicago. In: C. W. Taylor and F. Barron (Eds.), *Scientific Creativity*, pp. 251–64. New York: Wiley.
Blume, F. (Ed.) (1949–68) *Die Musik in Geschichte und Gegenwart* (14 vols.). Kassel: Bareureiter.

Boden, M. (1993) *The Creative Mind: Myths and Mechanisms*. London: Weidenfeld and Nicolson.

Bogen, J., De Zune, R., Ten Houten, W. & Marsh, J. (1972) The other side of the brain. IV: The A/P ratio. *Bulletin of the Los Angeles Neurological Society*, **37**, 49–59.

Bollobas, B. (1988) Ramanujan – a glimpse of his life and his mathematics. *Cambridge Review*, June, pp. 76–80.

Bonfield, W. A. (1953) The occurrence of clustering: the result of randomly arranged associates. *Journal of General Psychology*, **49**, 229–40.

Bonner, J. (1957) Going beyond the information given. In: J. Booner (Ed.), *Contemporary Approaches to Cognition: The Colerado Symposium*, pp. 41–69. Cambridge: Harvard University Press.

Bonville, J. (1981) *Kepler*. London: Secker and Warburg.

Bosch, van der, R. J. (1984) Eye tracking impairment: Attentional and psychometric correlates in psychiatric patients. *Journal of Psychiatric Research*, **18**, 277–86.

Botwinick, J. (1977) Intellectual abilities. In: J. E. Birren and K. W. Schaie (Eds.), *Handbook of the Psychology of Aging*. New York: Van Nostrand Reinhold.

Bouchard, T. J., Lykken, D. T., McGue, M., Segal, N. L. & Tellegen, D. (1990) Sources of human psychological differences: The Minnesota Study of Twins Reared Apart. *Science*, **250**, 223–8.

Bousfield, W. A. & Sedgewick, C. (1944) An analysis of sequences of restricted associative responses. *Journal of Genetic Psychology*, **30**, 149–65.

Bouthilet, L. (1948) The measurement of intuitive thinking. *Unpublished Ph.D. thesis, University of Chicago. Quoted by Westcott, 1968.*

Bower, G. H. (1970) Analysis of mnemonic devices. *American Scientist*, **58**, 496–510.

Bowers, K. S., Regehr, G., Balthasard, C. & Parker, N. (1990) Intuition in the context of discovery. *Cognitive Psychology*, **22**, 71–110.

Bown, W. & Mehta, R. (1993) The seamy side of swing bowling. *New Scientist*, **139**, 20–4.

Boyatzin, R. E. (1982) *The Competent Manager*. New York: Wiley.

Bradley, R. & Swartz, N. (1979) *Possible Worlds: An Introduction to Logic & Philosophy*. Oxford: Blackwell.

Braff, D. L. (1981) Impaired speech of information processing in non-mediated schizotypal patients. *Schizophrenia Bulletin*, **7**, 499–508.

Braff, D. L. & Saccuzzo, D. P. (1981) Information processing dysfunction in schizophrenia: a two factor deficit theory. *American Journal of Psychiatry*, **138**, 1051–6.

Braff, D. L. & Saccuzzo, D. P. (1982) Effects of antipsychotic medication on speed of information processing in schizophrenic patients. *American Journal of Psychiatry*, **139**, 1127–30.

Brain, R. (1960) *Some Reflections on Genius*. London: Pitman.

Bramwell, B. S. (1948) Galton's 'Hereditary Genius' and closest following generation since 1869. *Eugenics Review*, **39**, 146–53.

Brandon, S., Cowley, P., McDonald, C., Neville, P., Palmer, R. & Wellstood-Eason, S. (1985) Leicester ECT trial: Results in schizophrenia. *British Journal of Psychiatry*, **146**, 177–83.

Bringmann, W. G. & Tweney, R. D. (1980) *Wundt Studies*. Toronto: Hogrefe.

Broad, W. & Wade, N. (1982) *Betrayers of the Truth: Fraud and Deceit in Science*. Oxford: University Press.

Broadbent, D. E. (1971) *Decision and Stress*. London: Academic Press.

Broadbent, D. (1977) The hidden pre-attentive processes. *American Psychologist*, **32**, 109–18.

Brock, I. D. (1988) *Robert Koch: A Life in Medicine and Bacteriology*. New York: Springer Verlag.

Brockington, J. F., Kendell, R. E. & Leff, J. P. (1978) Definition of schizophrenia: concordance and prediction of outcome. *Psychological Medicine*, **8**, 387–98.

Brockington, J. F., Kendell, R. E. & Wainwright, S. (1980) Depressed patients with

schizophrenia or paranoid symptoms. *Psychological Medicine*, **10**, 665–75.
Brody, N. (1992) *Intelligence* (Second edition). New York: Academic Press.
Brody, R. A. and Page, B. I. (1975) The impact of events on presidential popularity: The Johnson and Nixon administrations. In: A. Wildavsky (Ed.), *Perspectives on the Presidency*. Boston: Little, Broden.
Bronowski, J. (1956) *Science and Human Values*. London: Hutchinson.
Brooks, P. (1984) Schizotypy and hemisphere function – II. Performance asymmetry as a verbal divided visual-field task. *Personality and Individual Differences*, **5**, 649–56.
Brown, R. T. (1989) Creativity: What are we to measure? In: J. A. Glover, R. P. Ronning & C. R. Reynolds (Eds.), *Handbook of Creativity*, pp. 3–32. New York: Plenum Press.
Brown, R. & McNeill, D. (1966) The 'tip-of-the-tongue' phenomenon. *Journal of Verbal Learning and Verbal Behaviour*, **5**, 325–37.
Brugger, P., Gamma, A., Muri, R., Schafer, M. & Taylor, K. (1993) Functional hemispheric asymmetry and belief in ESP. *Personality and Individual Differences* (in press).
Bruner, J. (1957) Going beyond the information given. In: J. Bruner (Ed.), *Contemporary Approaches to Cognition*: The Colarado Symposium, pp. 41–69. Cambridge: Harvard University Press.
Bryan-Tuckett, R. & Silverman, L. N. (1984) Effects of subliminal stimulation of symbiotic fantasies on the academic performance of emotionally handicapped students. *Journal of Conselling Psychology*, **31**, 295–305.
Buchsbaum, M. S., Coursey, R. D. & Murphy, P. L. (1976) The biochemical high-risk paradigm: behavioral and familial correlates of low platelet monoamine oxidase activity. *Science*, **194**, 339–41.
Bullen, J. G. & Hemsley, D. R. (1986) Schizophrenic: A failure to control the content of consciousness. *British Journal of Clinical Psychology*, **26**, 25–33.
Bullough, V., Bullough, B. & Munro, M. (1981) History and creativity: Research problems and some possible solutions. *The Journal of Creative Behaviour*, **15**, 102–16.
Bunge, M. (1967) *Scientific Research* (Vol. 1). New York: Springer.
Burks, B., Jensen, D. & Terman, L. M. (1930) *The Promised Youth*. Stanford: University Press.
Burt, C. (1943) Ability and insane. *British Journal of Educational Psychology*, **13**, 83–98.
Burt, C. (1963) Is intelligence distributed normally? *British Journal of Statistical Psychology*, **16**, 175–90.
Byerley, W. *et al.* (1990) D_2 dopamine receptor gene linked to manic-depression in three families. *Psychiatric Genetics*, **1**, 55–62.
Callahan, L. A. & Saccuzzo, D. P. (1986) Associative intrusions in the rest of behavior of the first-degree relatives of adult schizophrenics: A preliminary study. *Journal of Nervous and Mental Diseases*, **174**, 240–2.
Cameron, N. (1939) Deterioration and regression in schizophrenic thinking. *Journal of Abnormal and Social Psychology*, **34**, 265–70.
Cameron, N. (1947) *The Psychology of Behavior Disorders*. Boston: Houghton Mifflin.
Cameron, N. & Magaret, A. (1950) Experimental studies in thinking: I. Scattered speech in the response of normal subjects to incomplete sentences. *Journal of Experimental Psychology*, **39**, 617–27.
Cameron, N. & Magaret, A. (1951) *Behavior Pathology*. Boston: Houghton Mifflin.
Cammer, L. (1970) Schizophrenic children of manic-depressive parents. *Diseases of the Nervous System*, **31**, 177–80.
Campbell, D. T. (1960) Blind variations and selective retention in creative thought as in other knowledge processes. *Psychological Review*, **67**, 380–400.
Campbell, D. T. (1974) Unjustified variation and selective retention in scientific discovery. In: F. J. Ayala & T. Dobzhansky (Eds.) *Studies in the Philosophy of Biology*, pp. 131–49. London: Macmillan.
Campbell, D. T. & Tansofer, H. (1966) Schopenhauer (?), Sequin, Lubinoff and Zefender in

anticipation of Ememert's Law: With comments on the uses of eponymy. *Journal of the History of the Behavioral Sciences*, **2**, 58–63.

Canavan, A. G. M., Dunn, G. & McMillan, T. (1986) Principal components of the WAIS-R. *British Journal of Clinical Psychology*, **25**, 81–5.

Cangelosi, D. M. & Schaefer, C. E. (1991) A twenty-five year follow-up study of ten exceptionally creative adolescent girls. *Psychological Reports*, **68**, 307–11.

Cannell, D. M. (1993) *George Green, Mathematician and Physicist, 1793–1841: The Background to his life and work*. London: Athlone Press.

Canter, S. (1973) Some aspects of cognitive function in twins. In: G. Claridge, S. Canter & W. I. Hume (Eds.), *Personality Differences and Biological Variation: A Study of Twins*. Oxford: Pergamon Press.

Carlsson, A. (1988) The current status of the dopamine hypothesis of schizophrenia. *Neuropsychopharmacology*, **1**, 178–203.

Carroll, J. B. (1941) A factor analysis of verbal abilities. *Psychometrika*, **6**, 279–308.

Carroll, J. B. (1988) Individual differences in cognitive functioning. In: *Handbook of Experimental Psychology*, pp. 813–62. New York: Wiley.

Carroll, J. B. (1993) *Human Cognitive Abilities*. Cambridge: Cambridge University Press.

Caspar, M. (1959) *Johannes Kepler*. London: Abelard-Schuman.

Cattell, J. McK. (1903a) A statistical study of eminent men. *Popular Science Monthly*, **62**, 359–77.

Cattell, J. McK. (1903b) Statistics of American psychologists. *American Journal of Psychology*, **14**, 310–328.

Cattell, J. McK. & Brimhall, D. R. (1921) *American Men of Science*. Garrison: Science Press.

Cattell, R. B. (1934) Temperament Tests: II. Tests. *British Journal of Psychology*, **24**, 20–49.

Cattell, R. B. (1963) Personality of the researcher from measurements and biography. In: C. W. Taylor & F. Barron (Eds.), *Scientific Creativity: Its Recognition and Development*. New York: Wiley.

Cattell, R. B. (1971) *Abilities: Their Structure, Growth and Action*. Boston: Houghton-Mifflin.

Cattell, R. B. & Drevdahl, J. E. (1955) A comparison of the personality profile (16 P.F.) of eminent researchers with that of eminent teachers and administrators, and of the general populations. *British Journal of Psychology*, **46**, 248–61.

Cattell, R. B., Eber, H. W. & Tatsuoka, M. M. (1970) *Handbook for the 16 PF*. Champaign, Illinois: Institute for Personality and Ability Testing.

Ceci, S. J. (1990) *On Intelligence – More or Less*. Englewood Cliffs, N.J.: Prentice Hall.

Ceci, S. J. & Bronfenbrenner, U. (1985) Don't forget to take the cake out of the oven: Strategic time-monitoring, prospective memory, and context. *Child Development* , **56**, 175–90.

Ceci, S. J. & Liker, J. (1986) A day at the races: A study of IQ, expertise, and cognitive complexity. *Journal of Experimental Psychology: General*, **115**, 255–66.

Chambers, J. A. (1964) Relating personality and biographical factors to scientific creativity. *Psychological Monograph: General and Applied*, **78**, Whole No. 584.

Chand, I. & Runco, M. A. (1993) Problem finding skills as components in the creative process. *Personality and Individual Differences*, **14**, 155–62.

Chapman, L. J. (1956) Distractibility in the conceptual performance of schizophrenics. *Journal of Abnormal and Social Psychology*, **53**, 286–91.

Chapman, L. J. (1958) Intrusion of associative responses into schizophrenic conceptual performance. *Journal of Abnormal and Social Psychology*, **53**, 374–91.

Chapman, L. J. & Chapman, J. P. (1980) Scales for rating psychotic and psychotic-like experiences as continua. *Schizophrenia Bulletin*, **6**, 476–89.

Chapman, L. J., Chapman, J. P. & Miller, E. V. (1982) The reliabilities and intercorrelations of eight measures of proneness to psychosis. *Journal of Consulting and Clinical Psychology*, **50**, 187–95.

Chapman, L. J., Chapman, J. P. & Raulin, M. L. (1976) Scales for physical and social

anhedonia. *Journal of Abnormnal Psychology*, **85**, 374–82.
Chapman, L. J., Chapman, J. P. & Raulin, M. L. (1978) Body image aberration in schizophrenics. *Journal of Abnormal Psychology*, **87**, 399–407.
Chapman, L. J., Chapman, J. P., Numbers, J. S., Edell, W. S., Carpenter, B. N. & Beckfield, D. (1984) Impulsive non-conformity as a trait contributing to the prediction of psychotic-like and schizotypal symptoms. *Journal of Nervous and Mental Diseases*, **172**, 681–91.
Chapman, L. J., Edell, W. S. & Chapman, J. P. (1980) Physical anhedonia, perceptual aberration and psychosis proneness. *Schizophrenia Bulletin*, **6**, 635–53.
Chapman, L. J. & Taylor, J. (1957) Breadth of deviate concepts used by schizophrenics. *Journal of Abnormal and Social Psychology*, **54**, 118–23.
Charlton, S. & Bakan, P. (1988–9) Cognitive complexity and creativity. *Cognition and Personality*, **8**, 315–22.
Chase, W. G. & Ericson, K. A. (1982) Skill and working memory. In: G. W. Bower (Ed.), *The Psychology of Learning and Motivation*, Vol. 16. New York: Academic Press.
Checkley, S. A. (1980) Neuroendocrine tests of monoamine function in man: a review of basic theory and its application to the study of depressive illness. *Psychological Medicine*, **10**, 35–53.
Ciarlo, D. D., Lidz, T. & Ricci, J. (1967) Word meaning in parents of schizophrenics. *Archives of General Psychiatry*, **17**, 470–77.
Cicero, M. T. (1943) *Tuscularum Disputationes*. Cambridge: Harvard University Press.
Claridge, G. S. (1972) The schizophrenics as nervous types. *British Journal of Psychiatry*, **121**, 1–12.
Claridge, G. (1981) Psychoticism. In: R. Lynn (Ed.), *Dimensions of Personality*, pp. 79–109. London: Pergamon Press.
Claridge, G. (1983) The Eysenck Psychoticism Scale. In: J. V. Butcher & C. D. Spielberger (Eds.), *Advances in Personality Assessment*, Vol. 2, pp. 71–114. Hillsdale, N.J.: Lawrence Erlbaum.
Claridge, G. (1985) *Origins of Mental Illness*. Oxford: Blackwell.
Claridge, G. & Broks, P. (1984) Schizotypy and hemisphere function. I. Theoretical considerations and the measurement of schizotypy. *Personality and Individual Differences*, **5**, 633–48.
Claridge, G. S. & Chappa, H. J. (1973) Psychoticism: A study of its biological basis in normal subjects. *British Journal of Social and Clinical Psychology*, **12**, 175–87.
Claridge, G. S., Robinson, D. L. & Birchall, P. M. A. (1985) Psychophysiological evidence of 'psychoticism' in schizophrenics' relatives. *Personality and Individual Differences*, **6**, 1–10.
Clark, P. W. (1973) *Einstein: The Life and Times*. London: Hodder & Stoughton.
Clark, R. D. & Rice, G. A. (1982) Family constellations and eminence: 76 birth orders of Nobel Prize winners. *The Journal of Psychology*, **110**, 281–7.
Cloninger, C. D. (1987) A systematic method for clinical description and classification of personality. *Archives of General Psychiatry*, **44**, 573–88.
Cloninger, C. D. Martin, R. L., Guze, S. B. & Clayton, P. J. (1985) Diagnosis and prognosis in schizophrenics. *Archives of General Psychiatry*, **42**, 15–25.
Close, S. & Suter, S. (1903) Drawing and the cerebral hemispheres: Bilateral EEG alpha. *Biological Psychology*, **16**, 15–27.
Coccaro, E., Siever, L., Klar, H., Maurer, G., Cochrane, K., Cooper, T., Mahs, R. & Davis, K. (1989) Serotonergic studies in patients with affective personality disorders. *Archives of General Psychiatry*, **46**, 587–99.
Cohen, B. (1985) *Revolution of Science*. Cambridge: Harvard University Press.
Cohen, B. D. (1987a) Self-editing deficits in schizophrenia. *Journal of Psychiatric Research*, **14**, 267–73.
Cohen, B. D. (1987b) Referent communication disturbances in schizophrenia. In: S. Schwartz (Ed.), *Language and Cognition in Schizophrenia*, pp. 1–34. Hillsdale, N.J.: Erlbaum.

Cohen, M. R. & Nagel, N. (1936) *An Introduction to Logical Scientific Method.* New York: Harcourt, Brace & Co.

Cole, J. & Zuckerman, H. (1987) Marriage, motherhood and research performance in science. *Scientific American*, **255**, 115–25.

Cole, S. (1979) Age and scientific performance. *American Journal of Sociology*, **84**, 958–77.

Coleman, V. (1988) *The Health Scandal.* London: Sidgewick & Jackson.

Coleridge, S. T. (1895) *The Poetical Works of Samuel Taylor Coleridge.* London: Macmillan.

Collins Dictionary and Thesaurus. (1988) London: Collins.

Collins, A. M. & Loftus, E. F. (1975) A spreading-activation theory of semantic processing. *Psychological Review*, **82**, 407–28.

Collins, A. M. & Quillian, M. R. (1970) Facilitating retrieval from semantic memory: The effect of repeating part of an inference. *Actal Psychologica*, **33**, 304–14.

Commings, D. E. (1991) The dopamine D_2 receptor locus as a modifying gene in neuropsychiatric disorders. *Journal of the American Medical Association*, **266**, 1793–800.

Condon, T. J. & Allen, G. J. (1980) Role of psychoanalytic merging fantasies in systematic desensitization: A rigorous methodological examination. *Journal of Abnormal Psychology*, **89**, 437–43.

Constantinople, A. (1973) Masculinity–femininity: An exception to the famous dictum? *Psychological Bulletin*, **80**, 389–407.

Cooper, J. E. (1967) Diagnostic change in a longitudinal study of psychiatric patients. *British Journal of Psychiatry*, **113**, 129–42.

Cooper, J. E., Kendell, R. E., Gurland, B. J., Sharpe, L., Copeland, J. R. & Simon, R. (1972) *Psychiatric Diagnosis in New York and London.* Oxford University Press: Maudsley Monograph, No. 20.

Coren, S. (Ed.) (1990) *Left-handedness: Behavioral Implications and Anomalies.* Amsterdam: North-Holland.

Cox, C. (1926) *The Early Mental Traits of Three Hundred Geniuses.* Stamford: University Press.

Cramer, P. (1968) *Word Associations.* New York: Academic Press.

Cranberg, L. D. & Albert, M. I. (1988) The chess mind. In: L. H. Obler and D. Fein (Eds.), *The Exceptional Brain: Neuropsychology of Talent and Special Abilities.* New York: Guilford Press.

Creese, I., Burt, D. & Snyder, S. (1977) Dopamine receptor binding enhancement accompanies lesion-induced behavioral supersensitivity. *Science*, **197**, 596–8.

Crockett, W. (1965) Cognitive complexity and impress formation. In: B. A. Maher (Ed.), *Progress in Experimental Personality Research.* New York: Academic Press.

Crider, A., Solomon, P. R. & McMahon, M. A. (1982) Disruption of selective attention in the rat following d-amphetamine administration. Relationship to schizophrenic attention disorder. *Biological Psychiatry*, **17**, 351–61.

Cropley, A. J. (1992) *Fostering Creativity.* Norwood, N.J.: Ablex.

Croughan, J. L., Welner, A. & Robinson, E. (1974) The group of schizo-affective and related psychoses – critique, record, follow-up and family studies. *Archives of General Psychiatry* , **31**, 632–7.

Crow, T. J. (1980) Molecular pathology of schizophrenia: more than one disease process? *British Medical Journal*, **280**, 66–8.

Crow, T. J. (1986) The continuum of psychosis and its implication for the structure of the gene. *British Journal of Psychiatry*, **149**, 419–29.

Crow, T. J. (1987) (Ed.) Recurrent and Chronic Psychoses. *British Medical Bulletin* , **43**, 3.

Crow, T. J. (1989) Pseudoautosomal locus for the cerebral dominance gene. *Lancet*, **II**, 339–40.

Crow, T. J. (1990) The continuum of psychosis and its genetic origins. *British Journal of Psychiatry*, **156**, 788–97.

Crow, T. J. & Cooper, S. J. (1986) *The genetics of the unitary psychosis.* Proceedings of the

IVth World Congress of Biological Psychiatry. Amsterdam: Elsevier.
Csikszentmihalyi, M. & Beattie, O. U. (1979) Life themes: A theoretical and empirical exploration of their origins and effects. *Journal of Humanistic Psychology*, **19**, 45–63.
Csikszentmihalyi, M. & Getzels, J. W. (1973) The personality of young artists: an empirical and theoretical exploration. *British Journal of Psychology*, **64**, 91–104.
Dalby, J. T., Morgan, D. & Lee, M. C. (1986) Single-case studies: schizophrenia and mania in identical twins. *Journal of Nervous and Mental Diseases*, **174**, 308–14.
Danzinger, D. (1990) *Constructing the Subject: Historical Origin of Psychological Research*. Cambridge: Cambridge University Press.
Davis, G. (1975) In frumious pursuit of the creative process. *Journal of Creative Behavior*, **9**, 76–87.
Davis, H. (1974) What does the P scale measure? *British Journal of Psychiatry*, **125**, 161–7.
Davis, R. A. (1987) Creativity in neurological publications. *Neurosurgery*, **20**, 652–63.
Deary, I. & Caryl, P. (1993) Intelligence, EEG, and Evoked Potentials. In: P. A. Vernon (Ed.), *Biological Approaches to Human Intelligence*. New Jersey: Ablex.
Decina, P., Luscan, L. R. & Linder, J. R. (1989) Parent–child pairs with major psychiatric diseases. 142nd meeting of the American Psychiatric Association, San Francisco, May 1989. Abstract No. 105. Washington, D. C. *American Psychiatric Association. Cited by Crow, 1990*.
Dellas, M. & Gaier, E. L. (1970) Identification of creativity: The individual. *Psychological Bulletin*, **73**, 55–73.
Delva, N. J. & Letemendia, F. J. (1982) Lithium treatment in schizophrenia and schizoaffective disorders. *British Journal of Psychiatry*, **141**, 387–400.
Delva, N. J. & Letemendia, F. J. (1986) Lithium treatment – schizophrenia and schizoaffective disorders. In: A. Keer & R. P. Snaith (Eds.), *Contemporary Issues in Schizophrenia*. London: Gaskell.
Dennis, W. (1954) Productivity among American psychologists. *American Psychologist*, **9**, 191–4.
Dennis, W. (1955) Variation in productivity among creative workers. *Scientific Monthly*, **80**, 277–8.
Dennis, W. (1966) Creative productivity between the ages of 20 and 80 years. *Journal of Gerontology*, **21**, 1–8.
Dentler, R. A. & Mackler, B. (1964) Originality: Some social and personal determinants. *Behavioral Science*, **9**, 1–7.
Denvir, B. (1991) *Impressionism*. London: Studio Editions.
Depue, R. A. (1993) *Neurobehavioral Systems, Personality, and Psychopathology*. New York: Springer Verlag.
Depue, R. A. & Spoont, M. O. (1986) Conceptualizing a serotonin trait: A behavioral dimension of constraint. *Annals of the New York Academy of Sciences*, **487**, 47–62.
Depue, R. A. & Zald, D. (1993) Biological and environmental processes in non-psychotic psychopathology: A neurobehavioral system perspective. In: C. Costello (Ed.), *Basic Issues in Psychopathy*. New York: Guilford Press.
Depue, R. A., Luciana, M., Arkin, P. & Leon, A. (1994) Dopamine and the structure of personality: Relation of agonist-induced dopamine activity to positive emotionality. *Journal of Personality and Social Psychology* (in press).
Detterman, D. K. & Daniel, M. H. (1989) Correlations of mental tests with each other and with cognitive variables are highest for low IQ groups. *Intelligence* , **13**, 349–59.
Detterman, D. & Spry, K. C. (1988) Is it smart to play the horses? *Journal of Experimental Psychology: General*, **117**, 91–5.
Dewey, J. (1910) *How to Think*. Boston: Heath.
Dillon, J. T. (1982) Problem finding and solving. *Journal of Creative Behaviour*, **16**, 97–111.
Dixon, R. A., Kramer, D. A. & Baltes, P. B. (1985) Intelligence: A life-span developmental

perspective. In: B. B. Walman (Ed.), *Handbook of Intelligence*, pp. 301–52. New York: Wiley.
Dolan, R. J., Calloway, S. P. & Mann, A. H. (1985) Cerebral ventricular size in depressed subjects. *Psychological Medicine*, **15**, 873–8.
Domino, G. (1974) Assessment of cinematographic creativity. *Journal of Personality and Social Psychology*, **30**, 150–4.
Donahue, W. H. (1988) Kepler's fabricated figures. *Journal of the History of Astronomy*, **19**, 217–37.
Dorner, D. & Krenzig, H. (1983) Problemlosefahigkeit und Intelligenz. *Psychologische Rundschau*, **34**, 185–92.
Dorner, D., Reither, H. & Standel, T. (1985) *Lothausen: Vom Umgang mit Unbestimmtheit und Komplexitat*. Bern: Huber.
Drane, B. K. (1986) Clinical psychiatry and the physiodynamics of the limbic system. In: B. K. Drane & K. E. Livingstone (Eds.), *The Limbic System: Functional Organization and Clinical Disorders*, pp. 285–305.
Drevdahl, J. E. (1956) Factors of importance for creativity. *Journal of Clinical Psychology*, **12**, 21–6.
Drevdahl, J. E. (1964) Some developmental and environmental factors in creativity. In: C. W. Taylor (Ed.), *Widening Horizons in Creativity: The Proceedings of the Fifth Utah Creativity Research Conference*, pp. 170–202. New York: Wiley.
Drevdahl, J. E. & Cattell, R. B. (1958) Personality and creativity in artists and writers. *Journal of Clinical Psychology*, **14**, 107–11.
Dryden, J. (1681) *Absalom and Achitophel*, lines 163–4.
Dubos, R. (1950) *Louis Pasteur: Free Lance of Science*. New York: Charles Scribner.
Dunbar, D. (1993) Seeing biology through Aristotle's eyes. *New Scientist*, 20 Feb., 39–42.
Dykes, M. & McGhie, A. (1976) A comparative study of attentional strategies of schizophrenic and highly creative normal subjects. *British Journal of Psychiatry*, **128**, 50–6.
Eason, R. G. & White, C. T. (1961) Muscular tension, effort and tracking difficulty: studies of parameters which affect tension level and performance efficiency. *Perceptual and Motor Skills*, **12**, 331–72.
Easterbrook, J. A. (1959) The effect of emotion as one utilization and the organization of behavior. *Psychological Review*, **66**, 183–201.
Eaves, L. J., Eysenck, H. J. & Martin, N. G. (1989) *Genes, Culture and Personality: An Empirical Approach*. New York: Academic Press.
Eckblad, M. & Chapman, L. J. (1983) Magical ideation as an indicator of schizotypy. *Journal of Consulting and Clinical Psychology*, **51**, 215–75.
Edelman, G. (1992) *Bright Air, Brilliant Fire*. London: Allan Lane.
Edwards, B. (1979) *Drawing on the Right Side of the Brain*. Los Angeles: Trader.
Effron, E. (1984) *The Apocalyptics*. New York: Simon and Schuster.
Efron, R. (1990) *The Decline and Fall of Hemispheric Specialization*. Hillsdale, N. J.: Erlbaum.
Eisenman, R. (1965) Aesthetic preferences of schizophrenics. *Perceptual and Motor Skills*, **20**, 601–4.
Eisenman, R. (1966) The effect of disapproval on aesthetic preferences of schizophrenics. *The Journal of General Psychology*, **75**, 315–18.
Eisenman, R. (1968) Personality and demography in complexity–simplicity. *Journal of Counselling and Clinical Psychology*, **32**, 140–3.
Eisenman, R. (1969) Creativity, awareness, and liking. *Journal of Counselling and Clinical Psychology*, **33**, 157–60.
Eisenman, R. (1987) Creativity, birth order, and risk-taking. *Bulletin of the Psychodynamic Society*, **25**, 87–8.
Eisenman, R. (1990) Creativity, preference for complexity, and physical and mental illness. *Creativity Research Journal*, **3**, 231–6.

Eisenman, R. (1991) *From Crime to Creativity: Psychological and Social Factors in Deviance*. Iowa: Kendall Hunt.
Eisenman, R. (1992) Creativity in Prisoners: Conduct disorders and psychotics. *Creativity Research Journal*, **5**, 175–81.
Eisenman, R. (1994) *Contemporary Social Issues: Drugs, Crime, Creativity and Education*. Ashland, Ohio: Book Makers.
Eisenman, R. & Coffee, S. (1964) Aesthetic preferences of art students and mathematics students. *Journal of Psychology*, **58**, 375–8.
Eisenman, R. & Rappaport, J. (1967) Complexity preference and semantic differential ratings of complexity-simplicity and symmetry-asymmetry. *Psychometric Science*, **7**, 147–8.
Eisenman, R. & Robinson, N. (1967) Complexity–simplicity, creativity, intelligence and other correlates. *Journal of Psychology*, **67**, 331–4.
Eisenman, R., Borod, J. & Grossman, J. (1972) Sex differences in the inter-relationships of authoritarianism, anxiety, creative attitudes, preference for complex polygons, and the Barron–Welsh Art Scale. *Journal of Clinical Psychology*, **28**, 549–50.
Eisenstadt, J. M. (1978) Parental loss and genius. *American Psychologist*, **33**, 211–23.
Ellis, H. (1904) *A Study of British Genius*. Oxford: Hurst and Blackwell.
Ellis, H. (1926) *A Study of British Genius*. New York: Houghton-Mifflin.
Elsasser, G. (1952) *Die Nachkommen Geisteskranken Ehepaare*. Stuttgart: Thieme.
Encyclopaedia Brittanica (1974) (15th edition, 30 vols.) Chicago: Encyclopaedia Brittanica.
Endler, N. S., Rushton, J. P. & Roediger, H. L. (1978) Productivity and scholarly impact (citations) of British Canadian and U.S. Departments of Psychology. *American Psychologist*, **33**, 1064–82.
Epstein, S. (1953) Overinclusive thinking in a schizophrenic and a control group. *Journal of Counseling Psychology*, **17**, 384–8.
Erdelyi, M. H. (1992) Psychodynamics and the unconscious. *American Psychologist*, **47**, 784–7.
Ertel, S. (1982) Sonnentar tig Keit und Menschengeschichte: Veberprufung der Zusammen hangshypothese. *Lecture delivered at the University of Zurich, 22nd January, 1982*.
Ertel, S. (1988) Scientific discoveries: Testing for concomitance with extra-terrestrial events. *Paper presented at the 7th Annual Convention of the Society for Scientific Exploration, Cornell University, Ithaca, June 1988*.
Ertel, S. (1989) Synchronous bursts of creativity in independent cultures: Evidence for an extra-terrestrial connection. *Paper presented at the 8th Annual Meeting of the Society for Scientific Exploration, Boulder, Colorado, 15–17 June, 1989*.
Ertel, S. (1991) Patterns of scientific evolution: Short-term cycles and secular weaves. In: H. Best, E. Mochmann and M. Thaller (Eds.), *Computers in the Humanities and Social Sciences*. Muchen: K. G. Saur.
Ertel, S. (1992a) Kulturhistorischer Wandol unter Makrooekologischen Perspektive. *Lecture given at Gottingen University*, 5.2.1992.
Ertel, S. (1992b) Sonnentaetigkeit – ein Trigger fuer Revolutionen? Voberpruefung der Tschijewsky-Hypthese. *Private Publication: Institut fur Psychologie*, Goettingen, Gosslerstr. 14.
Everitt, B. S., Gourlay, A. J. & Kendell, R. E. (1971) An attempt at validation of traditional psychiatric syndromes by cluster analysis. *British Journal of Psychiatry*, **119**, 399–412.
Everson, R. D. & Fraumeni, J. F. (1975) Mortality among medical students and young physicians. *Journal of Medical Education*, **50**, 809–11.
Eysenck, H. J. (1939) Primary mental abilities. *British Journal of Educational Psychology*, **9**, 270–5.
Eysenck, H. J. (1941) 'Type' factors in aesthetic judgments. *British Journal of Psychology* , **31**, 262–70.
Eysenck, H. J. (1947) *Dimensions of Personality*. London: Routledge & Kegan Paul.
Eysenck, H. J. (1950) Criterion analysis: An application of the hypothetico-deductive method to factor analysis. *Psychological Review*, **57**, 38–53.

Eysenck, H. J. (1952a) *The Scientific Study of Personality*. London: Routledge & Kegan Paul.
Eysenck, H. J. (1952b) Schizothymia-cyclothymia as a dimension of personality. II. Experimental *Journal of Personality*, **20**, 345–84.
Eysenck, H. J. (1956) The questionnaire measurement of neuroticism and extraversion. *Revista di Psicologia*, **50**, 113–40.
Eysenck, H. J. (1957) *The Dynamics of Anxiety and Hysteria*. London: Routledge & Kegan Paul.
Eysenck, H. J. (1959) The Rorschach Test. In: O. Buros (Ed.), *The Fifth Mental Measurement Yearbook*, pp. 276–8. New Jersey: Gryphon Press.
Eysenck, H. J. (1960) The place of theory in psychology. In: H. J. Eysenck (Ed.), *Experiments in Personality*, **2**, 303–15.
Eysenck, H. J. (1961) Psychosis, drive and inhibition: A theoretical and experimental account. *American Journal of Psychiatry*, **118**, 198–204.
Eysenck, H. J. (Ed.) (1964) *Experiments in Motivation*, pp. 285–91. London: Pergamon Press.
Eysenck, H. J. (1965) *Smoking, Health and Personality*. New York: Basic Books.
Eysenck, H. J. (1967) *The Biological Basis of Personality*. Springfield: C. C. Thomas.
Eysenck, H. J. (1970a) *The Structure of Human Personality*. London: Methuen.
Eysenck, H. J. (1970b) A dimensional system of psychodiagnostics. In: A. R. Mahrer (Ed.). *New Approaches to Personality Classification*, pp. 169–208. New York: Columbia University Press.
Eysenck, H. J. (1972a) *Psychology is About People*. London: Allen.
Eysenck, H. J. (1972b) An experimental and genetic model of schizophrenia. In: A. R. Kaplan (Ed.). *Genetic Factors in Schizophrenia*. Springfield: C. C. Thomas.
Eysenck, H. J. (1972c) The development of aesthetic sensitivity in children. *Journal of Child Psychology and Psychiatry*, **13**, 1–10.
Eysenck, H. J. (1973a) Personality, learning and 'anxiety'. In: H. J. Eysenck (Ed.), *Handbook of Abnormal Psychology* (2nd Edition), pp. 390–419. London: Pitman.
Eysenck, H. J. (Ed.) (1973b) *Handbook of Abnormal Psychology*. London: Pitman.
Eysenck, H. J. (1976) *Sex and Personality*. London: Open Books.
Eysenck, H. J. (1979) *The Structure and Measurement of Intelligence*. New York: Springer Verlag.
Eysenck, H. J. (1980) *The Causes and Effects of Smoking*. London: Maurice Temple Smith.
Eysenck, H. J. (1981) Aesthetic preferences and individual differences. In: D. O'Hare (Ed.), *Psychology and the Arts*, pp. 76–101. Brighton: Harvester Press.
Eysenck, H. J. (Ed.) (1982) *A Model for Intelligence*. New York: Springer Verlag.
Eysenck, H. J. (1983a) The roots of creativity: Cognitive ability or personality trait? *Roeper Review*, **5**, 10–12.
Eysenck, H. J. (1983b) Is there a paradigm in personality research? *Journal of Research in Personality*, **17**, 369–76.
Eysenck, H. J. (1983c) Revolution dans la theorie et al mesure de l'intelligence. *Rerne Canadienne de Psycho-Education*, **12**, 3–17.
Eysenck, H. J. (1984) Meta-analysis: An abuse of research integration. *Journal of Special Education*, **18**, 41–59.
Eysenck, H. J. (1985a) The theory of intelligence and the psychophysiology of cognition. In: R. J. Sternberg (Ed.), *Advances in Research on Intelligence*, Vol. IV. Hillsdale: Erlbaum.
Eysenck, H. J. (1985b) The place of theory in a world of facts. *Annals of Theoretical Psychology*, **3**, 17–114.
Eysenck, H. J. (1986a) The theory of intelligence and the psychophysiology of cognition. In: R. J. Sternberg (Ed.), *Advances in the Psychology of Human Intelligence*. Vol. **3**. Hillside, N. J.: Lawrence Erlbaum.
Eysenck, H. J. (1986b) Intelligence: The new look. *Psychologische Beitrage*, **28**, 332–65.
Eysenck, H. J. (1987a) The definition of personality disorders and the criteria appropriate for their description. *Journal of Personality Disorders*, **1**, 211–19.

Eysenck, H. J. (1987b) Speed of information processing, reaction time, and the theory of intelligence. In: P. A. Vernon (Ed.), *Speed of Information-Processing and Intelligence*, pp. 21–67. Norwood, N. J.: Ablex.

Eysenck, H. J. (1988) The concept of 'Intelligence': Useful or Useless? (Guest Editorial). *Intelligence*, **12**, 1–16.

Eysenck, H. J. (1989a) Die Bewertung der Kreativitat mit Hilfe des Psychotizismus-Wertes. In: R. Lindner (Ed.), *Einfallsreiche Vernunft-Kreativ durch Wissen oder Gefuhl?* Zurich: Edition Interfrom.

Eysenck, H. J. (1989b) La regression a la moyenne et la genetique de l'intelligence. *La Revue Tires a Part*, **10**, 28–35.

Eysenck, H. J. (1990a) *Rebel With a Cause*. London: W. H. Allen.

Eysenck, H. J. (1990b) *Decline and Fall of the Freudian Empire*. London: Viking Press.

Eysenck, H. J. (1991a) *Smoking, Personality and Stress: Psychosocial Factors in the Prevention of Cancer and Coronary Heart Disease*. New York: Springer Verlag.

Eysenck, H. J. (1991b) The biological basis of intelligence. In: P. A. Vernon (Ed.), *Biological Approaches to the Study of Human Intelligence*. Norwood, N. J.: Ablex.

Eysenck, H. J. (1991c) Dimensions of personality: 16, 5 or 3? – Criteria for a taxonomic paradigm. *Personality and Individual Differences*, **12**, 8, 773–90.

Eysenck, H. J. (1992a) The definition and measurement of psychoticism. *Personality and Individual Differences*, **13**, 757–85.

Eysenck, H. J. (1992b) *Rebel With a Cause*. New York: Viking Press.

Eysenck, H. J. (1992c) Meta-analysis: Sense or nonsense? *Pharmaceutical Medicine*, **6**, 113–19.

Eysenck, H. J. (1993a) Creativity and Personality: Suggestion for a theory. *Psychological Inquiry*, **4**, 147–78.

Eysenck, H. J. (1993b) Creativity and personality: Word Association, Origence and Psychoticism. *Creativity Research Journal*, **7**, 209–16.

Eysenck, H. J. & Barrett, P. (1985) Psychophysiology and the measurement of intelligence. In: C. R. Reynolds & V. Wilson (Eds.), *Methodological and Statistical Advances in the Study of Individual Differences*. New York: Plenum Press.

Eysenck, H. J. & Barrett, P. (1993a) The nature of schizotypy. *Psychological Reports*, **73**, 59–63.

Eysenck, H. J. & Barrett, P. (1993b) Brain research related to giftedness. In: K. A. Heller, F. J. Monks and A. H. Passow (Eds.), *The International Handbook for Research on Giftedness and Talent*. London: Pergamon Press.

Eysenck, H. J. & Castle, M. (1970) A factor-analytic study of the Barron–Welsh Art Scale. *Psychological Research*, **20**, 523–5.

Eysenck, H. J. & Eysenck, M. W. (1985) *Personality and Individual Differences*. New York: Plenum Press.

Eysenck, H. J. & Eysenck, S. B. G. (1975) *Manual of the Eysenck Personality Questionnaire*. London: Hodder & Stoughton. San Diego: EDITS.

Eysenck, H. J. & Eysenck, S. B. G. (1976) *Psychoticism as a Dimension of Personality*. London: Hodder & Stoughton.

Eysenck, H. J. & Eysenck S. B. G. (1978) Psychopathy, personality and genetics. In: R. Hare and D. Schalling (Eds.), *Psychopathic Behavior*, pp. 197–223. London: Wiley.

Eysenck, H. J. & Eysenck, S. B. G. (1992a) *The Manual of the EPQ-R and the Impulsiveness, Venturesomeness and Empathy Scales*. London: Hodder & Stoughton.

Eysenck, H. J. & Eysenck, S. B. G. (1992b) Peer review: Advice to referees and contributors. *Personality and Individual Differences*, **13**, 393–9.

Eysenck, H. J. & Frith, C. D. (1977) *Reminiscence, Motivation and Personality*. New York: Plenum.

Eysenck, H. J. & Furnham, A. (1993) Personality and the Barron–Welsh Art Scale. *Psychological Reports*, **76**, 837–8.

Eysenck, H. J. & Gudjonsson, G. (1989) *The Causes and Cures of Criminality*. New York: Plenum Press.

Eysenck, H. J. & Nias, D. K. B. (1982) *Astrology – Science or Superstition*. London: Maurice Temple Smith.

Eysenck, H. J., Granger, G. W. & Brengelman, J. C. (1957) *Perceptual Process of Mental Illness*. London: Chapman & Hall.

Eysenck, H. J., Wakefield, J. & Friedman, A. (1983) Diagnosis and Clinical Assessment: The DSM-III. *Annual Review of Psychology*, **37**, 167–93.

Eysenck, H. J. & Wilson, G. D. (1973) *The Experimental Study of Freudian Theories*. London: Methuen.

Eysenck, H. J. & Wilson, G. (1979) *The Psychology of Sex*. London: Dent.

Eysenck, S. B. G., Eysenck, H. J. & Barrett, P. (1985) A revised version of the psychoticism scale. *Personality and Individual Differences*, **6**, 21–9.

Falcone, D. & Loder, M. (1984) A modified lateral eye movement measure, the right hemisphere and creativity. *Perceptual and Motor Skills*, **58**, 823–30.

Farmer, A. E., McGuffin, P. & Bebbington, P. (1988) The phenomena of schizophrenia. In: P. Bebbington and P. McGuffin (Eds.), *Schizophrenia: The Major Issues*, pp. 36–50. London: Heinemann.

Farmer, A. E., McGuffin, P., Harvey, I. & Williams, M. (1991) Schizophrenia: How far can we go in defining the phenotype? In: P. McGuffin & R. Murray (Eds.), *The New Genetics of Mental Illness*. London: Butterworth-Heinemann.

Farmer, C. (1961) An experimental approach to the measurement of intuition. *Unpublished MS, Teacher's College, Columbia. Quoted by Westcott, 1968.*

Farmer, E. W. (1974) Psychoticism and person-orientation as general personality characteristics of importance for different aspects of creative thinking. *Glasgow: Unpublished B.Sc. thesis. Quoted in Eysenck and Eysenck, 1976.*

Farmer, E. W., McGuffin, P. & Gottesman, I. (1987) Twin corcordance for DSM-III schizophrenia: scrutinizing the validity of the definition. *Archives of General Psychiatry*, **44**, 634–41.

Farnsworth, P. R. (1969) *The Social Psychology of Music*. 2nd Edn, p. 228. Amer: Iowa State University Press.

Feather, N. (1959) *Mass, Length and Time*. Edinburgh: University Press.

Fechner, G. T. (1897) *Vorschule de Aesthetik*. Leipzig: Breitkopf und Haertel.

Fisher, R. B. (1977) *Joseph Lister*. London: McDonal and Lane.

Flanders, T. (1993) Major meeting on new cosmologies. *Journal of Scientific Exploration*, **7**, 19–22.

Flashar, H. (1966) *Melancholie and Melancholiker*. Berlin: Springer.

Fletcher, R. (1991) *Science, Ideology, and the Media*. London: Transaction Publishers.

Flor-Henry, P. (1969) Psychoses and temporal lobe epilepsy: A controlled investigation. *Epilepsia*, **10**, 363–395.

Flor-Henry, P. (1974) Psychoses, neuroses and epilepsy. *British Journal of Psychiatry*, **124**, 144–50.

Flor-Henry, P. (1983) *Cerebral Basis of Psychopathology*. Boston: J. Wright.

Flor-Henry, P. & Gruzelier, J. (Eds.) (1983) *Laterality and Psychopathology*. New York: Elsevier.

Flynn, J. R. (1987) Massive IQ gains in 14 nations: What IQ tests really measure. *Psychological Bulletin*, **101**, 171–91.

Folgmann, E. (1933) An experimental study of composer-preference of four outstanding symphony orchestras. *Journal of Experimental Psychology*, **16**, 709–24.

Frank, P. G. (1957) (Ed.) *The Validation of Scientific Theories*. Boston: Beacon Press.

Frearson, W., Eysenck, H. J. & Barrett, P. T. (1990) The Furneaux model of human problem-solving: Its relationship to reaction time and intelligence. *Personality and Individual*

Differences, **11**, 239–57.

Frederickson, N. (1984) Implications of cognitive theory for instruction in problem solving. *Review of Educational Research*, **54**, 363–467.

Frenkel-Brunswick, E. (1949) Intolerance of ambiguity – as an emotional and perceptual personality variable. *Journal of Personality*, **18**, 108–43.

Freud, S. (1925) *An Autobiographical Study*. (1973 Edition). London: Hogarth Press.

Freud, S. (1938/1900) *The Interpretation of Dreams*. New York: Modern Library.

Frey, P. W. & Sears, R. J. (1978) Model of conditioning incorporating the Rescorla-Wagner associative axiom, a dynamic attention process, and a catastrophe rule. *Psychological Review*, **85**, 321–40.

Frith, C. D. (1979) Consciousness, information processing and schizophrenia. *British Journal of Psychiatry*, **134**, 225–35.

Frith, C. D. & Done, D. J. (1988) Towards a neurospychology of schizophrenia. *British Journal of Psychiatry*, **153**, 437–43.

Fromm, E. (1955) *The Sane Society*. New York: Rinehart.

Fromm, E. (1978) Primary and secondary process in waking and in altered states of consciousness. *Journal of Altered States of Consciousness*, **4**, 115–28.

Frosch, W. A. (1987) Moods, madness and music. I. Major affective disease and musical creativity. *Comprehensive Psychiatry*, **28**, 315–22.

Fruchter, B., (1948) The nature of verbal tests. *Educational and Psychological Measurement* , **8**, 33–47.

Fudin, R. (1986) Subliminal psychodynamic activation: Mammy and I are not yet one. *Perceptual and Motor Skills*, **63**, 1159–79.

Furneaux, W. D. (1960) Intellectual abilities and problem-solving behaviour. In: H. J. Eysenck (Ed.), *Handbook of Abnormal Psychology*, pp. 167–92. London: Pitman.

Furnham, A. (1992) Personality and learning style: A study of three instruments. *Personality and Individual Differences*, **13**, 429–38.

Furnham, A. & Yazdanpanaki, T. (1994) Personality differences and group persons individual brainstorming. *Personality and Individual Differences* (in press).

Galton, F. (1869, republished in 1978) *Hereditary Genius*. New York: Julian Friedmann.

Galton, FG. (1877) Typical laws of heredity. *Nature*, **15**, 492–5, 512–14, 532–3.

Gardner, H. (1982) *Art, Mind, and Brain*. New York: Basic Books.

Gardner, H. (1983) *Frames of Mind: The Theory of Multiple Intelligences*. New York: Basic Books.

Garnett, J. C. (1918) General ability, cleverness and purpose. *British Journal of Psychology*, **7**, 345–66.

Garskof, H. & Houston, J. P. (1963) Measurement of verbal relatedness: An ideographic approach. *Psychological Review*, **70**, 277–88.

Gattaz, W. F. (1981) H2A-B27 as a possible genetic marker of psychoticism. *Personality and Individual Differences* , **2**, 57–60.

Gattez, W. F., Ewald, R. W. & Beckman, W. (1980) The HLA system of schizophrenia. *Archiv fuer Psychiatrie und Nervenkrankheiten*, **278**, 205–11.

Gattaz, W. F., Seitz, M. & Beckmann, H. (1985) A possible association between H2A-B27 and vulnerability to schizophrenia. *Personality and Individual Differences* , **6**, 283–5.

Geary, R. C. (1947) Testing for normality. *Biometrika*, **43**, 1070–100.

Gedo, J. E. (1972) On the psychology of genius. *International Journal of Psychoanalysis*, **53**, 199–203.

Gershon, E. S. & Rieder, R. O. (1980) Are mania and schizophrenia genetically distinct? In: R. H. Belmaker & H. M. van Praeg (Eds.), *Mania: An evolving concept*. New York: Spectrum.

Gershon, E. S., Burney, W. E. & Leckman, J. F. (1976) The inheritance of affective disorders: a review of data and hypotheses. *Behavior Genetics*, **6**, 276–81.

Gerhson, E., Hamovit, J. H., Guroff, J. J. & Nurnberger, J. I. (1987) Birth cohort changes in manic and depressive disorders in relatives of bipolar and schizoaffective patients. *Archives of General Psychiatry*, **44**, 314–19.

Gershon, E. S., Hamovit, J., Guroff, J. J., Dibble, E., Leckmann, J. F., Sceery, W., Tarquin, S., Nurnberger, J. I., Goldin, L. R. & Bunney, W. E. (1982) A family study of schizo-affective, bipolar I, bipolar II, unipolar and normal control patients. *Archives of General Psychiatry*, **39**, 1157–67.

Getzels, J. W. & Csikzentmihalyi, M. (1976) *The Crative Vision: A Longitudinal Study of Problem Finding in Art*. New York: Wiley.

Getzels, J. W. & Jackson, P. W. (1962) *Creativity and Intelligence*. New York: Wiley.

Getzels, J. W. & Jackson, P. W. (1963) The highly intelligent and the highly creative adolescent: A summary of some research findings. In: C. W. Taylor and F. Barron (Eds.), *Scientific Creativity: Its Recognition and Development*, pp. 161–72. New York: Wiley.

Gewirtz, J. L. (1948) Studies in word fluency. I: Its relation to vocabulary and mental age in young children. *Journal of Genetic Psychology*, **72**, 165–76.

Ghiselin, B. (1952) *The Creative Process*. New York: Mentor.

Ghiselin, B. (Ed.) (1954) *The Creative Process: A Symposium*. Berkeley: University of California Press.

Gibbs, F. W. (1965) *Joseph Priestley*. London: Nelson.

Gide, A. (1932) *Oeuvres Complites*, Vol. **1**, p. 518. Paris.

Gidwani, D. G. S. (1971) The effect of previously learned habits and personality variables on verbal conditioning. University of London: *Unpublished Ph.D. thesis*.

Gilfillan, S. (1935) *The Sociology of Neuroticism*. Chicago: Follet.

Gillespie, C. C. (1960) *The Edge of Objectivity*. Princeton, N. J.: Princeton University Press.

Gleick, J. (1987) *Chaos*. London: Cardinal.

Gleick, J. (1992) *Genius: Richard Feynman and Modern Physics*. New York: Little, Brown & Co.

Gleitman, H., Nachmius, J. & Neisser, W. (1954) The S-R reinforcement theory of extinction. *Psychological Review*, **61**, 23–33.

Glover, J. A. (1979) Levels of questions asked in interview and reading sessions by creative and relatively non-creative college students. *Journal of Genetic Psychology*, **135**, 103–8.

Glover, J. A., Ronning, R. R. & Reynolds, C. R. (Eds.) (1989) *Handbook of Creativity*. New York: Plenum Press.

Goertzel, M., Goertzel, V. & Goertzel, T. (1978) *Three Hundred Eminent Personalities*. San Francisco: Jossey-Bass.

Goertzel, V. & Goertzel, M. G. (1962) *Cradles of Eminence*. London: Constable.

Goldberg, S. C., Schooler, V. R. & Mattson, N. (1968) Paranoid and withdrawal symptoms in schizophrenia: relationship to reaction time. *British Journal of Psychiatry*, **114**, 1161–5.

Goldstein, I. B. (1965) The relationship of muscle tension and autonomy activity to psychiatric disorders. *Psychosomatic Medicine*, **27**, 39–52.

Goodwin, F. & Jamison, K. R. (1990) *Manic-depressive Illness*. New York: Oxford University Press.

Gossop, M. R. & Eysenck, S. B. G. (1980) A further investigation into the personality of drug addicts in treatment. *British Journal of Addiction*, **75**, 305–11.

Goswami, A. (1990) Consciousness in quantum physics and the mind–body problem. *Journal of Mind and Behaviour*, **11**, 75–96.

Goswami, A. (1993) *The Self-Aware Universe: How Consciousness Creates the Material World*. New York: G. P. Putnam.

Gottesman, I. I. (1987) The psychotic hinterlands or the fringes of lunacy. *British Medical Bulletin* , **43**, 557–69.

Gottesman, I. I. & Bertelsen, A. (1991) Schizophrenia: Classical approaches with new twists and provocative results. In: P. McGuffin and R. Murray (Eds.), *The New Genetics of Mental Illness*, pp. 85–97. London: Butteworth-Heinemann.

Gottesman, I. I. & Shields, J. (1972) A polygenic theory of schizophrenia. *International Journal of Mental Health*, **1**, 107–15.

Gottesman, I. I. & Shields, J. (1972) *Schizophrenia and Genetics*. New York: Academic Press.

Götz, K. O. & Götz, K. (1973) Introversion–extraversion and neuroticism in gifted and ungifted art students. *Perceptual and Motor Skills*, **30**, 675–8.

Götz, K. O. & Götz, K. (1979a) Personality characteristics of successful artists. *Perceptual and Motor Skills*, **49**, 919–24.

Götz, K. O. & Götz, K. (1979b) Personality characteristics of professional artists. *Perceptual and Motor Skills*, **49**, 327–34.

Gouchie, C. & Kimura, D. (1990) The relationship between t levels and cognitive ability patterns. *Psychoneuroendocrinology*, **16**, 1–30.

Gough, H. G. (1957) *Manual for the Californian Psychological Inventory*. Palo Alto: Consulting Psychologists Press.

Gough, H. G. (1976) Studying creativity by means of Word Association tests. *Journal of Applied Psychology*, **61**, 348–53.

Gough, H. G. (1979) A creative personality scale for the Adjective Check List. *Journal of Personality and Social Psychology*, **37**, 1398–405.

Gough, H. G. & Heilbrunn, A. B. (1983) *The Adjective Check List Manual*. Palo Alto: Consulting Psychologists Press.

Gray, C. E. (1966) A measurement of creativity in Western civilization. *American Anthropologist*, **68**, 1384–417.

Gray, J. A. (1970) The psychophysiological basis of introversion–extraversion. *Behaviour Research and Therapy* , **8**, 249–66.

Gray, J. A. (1972) Learning theory, the conceptual nervous system and personality. In: V. D. Nebylitsyn & J. A. Gray (Eds.), *The Biological Basis of Individual Behavior*. New York: Academic Press.

Gray, J. A. (1975) *Elements of a Two-Process Theory of Learning*. New York: Academic Press.

Gray, J. A. (1976) Causal theories of personality. In: J. R. Royce (Ed.), *Multivariate Analysis and Psychological Theories*. New York: Academic Press.

Gray, J. A., Feldon, J., Rawlins, J. P., Hemsley, D. R. & Smith, A. D. (1991) The neuropsychology of schizophrenia. *Behavioral and Brain Sciences* , **14**, 1–84.

Gray, N., Pickering, A. & Gray, J. (1994) Psychoticism and dopamine D2 binding in the basal ganglia using single photon emission tomography. *Personality and Individual Differences* (in press).

Greenberg, N. (1977) The effects of subliminal neutral and aggressive stimuli on the thought processes of schizophrenics. *Canadian Journal of Behavioral Science*, **9**, 187–96.

Greenwald, A. G. (1992) Unconscious cognitions reclaimed. *American Psychologist* , **47**, 766–79.

Gribbin, J. (1992) Astronomers double the age of the universe. *New Scientist*, 4th January, p. 12.

Griesinger, W. (1861) *Die Pathologie und Therapie der Psychischen Krankheiten*. Stuttgart: Krabbe.

Griffin, L. & Eisenman, R. (1972) Auditory complexity and musical ability. *Perceptual and Motor Skills*, **35**, 43–6.

Griffin, D. & Tversky, A. (1992) The weighing of evidence and the determination of confidence. *Cognitive Psychology* , **24**, 411–35.

Griffiths, J. J., Mednick, S., Schulsinger, F. & Diderichsen, B. (1980) Verbal associative disturbance in children with a high risk for schizophrenics. *Journal of Abnormal Psychology*, **89**, 125–61.

Grove, M. S. & Eisenman, R. (1970) Personality correlates of complexity and simplicity. *Perceptual and Motor Skills*, **31**, 387–91.

Gruber, H. E. & Barrett, P. H. (1974) *Darwin on Man*. London: Wildwood House.

Gruzelier, J. H. (1973) Bilateral asymmetry of skin conductance orienting activity and levels in schizophrenia. *Biological Psychology*, **1**, 21–41.
Gruzelier, J. H. (1984) Hemispheric imbalance in schizophrenia. *International Journal of Psychology*, **1**, 227–40.
Gruzelier, J. H. & Liddiard, D. V. (1989) The neuropsychology of schizophrenia in the context of topographical mapping of electrocortical activity. In: K. Maurer (Ed.), *Topographic mapping of EEG and Evoked Potentials*. New York: Springer Verlag.
Gruzelier, J. H. & Venables, P. H. (1974) Bimodality and lateral asymmetry of skin conductance orienting activity in schizophrenics. Replication and evidence of lateral asymmetry in patients with depression and disorders of personality. *Biological Psychiatry*, **8**, 55–73.
Gruzelier, J. H., Seymour, K., Wilson, L., Jolley, A. & Hirsch, S. (1988) Impairments of neuropsychological tests of hippocampal function and word fluency in remitting schizophrenia and affective disorder. *Archives of General Psychiatry*, **45**, 623–9.
Guilford, J. P. (1950) Creativity. *American Psychologist*, **5**, 444–54.
Guilford, J. P. (1967) *The Nature of Human Intelligence*. New York: McGraw-Hill.
Guilford, J. P. (1981) Potentiality for creativity. In: J. C. Gowan, J. Khatena & E. P. Torrance (Eds.), *Creativity: Its Educational Implications* (2nd edition, pp. 1–5). Dubuque: Kendall, Hunt.
Guilford, J. P. & Christensen, P. R. (1975) The one-way relation between creative potential and IQ. *Journal of Creative Behaviour* , **7**, 247–52.
Guiora, A., Bolin, R., Dutton, C. & Meere, B. (1965) Intuition: A preliminary statement. *Psychiatric Quarterly Supplement*, **39**, Part 1, 110–22.
Guislain, J. (1833) *Traite des Phrenopathies*. Bruxelle: Etablissement Encyclographique.
Hadamard, J. (1945) *An Essay on the Psychology of Invention in the Mathematical Field*. Princeton: University Press.
Haecker, V. & Ziehen, T. (1931) Beitrag zur Lehre von der Vererbung und Analyse der zeichnerischen und mathematischen Begabung, insbesondere mit Bezug auf die Korrelation zur musikalischen Begabug. *Zeitschrift fur Psychologie*, **120**, 1–45; **121**, 1–103.
Hagstrom, W. O. (1974) Competition in science. *American Sociological Review*, **39**, 1–18.
Hall, C. S. (1953) *The Meaning of Dreams*. New York: Harper.
Halpern, D. E. (1992) *Sex Differences in Cognitive Abilities*. Hillsdale, N. J.: Laurence Erlbaum.
Hammer, M. & Zubin, J. (1968) Evolution, culture and psychopathology. *Journal of General Psychology*, **78**, 154–75.
Hampson, E. & Kimura, D. (1988) Reciprocal effects of hormonal fluctuation on human motor and perceptuo-spatial skills. *Behavioural Neuroscience*, **102**, 456–9.
Hampson, H. (1990) Estrogen-related variations in human spatial and articulatory-motor skills. *Psychoneuroendocrinology*, **15**, 97–111.
Hankoft, L. D. (1975) The hero as madman. *Journal of the History of the Behavioral Sciences*, **11**, 3215–333.
Happe, K. (1988) Hemispheric specialization and creativity. In: K. Happe, *Hemispheric Specialization*, pp. 303–15. Philadelphia: Saunders.
Happe, K. & Kyle, N. (1990) Dual brain, creativity and health. *Creativity Research Journal*, **3**, 150–7.
Hardy, G. H. (1940) *Ramanujan*. Cambridge: Cambridge University Press.
Hardy, G. H. (1979) *Collected Papers*. Oxford: Clarendon Press.
Hare, E. H. (1978) Variation in the seasonal distribution of births of psychotic patients in England and Wales. *British Journal of Psychiatry*, **132**, 155–8.
Hare, E. H. (1987a) Epidemiology of schizophrenia and affective psychoses. *British Medical Bulletin* , **43**, 514–30.
Hare, E. H. (1987b) Psychopathy and perceptual asymmetry during verbal dichotic listening.

Journal of Abnormal Psychology, **93**, 141–9.
Hare, E. H. & Walter, S. D. (1978) Seasonal variations in admission of psychiatric patients and its relation to seasonal variations in their birth. *Journal of Epidemiology and Community Health*, **37**, 47–52.
Hare, E. H., Price, J. S. & Slater, E. (1974) Mental disorder and season of birth. *British Journal of Psychiatry*, **124**, 81–6.
Hare, R. D. (1980) A research scale for the assessment of psychopathy in criminal populations. *Personality and Individual Differences*, **1**, 111–20.
Hare, R. D. (1985) Comparison of procedures for the assessment of psychopathy. *Journal of Consulting and Clinical Psychology*, **53**, 7–16.
Hare, R. D. & McPherson, L. M. (1984) Psychopathy and perceptual asymmetry during verbal dichotic listening. *Journal of Abnormal Psychology*, **93**, 141–9.
Hargreaves H. L. (1927) The 'faculty' of imagination. *British Journal of Psychology*. Monograph Suppl., **10**, 74.
Hanard, S. (1972) Creativity, lateral saccocles and the non-dominant hemisphere. *Perceptual and Motor Skills* , **34**, 653–4.
Harpaz, I. (1990) Asymmetry of hemispheric functions and creativity: An empirical examination. *Journal of Creative Behavior*, **24**, 161–70.
Hart, M. (1993) *The 100: A Ranking of the Most Influential People in History*. London: Simon & Schuster.
Hasenfus, N. & Magaro, P. (1976) Creativity and schizophrenia: An equality of empirical constructs. *British Journal of Psychiatry*, **129**, 346–9.
Hassler, M., Birbaumer, N. & Nieschlag, E. (1992) Creative musical behavior and sex hormones: Musical talent and spatial ability in the two sexes. *Psychoneuroendocrinology*, **97**, 55–70.
Hattie, J. & Fitzgerald, D. (1983) Do left-handers tend to be more creative? *Journal of Creative Behavior*, **17**, 269.
Hawking, S. (1988) *A Brief History of Time*. London: Bantam Press.
Heansley, P. & Reynolds, C. R. (1989) Creativity and intelligence. In: J. A. Glover, R. R. Ronning & C. R. Reynolds (Eds.), *Handbook of Creativity*, pp. 111–32. New York: Plenum Press.
Hearnshaw, L. S. (1979) *Cyril Burt, Psychologist*. London: Hodder & Stoughton.
Hellige, J. (1993) *Hemispheric Asymmetry*. Cambridge: Cambridge University Press.
Helmholtz, H. (1896) *Vortrage und Reden*. Braunschweig: Vieweg.
Helson, R. (1971) Women mathematicians and the creative personality. *Journal of Consulting and Clinical Psychology*, **36**, 210–11; 217–20.
Helson, R. & Crutchfield, R. S. (1970) Creative types in mathematics. *Journal of Personality*, **38**, 177–97.
Hemsley, D. R. (1975) A two-stage model of attention in schizophrenia. *British Journal of Social and Clinical Psychology*, **14**, 81–9.
Hemsley, D. R. (1976) Attention and information processing in schizophrenia. *British Journal of Social and Clinical Psychology*, **15**, 199–209.
Hemsley, D. R. (1987) An experimental psychological model for schizophrenia. In: R. Hafner, W. F. Gattaz & W. Janzanik (Eds.), *Search for the Causes of Schizophrenia*, pp. 139–88. Heidelberg: Springer Verlag.
Hemsley, D. R. (1991) What have cognitive defects to do with schizophrenia? In: G. Huber (Ed.), *Idiopathische Psychosen*. Stuttgart: Schattover.
Hendron, G. (1989) Using sign language to access right hand communication: A tool for teachers. *Journal of Creative Behaviour*, **23**, 116–20.
Henry, E. R. (1967) *Conference on criteria for research and application of research in management, leadership and creativity*. Greenboro, N. C.: Smith Richardson Foundation.
Heston, I. I. (1966) Psychiatric disorders in foster-home-reared children of schizophrenic mothers. *British Journal of Psychiatry*, **112**, 819–25.

Heston, C. C. (1970) The genetics of schizophrenic and schizoid diseases. *Science*, **167**, 249–56.
Hines, T. (1991) The myth of right hemisphere creativity. *Journal of Creative Behavior* , **25**, 223–7.
Hinton, J. & Craske, B. (1976) A relationship between Eysenck's P scale and change in muscle action potentials with attention to perceptual tasks. *British Journal of Psychology*, **67**, 461–6.
Hollingsworth, L. (1942) *Children above 180 IQ*. New York: World Books Co.
Hollister, L. E., Overall, J. E., Shelton, J., Pennington, V., Kimbell, I. & Johnson, M. (1967) Drug therapy of depression: amitriptyline, phenothenapine and their combination in different syndromes. *Archives of General Psychiatry*, **17**, 486–93.
Holzinger, K. J. (1934) *Preliminary report on Spearman Holzinger Unitary Trait Study, I & II.* Chicago: Statistical Laboratory, Department of Education.
Holzinger, K. J. (1935) *Preliminary Report on Spearman Holzinger Unitary Trait Study, III, IV, V and VI.* Chicago: Statistical Laboratory, Department of Education.
Holzman, P. S., Proctor, L. R. & Hughes, D. W. (1973) Eye tracking patterns in schizophrenia. *Science*, **181**, 179–81.
Holzman, P. S., Proctor, L. R., Levy, D. L., Yasillo, N. J., Metzer, H. Y. & Hart, S. W. (1974) Eye tracking dysfunction in schizophrenic patients and their relatives. *Archives of General Psychiatry*, **31**, 143–51.
Horley, J. M., Orley, J. & Teasdale, J. D. (1986) Levels of expressed emotion and relapse in depressed patients. *British Journal of Psychiatry*, **148**, 642–7.
Horn, J. L. & Cattell, R. B. (1967) Age differences in fluid and crystallized intelligence. *Acta Psychologica*, **26**, 107–29.
Horton, D. L., Marlowe, D. & Crowne, D. P. (1963) The effect of instructional set and need for social approval on commonality of word association responses. *Journal of Abnormal and Social Psychology*, **66**, 67–72.
Houndshell, D. & Smith, J. (1988) *Science and Corporate Strategy*. New York: Cambridge University Press.
Hovecar, D. (1979) Ideational fluency as a confounding factor in the measurement of originality. *Journal of Educational Psychology*, **71**, 191–6.
Hovecar, D. (1980) Intelligence, divergent thinking, and creativity. *Intelligence*, **4**, 25–40.
Hovecar, D. (1981) Measurement of creativity: Review and critique. *Journal of Personality Assessment*, **45**, 450–69.
Hovecar, D. & Bachelor, P. (1989) A taxonomy and critique of measurements used in the study of creativity. In: J. A. Glover, R. R. Ronning and C. R. Reynolds (Eds.), *Handbook of Creativity*, pp. 53–75. New York: Plenum Press.
Howe, M. J. A. (1989) *Fragments of Genius: The Strange Face of Idiots Savants*. London: Routledge.
Howieson, N. (1981) A longitudinal study of creativity: 1965–1975. *Journal of Creative Behavior*, **15**, 117–35.
Howieson, N. (1984) *The prediction of creative achievement from childhood measures: A longitudinal study in Australia, 1960–1983*. Unpublished doctoral dissertation, quoted by Torrance (1988).
Hoyle, F. (1962) *Astronomy*. London: Macdonald.
Hoyle, F. (1975) *Astronomy and Cosmology*. San Francisco: Freeman.
Hull, C. L. (1920) Quantitative aspects of the evolution of concepts. *Psychological Monographs*, **28**, No. 1 (Whole No. 1213).
Hull, C. L. (1943) *Principles of Behavior*. New York: Appleton-Century-Crofts.
Hull, C. L. (1951) *Essentials of Behavior*. New Haven: Yale University Press.
Hull, C. L. (1952) *A Behavior System*. New Haven: Yale University Press.
Hull, C. L. (1970) Quantitative aspects of the evolution of concepts. *Psychological Monographs*, **28**, No. 1 (Whole No. 123).
Hundal, P. S. & Upmanyu, V. V. (1981) Nature of emotional interaction elicited by Kent–Rosanoff Word Association Test: An empirical study. *Personality Study and Group Behavior*,

1, 50–61.
Hunt, W. A., Wittson, C. L. & Hunt, E. B. (1953) Theoretical and practical analysis of the diagnostic process. In: P. M. Hoch and J. Zubin (Eds.), *Current Problems in Psychiatric Diagnosis*. New York: Grune & Stratton.
Hunter, I. M. L. (1977) An exceptional memory. *British Journal of Psychology*, **68**, 155–64.
Hunter, J. E. & Hunter, R. F. (1984) Validity and utility of alternative predictors of job performance. *Psychological Bulletin*, **96**, 72–98.
Huntington, E. (1938) *Season of Birth: The Relation to Human Abilities*. New York: Wiley.
Huxley, J., Mayo, E., Hoffer, H. & Osmond, A. (1964) Schizophrenia as a genetic morphism. *Nature* , **204**, 220–21.
Iacono, W. G. & Lykken, D. T. (1979) Eye tracking and psychopathology. New procedures applied to a sample of normal monozygotic twins. *Archives of General Psychiatry*, **36**, 1361–9.
Iacono, W. G., Peloquin, L. J., Lumry, A., Valentine, B. H. & Tuason, V. B. (1982) Eye tracking in patients with unipolar and bipolar affective disorder in remission. *Journal of Abnormal Psychology*, **91**, 35–44.
Illing, R. (1963) *Pergamon Dictionary of Musicians and Music*. New York: Pergamon Press.
Illingworth, R. S. & Illingworth, C. M. (1969) *Lessons from Childhood: Some Aspects of Early Life of Unusual Men and Women*. Edinburgh: Livingstone.
Iwahara, S. (1957) Hulls concept of inhibition: A revision. *Psychological Reports*, **3**, 9–10.
Jablensky, A. (1988) Epidemiology of schizophrenia. In: P. Bebbington & P. MacGuffin (Eds.), *Schizophrenia: The Major Issues*, pp. 19–35. London: Heinemann.
Jackson, D. (1984) *Multidimensional Aptitude Battery*. Fort Huron: Research Psychologists Press.
Jacob, H. & Beckmann, N. (1986) Prenatal developmental disturbances in the limbic allocortex in schizophrenics. *Journal of Neural Transactions*, **65**, 303–26.
James, W. (1890) *Principles of Psychology*. London: Macmillan.
Jamison, R. (1989) Mood disorders and patterns of creativity in British writers and artists. *Psychiatry*, **52**, 125–34.
Jamison, R. (1993) *Touched with Fame*. New York: The Free Press.
Janos, P. M. (1987) A fifty-year follow-up of Terman's youngest college students and IQ-matched agemates. *Gifted Child Quarterly*, **31**, 55–8.
Jarvik, I. R. & Chadwick, S. B. (1973) Schizophrenia and survival. In: M. Hammer, K. Salzinger and S. Sutton (Eds.), *Psychopathology*, New York: Wiley.
Jenkins, J. (1959) Effects on word-association of the set to give popular responses. *Psychological Reports*, **5**, 94.
Jensen, A. R. & Sirka, S. W. (1993) Physical correlates of human intelligence. In: P. A. Vernon (Ed.), *Biological Approaches to the Study of Human Intelligence*. Norwood, N. J.: Ablex.
Jensen, A. R. & Reynolds, C. R. (1983) Sex differences on the WISC-R. *Personality and Individual Differences*, **4**, 223–6.
Jeste, D. V. & Lohr, J. B. (1989) Hippocampal pathologic findings in schizophrenia. *Archives of General Psychiatry*, **46**, 1019–29.
John, O. P. (1990) The 'Big Five' factor taxonomy: Dimensions of personality in the natural language and in questionnaires. In: L. A. Pervin (Ed.), *Handbook of Personality: Theory and Research*, pp. 66–100. New York: Guilford Press.
Johnson, D. M. (1955) *The Psychology of Thought and Judgment*. New York: Harper & Row.
Johnson, D. M. & Reynolds, F. (1941) A factor analysis of verbal ability. *Psychological Record*, **4**, 183–95.
Jones, E. E. & Nisbett, R. F. (1972) The actor and the observer: Divergent perceptions of the causes of behavior. In: E. E. Jones (Ed.), *Attribution: Perceiving the Causes of Behaviour*. Morristown, N. J.: General Learning Press.
Jones, G. V. (1978) Recognition failure and dual mechanisms in recall. *Psychological Review*, **85**, 464–9.

Jones, H. G. (1958) The status of inhibition in Hull's system: A theoretical revision. *Psychological Review*, **65**, 179–82.
Jones, H. G. (1960) Individual differences in inhibitory potential. *British Journal of Psychology*, **60**, 220–5.
Jones, R. M. (1970) *The New Psychology of Dreaming*. London: Penguin.
Jones, S. H. (1989) The Kamin blocking effect, incidental learning and choice reaction time in acute and chronic schizophrenics. *London: Unpublished Ph.D. thesis*.
Jones, S. H., Gray, J. A. & Hemsley, D. R. (1990) The Kamin Blocking Effect, incidental learning and psychoticism. *British Journal of Psychology*, **81**, 95–110.
Jones, S. H., Gray, J. A. & Hemsley, D. (1992a) The Kamin Blocking Effect, incidental learning and schizotypy (reanalysis). *Personality and Individual Differences*, **13**, 57–60.
Jones, S. H., Gray, J. A. & Hemsley, D. R. (1992b) Loss of the Kamin Blocking Effect in acute but not chronic schizophrenics. *Biological Psychiatry*, **32**, 739–55.
Joseph, M. H., Frith, C. D. & Waddington, J. L. (1979) Dopaminergic mechanisms and cognitive deficit in schizophrenia. A neurobiological model. *Psychopharmacology*, **63**, 273–80.
Joynson, R. B. (1989) *The Burt Affair*. London: Routledge.
Juda, A. (1949) The relationship between highest mental capacity and psychic abnormalities. *American Journal of Psychiatry*, **106**, 296–307.
Judd, L., Hubbard, B., Janowsky, D., Huey, L. & Attewell, P. (1977) The effects of lithium carbonate on affect, mood, and personality in normal subjects. *Archives of General Psychiatry*, **34**, 346–51.
Jung, C. G. (1921) *Psychologische Typen*. Zurich: Rascher & Cie.
Jung, C. (1926) *Psychological Types*. London: Routledge & Kegan Paul.
Jutai, J. W. (1988) Spatial attention in hypothetically psychosis-prone college students. *Psychiatry Research*, **27**, 207–15.
Kac, M. (1985) *Enigmas of Chance*. New York: Harper and Row.
Kaduson, H. G. & Schaefer, C. E. (1991) Concurrent validity of the creative personality scale of the adjective check-list. *Psychological Reports*, **69**, 601–2.
Kallman, F. J. (1938) *The Genetics of Schizophrenia*. New York: Augustin.
Kamin, L. J. (1969) Predictability, surprise, attention and conditioning. In: B. A. Campbell & R. M. Church (Eds.), *Punishment and Aversive Behavior*. New York: Appleton-Century-Crofts.
Kanigel, R. (1991) *The Man who Knew Infinity: A Life of the Genis Ramanujan*. New York: Charles Scribner.
Kant, O. (1942) The incidence of psychoses and other mental abnormality in the families of recovered and deteriorated schizophrenic patients. *Psychiatric Quarterly*, **16**, 176–84.
Karlsson, J. I. (1968) Genealogic studies of schizophrenia. In: D. Rosenthal & S. S. Kety (Eds.), *The Transmission of Schizophrenia*. Oxford: Pergamon Press.
Karlsson, J. L. (1970) Genetic association of giftedness and creativity with schizophrenia. *Hereditas*, **66**, 177–82.
Karlsson, J. (1974) Inheritance of schizophrenia. *Acta Psychiatrica Scandinavia*, Seygel, p. 247.
Kasanin, J. (1933) The acute schizo-affective psychoses. *American Journal of Psychiatry*, **3290**, 97–126.
Kasperson, C. J. (1978) Psychology of the scientists: XXXVII, Scientific creativity: A relationship with information channels. *Psychological Reports*, **42**, 6491–4.
Katz, A. (1980) Do left-handers tend to be more creative? *Journal of Creative Behavior*, **14**, 271.
Katz, A. (1983) Creativity of individual differences in asymmetric hemispheric functioning. *Empirical Studies of the Arts*, **1**, 3–16.
Katz, A. (1986) The relationship between creativity and cerebral hemisphericity for creative architects, scientists, and mathematicians. *Empirical Studies of the Arts*, **4**, 97–108.
Katz, A. N. (1994) Creativity of the cerebral hemispheres. In: M. Runco (Ed.), *Creativity*

Research Handbook. Cresskill, N. J.: Hampton Press.

Kaulins, A. (1979) Cycles in the birth of eminent humans. *Cycles*, **30**, 9–15.

Keating, D. P. (1984) The emperor's new clothes: The 'new look' in intelligence research. In: R. J. Sternberg (Ed.), *Advances in the Psychology of Human Intelligence*, Vol. 2, pp. 1–45. London: Erlbaum.

Keehn, R. J. (1974) Probability of death related to previous army rank. *Lancet*, **2**, 170.

Keele, W. (1965) *William Harvey*. London: Nelson.

Keele, S. W. & Neill, W. T. (1978) Mechanisms of attention. In: E. C. Carberette (Ed.), *Handbook of Perception*, Vol. 9, pp. 3–47. New York: Academic Press.

Kelley, M. P. & Coursey, R. D. (1994). Factor structure of schizotypy scales. In: *Personality and Individual Differences*. Baltimore: Johns Hopkins University.

Kemp, E. (1985) Psychological androgyny in musicians. *Bulletin of the Council for Research in Music Education*, **85**, 102–8.

Kemp, M. (1989) The 'Super-artist' as genius: The sixteenth–century view. In: P. Murray (Ed.), *Genius: The History of an Idea*, pp. 32–53. Oxford: Basil Blackwell.

Kendell, R. E. (1974) The stability of psychiatric diagnoses. *British Journal of Psychiatry*, **124**, 352–6.

Kendell, R. E. (1987) Diagnosis and classification of functional psychoses. *British Medical Bulletin*, **43**, 499–513.

Kendell, R. E. & Brockington, I. F. (1980) The identification of disease entities and the relationship between schizophrenic and affective psychoses. *British Journal of Psychiatry*, **137**, 324–31.

Kendell, R. E. & Gourley, J. (1978) The clinical distribution between the affective psychoses and schizophrenia. *British Journal of Psychiatry*, **117**, 261–6.

Kendler, K. S. & Gruenberg, A. M. (1984) An independent analysis of the Parrish adoption study of schizophrenia. IV. The relationship between psychiatric disorders as defined by DSM-III in the relatives and adoptees. *Archives of General Psychiatry*, **41**, 555–68.

Kendler, K. S. & Hewitt, J. (1992) The structure of self-report schizotypy twins. *Journal of Personality Disorders*, **6**, 1–17.

Kendler, K. S., Gruenberg, A. M. & Strauss, J. S. (1981) An independent analysis of the Copenhagen sample of the Danish adoption study of schizophrenia. *Archives of General Psychiatry*, **38**, 973–87.

Kendler, K. S., Gruenberg, A. M. & Tsuang, M. (1989) Psychiatric illness in the first degree relatives of schizophrenic and surgical control patients: A family study using DSM-III criteria. *Archives of General Psychiatry*, **42**, 770–9.

Kendrick, D. C. (1958a) Inhibition with reinforcement (conditions of inhibition). *Journal of Experimental Psychology*, **56**, 313–18.

Kendrick, D. C. (1958b) $_{S}I_{R}$ and drive level: A reply to Keehn and Sabbagh. *Psychological Reports*, **4**, 646.

Kendrick, D. C. (1959) Inhibition of, or with reinforcement? *Psychological Reports*, **5**, 639–40.

Kendrick, D. C. (1960) III. Effects of drive and effort on inhibition with reinforcement. *British Journal of Psychology*, **51**, 211–19.

Kent, G. H. & Rosanoff, A. J. (1910) A study of association in insanity. *American Journal of Psychiatry*, **67**, 37–96.

Kerwin, R. W., Patel, S., Meldrum, B. S., Czudock, C. & Reynolds, G. P. (1988) Asymmetrical loss of a glutamate receptor subtype in left hyppocampus in post-mortem schizophrenic brain. *Lancet*, (**1**), 583–4.

Kessel, N. (1989) Genius and mental disorder: A history of ideas concerning their conjunction. In: P. Murray (Ed.), *Genius: The History of an Idea*, pp. 196–212. Oxford: Basil Blackwell.

Kety, S. S., Rosenthal, D., Wender, P. H. & Schulsinger, F. (1968) The types and prevalence of mental illness in the biological adoptive families of adopted schizophrenics. In: D. Rosenthal and S. S. Kety (Eds.), *The Transmission of Schizophrenia*, pp. 345–62. London: Pergamon Press.

Khatena, J. & Torrance, E. P. (1973) *Thinking Creativity with Sounds and Words*. Bensenville: Scholastic Testing Service.

Kidd, R. T. & Powell, G. E. (1993) Raised left hemispheric activation in the non-clinical schizotypal personality. *Personality and Individual Differences*, **14**, 723–31.

Kidner, D. W. (1978) Personality and conceptual structure: An integrative model. *London: Unpublished Ph.D. thesis.*

Kihlstrom, J. F. (1987) The cognitive unconscious. *Science*, **237**, 1445–52.

Kihlstrom, J. F. (1990) The psychological unconscious. In: L. Pervin. (Ed.), *Handbook of Personality*, pp. 445–64. New York: Guilford Press.

Kihlstrom, J. F., Barnhardt, T. M. & Tataryn, D. J. (1992) The psychological unconscious: Found, lost, and regained. *American Psychologist*, **47**, 788–91.

Kilminster, C. (1989) Genius in mathematics. In: P. Murray (Ed.), *Genius: The History of an Idea*, pp. 181–95. Oxford: Basil Blackwell.

Kimble, G. (1961) *Hilgard and Marquis? Conditioning and Learning*. New York: Appleton-Century-Crofts.

Kimura, D. (1966) Dual functional asymmetry of the brain in animal perception. *Neuropsychologia*, **4**, 275–85.

Kinsbourne, M. (1974) Direction of gaze and distribution of cerebral thought processes. *Neuropsychologia*, **12**, 279–81.

Kipling, R. (1918) *Something of Myself*. London: Macmillan.

Kirk, S. A. & Kutchins, H. (1992) *The Selling of DSM: The Rhetoric of Science in Psychiatry*. New York: Aldine & Gruyter.

Kirsch, B. (1965) *Scales and Weights: A Historical Outline*. New Haven: Yale University Press.

Kitagava, E. M. & Hauser, P. M. (1973) *Differential Mortality in the United States: a Study in Socioeconomic Epidemiology*. Cambridge, M. A.: Harvard University Press.

Kitterle, F. L. (Ed.) (1991) *Cerebral Laterality: Theory and Research. The Toledo Symposium*. Hillsdale, N. J.: Erlbaum.

Kleinmutz, B. (1990) Why we still use our heads instead of formulas: Towards an integrative approach. *Psychological Bulletin*, **107**, 296–310.

Klerman, G. L. & Cole, J. O. (1965) Clinical pharmacology of imipramine and related antidepressant compounds. *Pharmacological Review*, **17**, 101–41.

Klibarsky, R., Panofsky, E. & Saxl, F. (1964) *Saturn and Melancholy*. London: Methuen.

Kliegl, R., Smith, J. & Baltes, P. B. (1986) Testing the limits, expertise and memory in adulthood and old age. In: F. Klix and H. Hagendorf (Eds.), *Proceedings of Symposium in Memoriam Hermann Ebbinghaus*. Amsterdam: North Holland.

Kliegl, R., Smith, J. & Baltes, P. B. (1989) Testing – the limits and the study of adult age differences in cognitive plasticity of a mnemonic skill. *Developmental Psychology*, **25**, 247–56.

Kliegl, R., Smith, J., Heckhausen, J. & Baltes, P. (1987) Mnemonic training for the acquisition of skilled digit memory. *Cognition and Instruction*, **4**, 203–23.

Kline, P. (1981) *Fact and Fantasy in Freudian Theory*. London: Methuen.

Klinteberg, B., Schalling, D., Edman, G., Oreland, L. & Aesberg, M. (1987) Personality correlates of platelet monoamine oxidase (MAO) activity in female and male subjects. *Neuropsychobiology*, **18**, 89–96.

Klugman, S. F. (1950) Speed of vocational interest and general adjustment. *Journal of Counselling Psychology*, **34**, 108–14.

Koch, S. (1954) Clark L. Hull. In: W. K. Estes (Ed.), *Modern Learning Theory*. New York: Appleton-Century-Crofts.

Koeher, A. L. (1944) *Configurations of Cultural Growth*. Berkeley: University of California Press.

Koenigsberger, L. (1965) *Hermann von Helmhultz*. New York: Dover.

Koestler, A. (1964) *The Art of Creation*. New York: Macmillan.

Kolb, D. (1976) *Learning Style Inventory: Technical Manual*. Boston, MA: Metzer.

Kolb, D. (1984) *Experimental Learning*. Englewood Cliffs, N. J.: Prentice Hall.

Koob, G. F. & Bloom, F. E. (1988) Cellular and molecular mechanisms of drug dependence. *Science*, **242**, 115–23.

Korn, J. H., Davis, R. & Davis, S. F. (1991) Historians' and chairpersons' judgments of eminence among psychologists. *American Psychologists*, **46**, 789–92.

Kraepelin, E. (1897) *Psychiatrie* (6th Edition). Leipzig: Barth.

Kraepelin, E. (1920) Die Goscheiruregsformen des Irreseins. *Zeitschrift fur die gesamte Neurologie und Psychiatrie*, **62**, 1–29.

Kramer, M. (1961) Some problems for international research suggested by observation on different first admission rates to the mental hospitals of England and Wales and of the United States. *Proceedings of the Third World Congress of Psychiatry*, Vol. 3, pp. 153–60. Montreal.

Krantz, D. H., Atkinson, R. C., Luce, R. D. & Suppes, P. (Eds.) (1974) *Contemporary Developments in Mathematical Psychology*, (Vol. 1). San Francisco: Freeman.

Krapf, D. & Muller-Oerlinghaussen, B. (1979) Changes in learning, memory, and mood during lithium treatment: Approach in a research strategy. *Acta Psychiatric Scandinavica*, **59**, 97–124.

Krause, W. (1932) Experimentelle Untersuchungen uber die Vererbung der zeichnerischen Begabung. *Zeitschrift fur Psychologie*, **126**, 86–145.

Krechevsky, P. (1932) 'Hypotheses' in rats. *Psychological Review*, **39**, 516–32.

Kreitman, N. (1961) The reliability of psychiatric diagnosis. *Journal of Mental Science*, **107**, 876–87.

Kreitman, N., Sainsbury, P., Morrissey, J., Towers, J. & Scrivener, J. (1961) The reliability of psychiatric assessment: An analysis. *Journal of Mental Science*, **107**, 887–908.

Kretschmer, E. (1931) *The Psychology of Men of Genius*. London: Kegal Paul.

Kretschmer, E. (1946) *Medizinische Psychologie*. Leipzig: Thieme.

Kretschmer, E. (1948) *Koperbau und Charakter*. Berlin: Springer.

Kretschmer, E. & Enke, W. (1936) *Die Personlichkeit der Athletiker*. Leipzig: Thieme.

Krige, J. (1980) *Science, Revolution and Discontinuity*. Sussex: Harvester Press.

Kris, E. (1952) *Psychoanalytic Explorations in Art*. New York: International Universities Press.

Kroeber, A. (1917) The superorganic. *American Anthropologist*, **19**, 163–214.

Kroeber, A. (1944) *Configurations in Culture Growths*. Berkeley: University of California Press.

Kubie, L. S. (1954) *Neurotic Distortion of the Creative Process*. Kansas: Kansas University Press.

Kuhn, T. S. (1957) *The Capernium Revolution*. Cambridge, Mass., Harvard University Press.

Kuhn, T. S. (1976) *The Structure of Scientific Revolutions*. Chicago: University of Chicago Press.

Laffel, J. & Feldman, S. (1962) The structure of single word and continuous associations. *Journal of Verbal Learning and Verbal Behavior*, **1**, 54–61.

Lakatos, I. (1970) Falsification and the methodology of scientific research programmes. In: I. Lakatos and A. Musgrave (Eds.), *Criticism and the Growth of Knowledge*. Cambridge: University Press.

Lakatos, I. (1978) *Philosophical Papers. Vol. 1: The Methodology of Scientific Research Programmes*, J. Worrall & G. Currie (Eds.). Cambridge: Cambridge University Press.

Lakatos, I. & Musgrave, A. (Eds.) (1970) *Criticism and the Growth of Knowledge*. Cambridge: University Press.

Lamartine. *Quoted in W. Lange-Eichbaum*.

Laming, D. (1973) *Mathematical Psychology*. New York: Academic Press.

Lange-Eichbaum, W. (1956) *Genie, Irrsein und Ruhm*. Munich: Ernst Reinhardt.

Langley, P., Simon, H. D. & Zytkow, J. (1987) *Scientific Discovery*. Cambridge, Mass.: MIT Press.

Lashley, K. S. (1929) *Brain Mechanisms and Intelligence: A Quantitative Study of Injuries to the Brain*. Chicago: University of Chicago Press.

Launay, G. & Slade, P. (1981) The measurement of hallucinatory predisposition in male and female prisoners. *Personality and Individual Differences*, **2**, 221–34.

Lave, J., Murtaugh, M. & Roche, D. (1984) The Dialectic of Arithmetic in grocery shopping. In: B. Roguff & J. Lave (Eds.), *Everyday Cognition: Its Development in Social Context*. Cambridge, M. A.: Harvard University Press.

Lawrence, D. H. (1949) Acquired distinctiveness of cues: I. Transfer between discrimination on the basis of familiarity with the stimulus. *Journal of Experimental Psychology*, **39**, 770–84.

Lawley, D. N. (1949) Acquired distinctiveness of cues: I. Transfer between discrimination on the basis of familiarity with the stimulus. *Journal of Experimental Psychology*, **39**, 770–84.

Lehman, H. C. (1953) *Age and Achievement*. Princeton: University Press.

Lehman, H. C. (1966) The psychologist's most creative years. *American Psychologist*, **21**, 363–9.

Lehman, H. C. & Witty, P. A. (1931) Scientific eminence and church membership. *Scientific Monthly*, **33**, 544–9.

Levey, A. B. & Martin, E. (1990) Evaluative conditioning: Overview and further options. *Cognition and Emotion*, **4**, 31–7.

Levey, A. B. & Martin, I. (1975) Classical conditioning of human 'evaluative' responses. *Behaviour Research and Therapy*, **13**, 221–6.

Levy, L. (1963) *Psychological Interpretation*. New York: Holt, Rinehart & Winston.

Ley, P. (1972) The reliability of psychiatric diagnosis: Some new thoughts. *British Journal of Psychiatry*, **121**, 41–3.

Lidberg, L., Modin, I., Oreland, L., Tuck, J. R. & Gillner, A. (1985) Platelet monoamine oxidase activity and psychopathy. *Psychiatry Research*, **16**, 339–43.

Lindgren, H. C. & Lindgren, F. (1965) Brainstorming and orneriness as facilitators of creativity. *Psychological Reports*, **16**, 572–83.

Lipton, R. B., Levy, D. L., Holtzman, P. S. & Levin, S. (1983) Eye movement dysfunction in psychiatric patients: a review. *Schizophrenic Bulletin*, **9**, 123–32.

Lisman, S. A. & Cohen, B. D. (1972) Self-editing deficits in schizophrenia: A word-association analogue. *Journal of Abnormal Psychology*, **79**, 181–8.

Litwack, T. R., Wiederman, C. F. & Yager, X. (1979) The fear of object loss, responsiveness to subliminal stimuli, and schizophrenic psychopathology. *Journal of Nervous and Mental Disease*, **167**, 79–90.

Loftus, E. F. & Klinger, M. R. (1992) Is the unconscious smart or dumb? *American Psychologist*, **47**, 761–5.

Lombroso, C. (1891) *The Man of Genius*. London: Walter Scott.

Lombroso, C. (1901) *The Man of Genius* (6th Edition). New York: Charles Scribner's Sons.

Lorr, M., Klett, C. J. & McNair, D. M. (1963) *Syndromes of Psychosis*. Pergamon Press.

Lorr, M., Klett, C. J., McNair, D. M. & Lasky, J. J. (1963) *Inpatient Multidimensional Psychiatric Scale Manual*. Palo Alto: Consulting Psychology Press.

Lotka, A. J. (1926) The frequency distribution of scientific productivity. *Journal of the Washington Academy of Sciences*, **16**, 317–23.

Lovibond, S. N. (1954) The Object Sorting Test and conceptual thinking in schizophrenia. *Australian Journal of Psychology*, **6**, 52–70.

Lubinski, D., Benbow, C. & Sanders, C. (1993) Reconceptualizing gender differences in achievement among the gifted: An outcome of contrasting attributes for personal fulfillment in the world of work. In: F. Hellen, and A. Passow (Eds.), *International Handbook for Research on Giftedness and Talent*. Oxford: Pergamon Press.

Lubinski, D. & Humphreys, L. (1990) A broadly based analysis of mathematical giftedness. *Intelligence*, **14**, 327–55.

Lubow, R. E. (1989) *Latent Inhibition and Conditions of Attention Theory*. New York: Cambridge University Press.

Lubow, R. E., Ingberg-Sacks, V., Zalstein-Orda, N. & Gewirtz, J. (1992) Latent inhibition in law and high 'psychotic-prone' normal subjects. *Personality and Individual Differences*, **13**, 563–72.

Lubow, R. E., Weiner, I., Schlossberg, A. & Baruch, I. (1987) Latent inhibition and schizophrenia. *Bulletin of the Psychonomic Society*, **25**, 464–7.

Ludwig, A. M. (1992) Creative achievement and psychopathology: Comparisons among professions. *American Journal of Psychotherapy*, **46**, 330–56.

Luria, A. R. (1975) *The Mind of a Mnemonist*. Harmondsworth: Penguin.

Lykken, D. T. (1982) Research with twins: The concept of emergenesis. *Psychophysiology*, **19**, 361–73.

Lykken, D. T., McGue, M., Tellegen, A. & Bouchard, T. J. (1992) Emergenesis. *American Psychologist*, **47**, 1565–77.

Lynn, J. (1968) Chronological age, professional age, and eminence in psychology. *American Psychologist*, **23**, 371–9.

Lynn, R. (1994) Sex differences in intelligence and brain size: a paradox resolved. *Personality and Individual Differences*, **17**, 257–72.

McCarthy, K. A. (1993) Indeterminacy and consciousness in the creative process: What quantum physics has to offer. *Creativity Research Journal*, **6**, 201–19.

McClelland, D. C. (1963) The calculated risk: An aspect of scientific performance. In: C. W. Taylor and F. Barron (Eds.), *Scientific Creativity: Its Recognition and Development*, pp. 184–92. New York: Wiley.

McClelland, D. C. (1973) Testing for competence rather than for 'intelligence'. *American Psychologist*, **28**, 1–14.

McConaghy, N. & Clancy, M. (1968) Familial relationships of allusive thinking in university students and their parents. *British Journal of Psychiatry*, **114**, 1079–87.

McCrae, R. R. (1987) Creativity, divergent thinking, and openness to experience. *Journal of Personality and Social Psychology*, **52**, 1258–65.

McCurdy, H. B. (1957) The childhood patterns of genius. *Journal of the Elista Mitchell Science Society*, **73**, 448–62. *Quoted by Ochse (1990)*.

McGuffin, P. (1979) Is schizophrenia an HLA-associated disease? *Psychological Medicine*, **9**, 721–8.

McGuffin, P. & Murray, R. (1991) *The New Genetics of Mental Illness*. London: Butterworth-Heinemann.

McGuffin, P., Murray, R. M. & Reveley, A. M. (1987) Genetic influence on the psychoses. *British Medical Bulletin*, **43**, 531–56.

McGuffin, P., Reveley, A. M. & Holland, A. (1982) Identical triplets: Non-identical psychosis? *British Journal of Psychiatry*, **140**, 1–6.

McKellar, P. (1957) *Imagination and Thinking*. New York: Basic Books.

McKenna, P. J. (1987) Pathology, phenomenology and the dopamine hypothesis of schizophrenia. *British Journal of Psychiatry*, **151**, 288–301.

McNeil, T. F. (1971) Prebirth and postbirth influence on the relationship between creative ability and recorded mental illness. *Journal of Personality*, **39**, 391–406.

McPherson, F. M., Presly, A. S., Armstrong, J. & Curtis, R. H. (1974) 'Psychoticism' and psychotic illness. *British Journal of Psychiatry*, **125**, 152–60.

MacKinnon, D. W. (Ed.) (1961) *The Creative Person*. Berkeley: University of California Press.

MacKinnon, D. W. (1962a) The nature and nurture of creative talent. *American Psychologist*, **17**, 484–95.

MacKinnon, D. W. (1962b) The personality correlates of creativity. In: G. S. Nielsen (Ed.), *Proceedings of the Fourteenth International Congress of Applied Psychology*, Vol. 2, pp. 11–39. Copenhagen: Munksgaard.

MacKinnon, D. W. (1965) Personality and the realization of reactive potential. *American Psychologist*, **20**, 273–81.

MacKinnon, D. W. (1978) *In Search of Human Effectiveness*. New York: Creative Education Foundation.

Mackintosh, N. J. (1975) A theory of attention: Variation in the associability of stimuli with reinforcement. *Psychological Revue*, **82**, 276–98.

Mackintosh, N. J. (1983) *Conditioning and Associative Learning*. Oxford: University Press.
Mackintosh, N. J. (1985) *Conditioning and Associative Learning*. Oxford: Clarendon Press.
Maddi, S. R. & Andrews, S. (1966) The need for variety in fantasy and self-description. *Journal of Psychiatry*, **34**, 610–25.
Magnusson, D. & Backteman, G. (1977) Longitudinal stability of person characteristics: Intelligence and creativity. *Reports from the Department of Psychology, The University of Stockholm, No. 511.*
Malmo, R., Shagass, L. & Davis, J. L. (1951) Electromyographic studies of muscular tension in psychiatric patients under stress. *Journal of Clinical and Experimental Psychopathology*, **12**, 45–66.
Mandler, G. (1975a) Consciousness: Respectable, useful and probably necessary. In: R. Salso (Ed.), *Information Processing and Cognition: The Loyola Symposium*. Hillsdale, N. J.: Erlbaum.
Mandler, G. (1975b) *Mind and Emotion*. New York: Wiley.
Mandler, G. (1985) *Cognitive Psychology: An Essay on Cognitive Science*. Hillsdale, N. J.: Erlbaum.
Manhattan Institute (1991) *Health, Lifestyle and Environment*. New York: Manhattan Institute.
Mannhaupt, H. R. (1983) Produktions normen fuer verbal Reaktionen zu 40 gelaufigen Kategorien. *Sprache Kognition*, **2**, 264–78.
Mannuzza, S., Spring, B., Gottlieb, M. D. & Kietzman, M. L. (1980) Visual detection threshold differences between psychiatric patients and normal controls. *Bulletin of the Psychosomic Society*, **15**, 69–72.
Mansfield, P. S. & Busse, T. V. (1981) *The Psychology of Creativity and Discovery*. Chicago: Nelson Hall.
Mansfield, P. S., Busse, T. V. & Krepelka, E. (1978) The effectiveness of creativity training. *Review of Educational Research*, **48**, 517–36.
Manuel, F. E. (1968) *A Portrait of Isaac Newton*. Cambridge, Mass.: Harvard University Press.
Maranell, G. M. (1970) The evaluation of presidents: An extension of the Schlesinger polls. *Journal of American History*, **57**, 104–13.
Marquardt, M. (1949) *Paul Ehrlich*. London: Heinemann.
Marshall, M. H., Neumann, C. P. & Robinson, M. (1979) Lithium, creativity and manic-depressive illness: Review and prospectus. *Psychosomatics*, **11**, 406–88.
Martin, I. & Levey, A. B. (1978) Evaluative Conditioning. *Advances in Behaviour Research and Therapy*, **1**, 57–101.
Martindale, C. (1972) Father's absence, psychopathology, and poetic eminence. *Psychological Reports*, **31**, 843–7.
Martindale, C. (1977) Creativity, consciousness and cortical arousal. *Journal of Altered States of Consciousness*, **3**, 69–87.
Martindale, C. (1981) *Cognition and Consciousness*. Homewood, Ill.: Dorsey.
Martindale, C. (1989) Personality, situation, and creativity. In: J. A. Glover, R. R. Ronning & C. R. Reynolds (Eds.), *Handbook of Creativity*. New York: Plenum.
Martindale, C. (1990) *The Clockwork Muse*. New York: Basic Books.
Martindale, C. (1991) *Cognitive Psychology: A Neural-Network Approach*. Pacific Grove, C. A.: Brooks/Cole.
Martindale, C. (1992) How can we measure a society's creativity? In: *Achievement Project Symposium*, Ashford, Kent.
Martindale, C. & Armstrong, J. (1974) The relationship of creativity to cortical activation and its operant control. *Journal of Genetic Psychology*, **124**, 311–20.
Martindale, C. & Greenough, J. (1973) The differential effect of increased arousal on creative and intellectual performance. *Journal of Genetic Psychology*, **123**, 329–35.
Martindale, C. & Hasenfus, N. (1978) EEG differences as a function of creativity, stage of the creative process and effort to be original. *Biological Psychology*, **6**, 157–67.

Martindale, C. & Hines, D. (1975) Creativity and cortical activation during creative, intellectual and EEG feedback tasks. *Biological Psychology*, **3**, 71–100.
Martindale, C., Hines, D., Mitchell, L. & Covello, E. (1984) EEG alpha asymmetry and creativity. *Personality and Individual Differences*, **5**, 79–86.
Maslow, A. (1976) Creativity in self-actuating people. In: A. Rothenberg & L. P. Hausman (Eds.), *The Creativity Question*, pp. 86–92. Durham, N. C.: Duke University Press.
Mason, S. T. (1984) *Catecholamines and Behaviour*. Cambridge: Cambridge University Press.
Masserman, J. H. & Carmichael, H. T. (1938) Diagnosis and prognosis in psychiatry. *Journal of Mental Science*, **84**, 853–946.
Matarazzo, J. D. (1972) *Wechsler's Measurement and Appraisal of Adult Intelligence*, 5th Edition. Baltimore: Williams and Wilkins.
Maudsley, H. (1909) Heredity variation and genius. *Medical and Legal Journal*, **27**, 117–29.
Maxwell, A. E. (1972) Difficulties in a dimensional description of symptomatology. *British Journal of Psychiatry*, **121**, 19–26.
Mednick, S. A. (1962) The associative basis of the creative process. *Psychological Review*, **69**, 220–32.
Mednick, S. A. & Mednick, M. T. (1964) An associative interpretation of the creative process. In: C. W. Taylor (Ed.), *Widening Horizons in Creativity*, pp. 54–68. New York: Wiley.
Mednick, S. A. & Schulsinger, F. (1968) Some premorbid characteristics related to breakdown in children with schizophrenic mothers. *Journal of Psychiatric Research*, Suppl. 1, **3**, 267–91.
Medvedev, Z. A. (1969) *The Rise and Fall of T. D. Lysenko*. New York: Columbia University Press.
Medvedev, Z. A. (1971) *The Medvedev Papers*. London: Macmillan.
Meehl, P. E. (1954) *Clinical versus Statistical Prediction: A Theoretical Analysis and a Review of the Evidence*. Minneapolis: University of Minnesota Press.
Meehl, P. E. (1962) Schizotaxia, schizotypy, schizophrenia. *American Psychologist*, **17**, 872–38.
Meehl, P. E. (1989) Schizotaxia revisited. *Archives of General Psychiatry*, **46**, 935–44.
Meeker, M. (1985) Towards a psychology of giftedness: A concept in search of measurement. In: B. B. Wolman (Ed.), *Handbook of Intelligence: Theories, Measurement and Applications*, pp. 787–99. New York: Wiley.
Meltzer, H. & Stahl, S. (1976) The dopamine hypothesis in schizophrenia. *Schizophrenia Bulletin*, **2**, 19.
Mendelssohn, K. (1976) *Science and Western Domination*. London: Thomas and Hudson.
Menninger, K., Ellenberger, H., Prayser, P. & Mayman, M. (1958) The unitary concept of mental illness. *Bulletin of the Menninger Clinic*, **22**, 4–12.
Merritt, R. D. & Balogh, D. W. (1984) The use of a backward masking paradigm to assess information-processing deficits among schizophrenics: a re-evaluation of Sheronko and Woods. *Journal of Nervous and Mental Disease*, **172**, 216–24.
Merten, T. (1992) Wortassoziation und Schizophrenie: Eine empirische Studie. *Der Nervenarzt*, **63**, 401–8.
Merten, T. (1993) Word association responses and psychoticism. *Personality and Individual Differences*, **14**, 837–9.
Merton, R. K. (1961a) The role of genius in scientific advance. *New Scientist*, **12**, 306–8.
Merton, R. K. (1961b) Singletons and multiples in scientific discovery: A chapter in the sociology of science. *Proceedings of the American Philosophical Society*, **105**, 470–86.
Merton, R. K. (1973) *The Sociology of Science*. Chicago: University of Chicago Press.
Metcalfe, J. & Wiehe, D. (1987) Intuition in insight and insight problem-solving. *Memory and Cognition*, **15**, 238–46.
Micceri, T. (1993) The unicorn, the normal curve, and other improbable creatures. *Psychological Bulletin*, **105**, 156–66.
Michael, W. B. & Wright, C. R. (1989) Psychometric issues in the assessment of creativity. In: J. A. Glover, R. R. Ronning and C. R. Reynolds (Eds.), *Handbook of Creativity*, pp. 33–52. New York: Plenum Press.

Middleton, W. E. K. (1966) *A History of the Thermometer*. Baltimore: The Johns Hopkins Press.
Miers, T. C. & Raulin, M. L. (1985) The development of a scale to measure to cognitive slippage. *Paper read at the Eastern Psychological Association Convention, Boston.*
Miller, A. I. (1992) Scientific creativity: A comparative study of Henri Poincaré and Albert Einstein. *Creativity Research Journal*, **5**, 385–418.
Miller, D. J. & Hersen, M. (1992) *Research Fraud in the Behavioural and Biomedical Sciences*. New York: Wiley.
Miller, E. N. & Chapman, L. J. (1983) Continued word association in hypothethically psychiatric-prone college students. *Journal of Abnormal Psychology*, **92**, 468–78.
Miller, G. A. (1956) The magical number seven, plus or minus two: Some limits on our capacity for processing information. *The Psychological Review*, **63**, 81–97.
Miller, G. A. (1962) *Psychology: The Science of Mental Life*. New York: Harper & Row.
Miller, J. G. (1942) *Unconsciousness*. New York: Wiley.
Miller, S., Saccuzzo, D. P. & Braff, D. L. (1979) Information processing deficits in remitted schizophrenia. *Journal of Abnormal Psychology*, **88**, 446–9.
Mjoen, J. A. (1925) Zur Erhanalyse der musikatischen Begabung. *Hereditas*, **7**, 109–28.
Moises, H. W. (1991) No linkage between D_2 dopamine receptor gene region and schizophrenia. *Archives of General Psychiatry*, **48**, 643–7.
Moles, A. (1968) *Information Theory and Esthetic Perception*. Urbana: University of Illinois Press. (Translated from German edition, published in 1958).
Moone, O. K. & Anderson, S. (1954) Search behavior and problem solving. *American Sociological Review*, **19**, 702–14.
Moran, L. J. (1953) Vocabulary knowledge and usage among normal and schizophrenic subjects. *Psychological Monograph*, **67** (Whole No. 370).
Mordell, L. J. (1941) Ramanujan. *Nature*, **148**, 642–7.
Moriyama, I. M., Krueger, P. E. & Stamler, J. (1971) *Cardiovascular Diseases in the United States*. Cambridge, MA.: Harvard University Press.
Morris, R. E. M. (Ed.) (1989) *Parallel Distributed Processing: Implications for Psychology and Neurology*. Oxford: Clarendon Press.
Most, G. (1989) The second Homeric renaissance: Allegorosis and genius in early modern poetics. In: P. Murray (Ed.), *Genius: History of an Idea*, pp. 54–75. Oxford: Basil Blackwell.
Moulin, L. (1955) The Nobel Prizes for the sciences from 1901–1950: An essay in sociological analysis. *British Journal of Sociology*, **6**, 246–63.
Munchard, M. D. & Gustafson, S. B. (1988) Creativity syndrome: Integration, application, and innovation. *Psychological Bulletin*, **103**, 27–43.
Muntaner, C., Garcia-Sevilla, L., Fernandos, A. & Torrubia, R. (1988) Personality dimensions, schizotypal and borderline personality traits and psychosis proneness. *Personality and Individual Difference*, **9**, 257–85.
Murray, H. A., MacKinnon, D. W., Miller, J. G., Fiske, D. W. & Haufmann, E. (1963) *Assessment of Men*. New York: Holt, Rinehart and Winston.
Murray, P. (1989a) *Genius: The History of an Idea*. Oxford: Basil Blackwell.
Murray, P. (1989b) Poetic genius and its classical origins. In: P. Murray (Ed.), *Genius: The History of an Idea*, pp. 9–31. Oxford: Basil Blackwell.
Murray, R. (1991) The neurodevelopmental basis of sex differences in schizophrenia. *Psychological Medicine*, **21**, 565–75.
Murray, R. H. (1925) *Science and Scientists in the Nineteenth Century*. London: Sheldon.
Murtaugh, M. (1985) The practice of arithmetic by American grocery shoppers. *Anthropology and Education. (Quoted by Ceci.)*
Myers, I. B. & McCaulley, M. H. (1985) *Manual: A Guide to the Development and Use of the Myers–Briggs Type Indicator*. Palo Alto, C. A.: Consulting Psychologists Press.
Neill, V. T. (1977) Inhibitory and facilitatory processes in selective attention. *Journal of Experimental Psychology: Human Perception and Performance*, **3**, 444–50.

Neisser, V. (1963) The multiplicity of thought. *British Journal of Psychology*, **54**, 1–14.

Neisser, V. (1967) *Cognitive Psychology*. New York: Appleton-Century-Crofts.

Neumann, H. (1859) *Lehrbuch der Psychiatrie*. Erlangen: Enke.

Newell, H., Shaw, J. C. & Simon, H. A. (1962) The processes of creative thinking. In: H. E. Gruber, G. Terrell and M. Wertheimer (Eds.), *Contemporary Approaches to Creative Thinking*, pp. 63–119. New York: Atherton Press.

Nichols, R. C. (1964) Parental attitudes of mothers of intelligent adolescents and creativity of their children. *Child Development*, **32**, 502–10.

Nicolson, H. (1947) The health of authors. *Lancet*, **II**, 709–14.

Nietzsche, F. (1927/1972) The birth of tragedy from the spirit of music. In: *The Philosophy of Nietzsche*. New York: Endon Library.

Nilsson, L. & Gardiner, J. M. (1991) Memory theory and the boundary conditions of the Tulviag–Wiseman Law. In: W. E. Hocksley and S. Levandowsky (Eds.), *Relating Theory and Data: Essays on Human Memory in Humor of Bennet B. Murdoch*. Hillsdale: Erlbaum.

Nisbett, R. E. & Wilson, T. D. (1977) Telling more than we can know: Verbal reports of mental processes. *Psychological Review*, **84**, 231–59.

Noble, E. P. (1993) The D_2 dopamine receptor gene: A review of association studies in alcoholism. *Behavior Genetics*, **23**, 119–29.

Noble, E. P., Blum, K. & Khalsa, H. (1992) Allelic association of the D_2 dopamine receptor gene in cocaine dependence. *18th Collegium Internationale Neuro-Psychopharmacologium, Nice, France, June 28–July 2. Quoted in Noble, 1993.*

Norman, D. A. (1970) *Memory and Attention*. New York: Wiley.

Norman M. F. (1972) *Markov Processes and Learning Models*. New York: Academic Press.

Norris, V. (1959) *Mental Illness in London*. London: Chapman and Hall.

Notcutt, B. (1943) Perservation and fluency. *British Journal of Psychology*, **33**, 200–208.

Nothen, M. M. (1992) Lack of association between D_1 and D_2 dopamine receptor genes and bipolar affective disorder. *American Journal of Psychiatry*, **149**, 199–201.

Nuechterlein, K. H. & Dawson, M. E. (1984) Information processing and attentional functioning in the developmental course of schizophrenic disorders. *Schizophrenia Bulletin*, **10**, 160–203.

Nyborg, H. (1979) Sex chromosome abnormalities and cognitive performance. V: Female sex hormones and discontinuous cognitive development. *Paper read at symposium on 'Cognitive Studies' at the Fifth Biennial Meeting of the International Society for the Study of Behavioural Development*, Lund, June 25–9.

Nyborg, H. (1983) Spatial ability in men and woman: Review and a new theory. *Advances in Human Research and Therapy*, **5** (whole issue), 39–40.

Nyborg, H. (1987) Individual differences or different individuals: That is the question. *The Behavioral and Brain Sciences*, **10**, 34–5.

Nyborg, H. (1988) Sex hormones and covarient body, brain and behavioral development. *Neuroendocrinology*, Letters (Abstract), **10**, 217.

Nyborg, H. (1991) Development of exceptional scientific creativity. *Paper presented at the XXII Congress of the International Society of Psychoneuroendocrinology, Sienna, June 17–20. 1991.*

Nyborg, H. (1993) *Sex, Body, Mind and Society: The Physiological Approach*. New York: Greenwood.

Nyborg, H., Nielsen, J., Naera, R. & Kastrup, K. W. (1992) Sex hormone therapy harmonizes body, brain, and ability development and restores visio-spatial ability in young girls with Turner's syndrome. *Paper presented at the 39th Annual Meeting of the American Academy of Child and Adolescent Psychiatry, Washington, D. C., October 20–5, 1992.*

O'Boyle, M. & Benbow, C. (1990) Handedness and its relationship stability and talent. In: S. Coren (Ed.), *Left-handedness: Behavioural Implications and Anomalies*, pp. 343–72. Amsterdam: North-Holland.

O'Callaghan, E. Gham, P. & Takis, N. (1991) Schizophrenia after prenatal exposure to 1951

A2 influenza epidemic. *Lancet*, **337**, 1248–50.
Oades, R. D. (1985) The role of norodrenaline in tuning and dopamine in switching between signals in the CNS. *Neuroscience and Biobehavioral Reviews*, **9**, 261–82.
Ochse, R. (1990) *Before the Gates of Excellence: The Determinants of Creative Genius.* Cambridge: Cambridge University Press.
Ochse, R. (1991) The relation between creative genius and psychopathology: An historical perspective and a new explanation. *South African Journal of Psychology*, **21**, 45–53.
Odegaard, O. (1966) An official diagnostic classification in an actual hospital practice. *Acta Psychiatrica Scandinavica*, **42**, 329–37.
Ogburn, W. K. & Thomason, D. (1922) Are inventions inevitable? A note on social evolution. *Political Science Quarterly*, **37**, 83–93.
Okuda, S. M., Runco, M. A. & Berger, D. (1991) Creativity and the finding and solving of real-world problems. *Journal of Psychoeducational Assessment*, **9**, 45–53.
Oltmans, T. F., Ohayan, J. & Neale, J. M. (1978) The effects of antipsychotic mediation and diagnostic criteria on distractibility in schizophrenia. *Journal of Psychiatric Research*, **14**, 81–91.
Oppenheimer, R. (1955) *The Open Mind.* New York: Simon and Schuster.
Osborn, A. F. (1953) *Applied Imagination.* New York: Scribner.
Osgood, C. E., Suci, G. J. & Tannenbaum, P. H. (1957) *The Measurement of Meaning.* Urbana: University of Illinois Press.
Over, D. (1980) Research productivity and impact of men and women in departments of psychology in the United Kingdom. *Bulletin of the British Psychological Society*, **33**, 385–6.
Over, R. (1982) The durability of scientific reputation. *Journal of the History of the Behavioral Sciences*, **18**, 53–61.
Overall, J. E. & Gorham, D. R. (1962) The Brief Psychiatric Rating Scale. *Psychological Reports*, **10**, 799–812.
Overbye, D. (1991) *Lonely Hearts of the Cosmos.* London: Macmillan.
Owen, W. A. (1969) Cognitive, non-cognitive and environmental correlates of mechanical ingenuity. *Journal of Applied Psychology*, **53**, 199–208.
Pahnatier, J. R. & Bornstein, P. H. (1980) Effects of subliminal stimulation of symbiatic merging fantasies on behavioral treatment of smokers. *Journal of Nervous and Mental Diseases*, **168**, 715–20.
Panofsky, E. (1962) Artist, Scientist, Genius: Notes on the Renaissance in Dammerung. In: *The Renaissance: Six Groups.* New York: Praeger.
Pareto, V. (1897) *Cours d'economie politique.* Paris: Armand.
Parry, J. B. (1959) The place of personality appraisal in vocational selection. *Occupational Psychology*, **33**, 147–56.
Pascal, B. (1925) *Pensees sur la verite de la religion chretienne.* Paris: Alcon.
Patai, R. (1971) *Tents of Jacob.* Englewood Cliffs, N. J.: Prentice-Hall.
Pato, C. N., Lander, E. S. & Schulz, S. C. (1989) Prospects for the genetic analysis of schizophrenia. *Schizophrenia Bulletin*, **15**, 365–72.
Patterson, T. (1987) Studies toward the subcortical pathogenesis of schizophrenia. *Schizophrenia Bulletin*, **13**, 555–76.
Pavy, D. (1968) Verbal behavior in schizophrenia. *Psychological Bulletin*, **70**, 164–78.
Payne, R. W. (1960) Cognitive abnormalities. In: H. J. Eysenck (Ed.), *Handbook of Abnormal Psychology*, pp. 193–261. London: Pitman.
Payne, R. W. & Hewlett, J. H. G. (1960) Thought disorder in psychotic patients. In: H. J. Eysenck (Ed.), *Experiments in Personality*, pp. 3–104. London: Routledge & Kegan Paul.
Payne, R. W., Matussek, P. & George, E. I. (1959) An experimental study of schizophrenic thought disorder. *Journal of Mental Science*, **105**, 627–52.
Pearce, J. M. & Hall, G. (1980) A model for Pavlovian learning: Variations in the effectiveness of conditioned but not of unconditioned stimuli. *Psychological Review*, **87**, 532–52.
Pearson, K. (1914) *The Life, Letters and Labours of Francis Galton.* Cambridge: Cambridge

University Press.
Pearson, R. (1991) *Race, Intelligence and Bias in Academe*. Washington, D. C.: Scott-Townsend Publishers.
Penrose, L. S. (1968) Critical survey of schizophrenic genetics. In: J. G. Howells (Ed.), *Modern Perspectives in World Psychiatry*. Edinburgh: Oliver & Boyd.
Perkins, D. (1992) The topography of invention. In: R. Weber & D. Perkins (Eds.) *Inventive Minds: Creativity: Technology*. Oxford: Oxford University Press.
Perris, C. (1974) A study of cycloid psychoses. *Acta Psychiatrica Scandinavica*, Supplement 253.
Peters, E., Pickering, A. & Hemsley, D. (1994) 'Cognitive Inhibition' and positive symptomatology in schizotypy. *British Journal of Clinical Psychology*, **33**, 33–48.
Pintner, R. & Farlano, G. (1943) Season of birth and mental differences. *Psychological Bulletin*, **40**, 25–35.
Planansky, K. (1972) Phenotypic boundaries and genetic specificity in schizophrenia. In: A. R. Kaplan (Ed.), *Genetic Factors in 'Schizophrenia'*, pp. 141–72. Springfield: C. C. Thomas.
Planck, M. (1949) *Scientific Autobiography*. New York: Philosophical Library.
Plomin, R., De Fries, J. C. & McClearn, G. E. (1990) *Behavioral Genetics* (Second edition). New York: W. H. Freeman and Co.
Plum, F. (1973) Neuropathological findings. In: F. O. Schmitt, G. Adelman & F. G. Worden (Eds.), *Neurosciences Research Symposium Summaries*, pp. 384–380. Cambridge: MIT Press.
Plutarch. Marcellus XVII 4–6. In: I. Thomas (Ed.), *Selections Illustrating the History of Greek Mathematics*. London: 1941, Vol. 2, 31–32.
Poincaré, H. (1908) *Science et Methode*. Paris: Ernest Flammarion.
Polanyi, M. (1958) *Personal Knowledge*. London: Routledge & Kegan Paul.
Polanyi, M. (1966) *The Tacit Dimension*. New York: Doubleday.
Pollock, H. M. & Malzberg, B. (1940) Hereditary and environmental factors in the causation of manic-depressive psychosis and dementia praecox. *American Journal of Psychiatry*, **96**, 1227–47.
Popper, K. R. (1959) *The Logic of Scientific Discovery*. London: Hutchinson.
Popper, K. R. (1979) *Conjectures and Refutations*. London: Routledge & Kegan Paul.
Porterfield, A. L. & Golding, S. L. (1985) Failure to find an effect of subliminal psychodynamic activation upon cognitive measures of pathology in schizophrenia. *Journal of Abnormal Psychology*, **94**, 630–9.
Posner, M. I. (1982) Cumulative development of attentional theory. *American Psychologist*, **37**, 168–79.
Post, F. (1994) Creativity and psychopathology: A study of 291 world famous men. *British Journal of Psychiatry*, **165**, 22–34.
Powell, A., Thomson, N., Hall, D. J. & Wilson, L. (1973) Parent–child concordance with respect to sex and diagnosis in schizophrenia and manic-depressive psychosis. *British Journal of Psychiatry*, **123**, 653–8.
Power, M. J. (1991) Cognitive science and behavioural psychotherapy: Where behaviour was, there shall cognition be. *Behavioural Psychotherapy*, **19**, 20–41.
Prentky, R. A. (1980) *Creativity and Psychopathology*. New York: Praeger.
Price, D. (1963) *Little Science, Big Science*. New York: Columbia University Press.
Pritchard, W. S. (1986) Cognitive event-related potential correlates of schizophrenia. *Psychological Bulletin*, **100**, 43–66.
Pritchard, W. S. (1993) The link between cigarette smoking and P: A serotonergic hypothesis. *Personality and Individual Differences* (in press).
Quillian, M. R. (1967) Word concepts: A theory and simulation of some basic semantic capabilities. *Behavioral Science*, **12**, 410–30.
Radford, J. (1990) *Child Prodigies and Exceptional Early Achievers*. New York: Harvester Wheatsheaf.
Raine, A. (1994) Schizotypal and borderline features in psychopathic criminals. In: *Personality*

and Individual Differences. Baltimore: Johns Hopkins University.
Rainoff, T. J. (1929) Wave-like fluctuation of creative productivity in the development of West-European physics in the eighteenth and nineteenth centuries. *Isis*, **12**, 287–319.
Rank, O. (1945) *Will Therapy and Truth and Reality*. New York: Alfred Knopf.
Rapaport, R. (1945) *Diagnostic Clinical Testing*. Chicago: Year Book Publishers.
Raskin, E. (1936) Comparison of scientific and literary ability: A biographical study of eminent scientists and men of letters of the nineteenth century. *Journal of Abnormal and Social Psychology*, **31**, 20–35.
Ratcliff, R. (1978) A theory of memory retrieval. *Psychological Review*, **852**, 59–103.
Raulin, M. L. (1984) Development of a scale to measure intense ambivalence. *Journal of Consulting and Clinical Psychology*, **52**, 63–72.
Raulin, M. L. & Wee, J. L. (1984) The development and initial validation of a scale of social fear. *Journal of Clinical Psychology*, **40**, 780–4.
Rawlings, D. (1983) *An inquiry into the nature of psychoticism as a personality dimension*. Oxford: Unpublished Ph.D. thesis.
Rawlings, D. (1985) Psychoticism, creativity and dichotic shadowing. *Personality and Individual Differences*, **6**, 737–42.
Rawlings, D. & Borge, A. (1987) Personality and hemispheric function: two experiments using the dichotic shadowing technique. *Personality and Individual Differences*, **8**, 483–8.
Rawlings, D. & Claridge, G. (1984) Schizotypal hemispheric function. III. Performance asymmetries on task of letter recognition and local-global processing. *Personality and Individual Differences*, **5**, 657–63.
Rawlings, D., Sherr, D. & Dempsey, A. (1994) The personality correlates of music preference. *Personality and Individual Differences* (in press).
Razran, G. H. S. (1954) The conditioned evocation of attitudes (cognitive conditioning?) *Journal of Experimental Psychology*, **48**, 278–82.
Rees, H. J. & Israel, H. V. (1935) An investigation of the establishment and operation of mental sets. *Psychological Monograph*, **46**, No. 6 (Whole No. 210).
Reich, T., Cloninger, C. R., Svarez, B. & Rice, J. (1982) Genetics of affective disorders. In: J. K. Wing and L. Wing (Eds.), *Handbook of Psychiatry*, 3. Cambridge: Cambridge University Press.
Reich, T., James, J. V. & Morris, C. A. (1972) The use of multiple thresholds in determining the mode of transmission of semi-continuous traits. *Annals of Human Genetics*, **36**, 163–84.
Reich, W. (1976) The schizophrenia spectrum: A genetic concept. *The Journal of Nervous and Mental Disease*, **162**, 3–12.
Reid, R. L. (1960) Intuition – Pavlov, Hull, Cupack. *British Journal of Psychology*, **51**, 226–32.
Rennert, H. (1982) Zum Modell 'Universalgenese der Psychosen' – Aspekte einer unkonventionellen Auffassung der psychiachen Krankheiten. *Fortschritte der Neurologie und Psychiatrie*, **50**, 1–29.
Reznikoff, M., Domino, G., Bridges, C. & Honeyman, M. (1973) Creative abilities in identical and fraternal twins. *Behavior Genetics*, **4**, 365–377.
Richards, E. (1991) *Vitamin C and Cancer: Medicine or Politics?* London: Macmillan.
Richards, R. L. (1981) Relationship between creativity and psychopathology: An evaluation and interpretation of the evidence. *Genetic Psychological Monograph*, **103**, 261–324.
Richards, R. L., Kinney, D. K., Benet, M. & Merzel, A. (1988a) Assessing everyday creativity: Characteristics of the lifeline creativity scales and validation with three large samples. *Journal of Personality and Social Psychology*, **54**, 476–85.
Richards, R. L., Kinney, D. E., Lunde, I., Benet, M. & Merzel, A. (1988b) Creativity in manic-depressives, cyclothymes, their normal and relatives, and control subjects. *Journal of Abnormal Psychology*, **97**, 281–9.
Rieder, R. O. (1979) Borderline schizophrenics: Evidence of its validity. *Schizophrenia Bulletin*, **5**, 39–46.
Roberts, D. & Claridge, G. (1991) A genetic model compatible with a dimensional view of

schizophrenia. *British Journal of Psychiatry*, **158**, 451–6.
Robinson, T. N. & Zahn, T. P. (1979) Co-variation of two-flash threshold and autonomic arousal for high and low scores on a measure of psychoticism. *British Journal of Social and Clinical Psychology*, **18**, 431–41.
Robinson, T. N. & Zahn, T. P. (1985) Psychoticism and arousal: Possible evidence for a linkage of P and psychopathy. *Personality and Individual Differences*, **6**, 47–66.
Roe, A. (1951a) A psychological study of eminent biologists. *Psychological Monographs: General and Applied*, **65**. Whole No. 331.
Roe, A. (1951b) A psychological study of physical scientists. *Genetic Psychology*, **43**, 121–239.
Roe, A. (1953) A psychological study of eminent psychologists and anthropologists, and a comparison with biological and physical scientists. *Psychological Monographs: General & Applied*, **67**, Whole No. 352.
Rogers, C. A. (1956) The orectic relations of verbal fluency. *Australian Journal of Psychology*, **8**, 27–46.
Rogers, C. R. (1976) Toward a theory of creativity. In: A. Rothenberg and L. P. Hausman (Eds.), *The Creativity Question*, pp. 296–305. Durham, N. C.: Duke University Press.
Rose, P. L. (1993) *Wagner*. London: Faber.
Rosenblatt, F. (1958) The Perception: A probabilistic model for information storage and organization in the brain. *Psychological Review*, **65**, 386–408.
Rosengren, K. E. (1985) Time and literary fame. *Poetics*, **14**, 157–72.
Rosenthal, D. (1970) *Genetic Theory and Abnormal Behavior*. New York: MacGraw-Hill.
Rosenthal, D., Wender, P. H., Key, S. S., Schulsinger, F., Welner, J. & Oestergaard, L. (1968) Schizophrenic offspring reared in adoptive homes. In: D. Rosenthal and S. Kety (Eds.), *The Transmission of Schizophrenia*, pp. 377–91. Oxford: Pergamon Press.
Rossman, J. (1931) *The Psychology of the Inventor: A Study of the Patentee*. Washington: The Inventor's Publishing Co.
Rost, D. (1993) *Lebensumweltanalyse Hoch Begabter Kinder*. Gottingen: Hogrefe.
Rothenberg, A. (1976) Homospatial thinking in creativity. *Archives of General Psychiatry*, **33**, 17–26.
Rothenberg, A. (1979) *The Emerging Goddess*. Chicago: University of Chicago Press.
Rothenberg, A. (1986) Artistic creation as stimulated by superimposed versus combined-composite visual images. *Journal of Personality and Social Psychology*, **50**, 370–81.
Rothenberg, A. & Sobel, R. S. (1980) Creation of literary metaphors as stimulated by superimposed versus separated visual images. *Journal ofMental Imagery*, **4**, 77–91.
Rothenberg, M. A. (1967) The effect of 'social' instructions on word-association behavior. *Journal of Verbal Learning and Verbal Behavior*, **6**, 298–300.
Routh, D. K. (1971) Instructional effects on word association commonality in high and low 'schizophrenic' college students. *Journal of Personality Assessment*, **35**, 139–47.
Routh, D. K. & Schneider, J. M. (1970) Word association and ink blot responses as a function of instructional sets and psychopathology. *Journal of Projective Technique and Personal Assessment*, **34**, 113–20.
Rudin, E. (1916) *Zur Vererbung und Neuentshehung der Dementia Praecox*. Berlin: Springer Verlag.
Runco, M. A. & Albert, R. S. (1986) The threshold theory regarding creativity and intelligence: An empirical test with gifted and non-gifted children. *Creative Child and Adult Quarterly*, **11**, 217–18.
Runco, M. A. & Charles, R. E. (1993) Judgments of originality and appropriateness as predictors of creativity. *Personality and Individual Differences* (in press).
Runco, M. A., Ebersole, P. & Mraz, W. (1991) Creativity and self-actualization. *Journal of Social Behavior and Personality*, **6**, 161–7.
Rushton, J. P. (1989) A ten-year scientometric revisit of British psychology departments. *The Psychologist*, **2**, 64–8.
Rushton, J. P. (1992) Cranial capacity related to sex, rank, and race in a stratified random

sample of 6,325 U.S. military personnel. *Intelligence*, **16**, 401–13.
Saccuzzo, D. P. & Schubert, D. L. (1981) Backward masking as a measure of slow processing in schizophrenic spectrum disorders. *Journal of Abnormal Psychology*, **90**, 305–12.
Sandifer, M. G., Hordern, A., Truberg, G. C. & Green, L. M. (1968) Psychiatric diagnosis: A comparative study in North Carolina, London and Glasgow. *British Journal of Psychiatry*, **114**, 1–9.
Sarkar, G. (1991) Direct sequencing of the dopamine D_2 receptor DRD2 in schizophrenics reveals three polymorphisms but no structure of changes: the receptor. *Genomics*, **11**, 8–14.
Sarton, G. (1927–48) *Introduction in the History of Science*, 5 vols. Washington, D. C.: Carnegie Institute of Washington, D.C.
Sass, L. (1992) *Madness and Modernism*. New York: Basic Books.
Schachter, S. (1963) Birth order, eminence and higher education. *American Sociological Review*, **28**, 757–68.
Schalling, D., Aesberg, M. & Edman, E. (1984) Personality and CSF monoamine metabolites. *Quoted by Zuckerman, 1991*.
Schalling, D., Edman, E. & Aesberg, M. (1987) Impulsive cognitive style and inability to tolerate boredom: psychobiological studies of temperamental vulnerability. In: M. Zuckerman (Ed.), *Biological Basis of Sensation-Seeking, Impulsivity and Anxiety*. Hillsdale, N. J.: Lawrence Erlbaum.
Schare, K. W. (1983) What can we learn from the longitudinal study of adult psychological development? In: K. W. Schare (Ed.), *Longitudinal Studies of Adult Psychological Development*, **18**, 383–90.
Schmajuk, N. A. (1987) Animal match of schizophrenia: The hippocampally lesioned animal. *Schizophrenia Bulletin*, **12**, 317–27.
Schmidt, H. & Fonda, C. (1956) The reliability of psychiatric diagnosis: A new look. *Journal of Abnormal and Social Psychology*, **52**, 262–7.
Schmorkler, J. (1966) *Invention and Economic Growth*. Cambridge, M. A.: Harvard University Press.
Schneider, W. A. & Schiffrin, R. M. (1977) Controlled and automatic human information processing: I. Detention, search and attention. *Psychological Review*, **84**, 1–66.
Scholes, P. A. (Eds.) (1955) *The Oxford Companion to Music*, (9th edition). London: Oxford University Press.
Schou, M. (1968) Special review of lithium in psychiatric therapy and prophylaxis. *Journal of Psychiatric Research*, **6**, 67–95.
Schou, M. (1979) Artistic productivity and lithium prophylaxis in manic-depressive illness. *British Journal of Psychiatry*, **135**, 97–103.
Schubert, D. S. (1973) Intelligence is necessary but not sufficient for creativity. *Journal of Genetic Psychology*, **122**, 45–7.
Schuldberg, D. (1990) Schizotypal and hypermanic traits, creativity and psychological health. *Creativity Research Journal*, **3**, 218–30.
Schuldberg, D. (1993) Personal resourcefulness: Cognitive aspects of functioning in high-risk research. *Psychiatry*, **56**, 137–52.
Schuldberg, D., French, C., Stone, B. L. & Heberle, J. (1988) Creativity and schizotypal traits: Creativity test scores and perceptual aberration, magical ideation, and impulsive non-conformity. *Journal of Nervous and Mental Disease*, **176**, 648–57.
Schulz, B. (1940) Kinder manisch-depressiver und auderer affektiv psychotischen Ehepaare. *Zeitschrift fur die gesamte Neurologie und Psychiatrie*, **169**, 311–412.
Schwartz, S. (1978a) Do schizophrenics give rare word associations? *Schizophrenic Bulletin*, **4**, 248–51.
Schwartz, S. (1978b) Language and cognition in schizophrenia: A review of synthesis. In: S. Schwartz (Ed.), *Language and Cognition in Schizophrenia*, pp. 237–76. Hillsdale, N. J.: Erlbaum.
Schartz, S. (1982) Is there a schizophrenic language? *Behavior and Brain Science*, **5**, 579–88.

Scribner, S. (1984) Studying working intelligence. In: B. Rogoft and J. Lane (Eds.), *Everyday Cognition: Its Development in Social Context*. Cambridge, M. A.: Harvard University Press.

Seltzer, C. C. & Jablon, S. (1977) Army rank and subsequent morality by cancer: twenty-three year follow-up. *American Journal of Epidemiology*, **105**, 559–66.

Sen, A. & Hagtvet, K. (1993) Correlations among creativity, intelligence, personality, and academic achievement. *Perceptual and Motor Skills*, **77**, 497–8.

Shaw, E., Mann, J. & Stokes, P. (1986) Effects of lithium carbonate on creativity in bipolar patients. *American Journal of Psychiatry*, **143**, 1166–9.

Shepherd, M., Brooke, E. M., Cooper, J. E. & Lin, T. (1968) An experimental approach to psychiatric illness. *Acta psychiatrica Scandinavica*, Suppl. 201.

Sherman, G. F. & Galaburda, A. M. (1985) Asymmetries in anatomy and pathology in the rodent brain. In: S. D. Elick (Ed.), *Cerebral Lateralization in Non-Human Species*, pp. 89–107. Orlando: Academic Press.

Shields, J., Heston, L. L. & Gottesman, I. I. (1975) Schizophrenia and the schizoid: The problem for genetic analysis. In: R. R. Fieve, D. Rosenthal & H. Brill (Eds.), *Genetic Research in Psychiatry*, pp. 176–97.

Shiflett, S. C. (1989) Validity evidence for the Myers–Briggs Type Indicator as a measure of hemisphere dominance. *Educational and Psychological Measurement*, **49**, 741–5.

Shockley, W. (1957) On the statistics of individual variation of productivity in research laboratories. *Proceedings of the Institute of Radio Engineers*, **45**, 279–90.

Shurkin, J. N. (1992) *Terman's Kinds: The Groundbreaking Study of How the Gifted Grow Up*. Boston: Little, Brown.

Shyrock, R. H. (1936) *The Development of Modern Medicine*. Philadelphia: University of Philadelphia Press.

Siddle, D. & Remington, B. (1987) Latent inhibition in human Pavlovian conditioning: Research and relevance. In: G. Davey (Ed.), *Cognitive Processes and Pavlovian Conditioning in Humans*, pp. 115–16. New York: Wiley.

Siever, L., Amin, F., Corcaro, E., Trestman, R., Silverman, J., Horwath, T., Makon, T., Kurtt, P., Albstiel, L. & Davidson, M. (1993) CSF homovanillic acid – schizotypal personality disorder. *American Journal of Psychiatry*, **150**, 149–51.

Siever, L. J., Haier, R. J., Coursey, R. D., Sostek, A. J., Murphy, D. L., Holtzman, P. S. & Bucksbaum, M. S. (1982) Smooth pursuit tracking impairment: Relation to other 'markers' of schizophrenia and psychologic correlates. *Archives of General Psychiatry*, **39**, 1001–5.

Silverman, A. (1969) The scanning control mechanisms and 'cognitive' filtering in paranoid and non-paranoid schizophrenia. *Journal of Consulting Psychology*, **28**, 385–93.

Silverman, L. H. (1983) The subliminal psychodynamic activation studies. *Journal of Abnormal Psychology*, **91**, 126–30.

Silverman, L. H., Candell, P., Pettit, T. F. & Blum, E. (1971) Further data on effects of aggressive activation and symbiotic merging on ego functioning of schizophrenics. *Perceptual and Motor Skills*, **32**, 93–4.

Silverman, L. H., Franks, S. G. & Dachinger, P. (1974) A psychoanalytic reinterpretation of the effectiveness of a systematic desensitization: Experimental data bearing on the rule of emerging fantasies. *Journal of Abnormal Psychology*, **83**, 313–18.

Silverman, L. H., Levinson, P., Mendelsohn, E., Angaro, R. & Bronstein, A. (1975) A clinical application of subliminal psychoanalytic activation. *Journal of Nervous and Mental Disease*, **161**, 379–92.

Silverman, L. H., Martin,. A., Nugaro, X. & Mendelsohn, E. (1978) Effect of subliminal stimulation of symbiotic fantasies on behaviour modification treatment of obesity. *Journal of Consulting and Clinical Psychology*, **46**, 432–44.

Silverman, L. H., Ross, D. I., Adler, J. M. & Lustig, D. A. (1978) Simple research paradigm for demonstrating subliminal psychodynamic activation: Effects of oedipal stimuli on dart throwing accuracy in college students. *Journal of Abnormal Psychology*, **87**, 341–57.

Silverman, L. H. & Silverman, D. K. (1964) A clinical experimental approach to the study of

subliminal stimulation. *Journal of Abnormal and Social Psychology*, **69**, 158–72.
Silverman, L. H. & Weinberger, J. (1985) Mammy and I are one: Implications for psychotherapy. *American Psychologist*, **40**, 1296–308.
Simon, B. (1978) *Mind and Madness in Ancient Greece*. Ithaca: Cornell University Press.
Simon, H. A. (1981) *Sciences of the Artificial*. Cambridge, M. A.: MIT Press.
Simons, R. F. & Katkin, W. (1985) Smooth pursuit eye movements in subjects reporting physical anhedonia and perceptual aberrations. *Psychiatry Research*, **14**, 275–89.
Simonton, D. L. (1975) Creativity, task complexity, and intuitive versus analytical problem solving. *Psychological Reports*, **37**, 351–4.
Simonton, D. K. (1976a) Biographical determinants of achieved eminence: A multivariate approach to the Cox data. *Journal of Personality and Social Psychology*, **33**, 218–26.
Simonton, D. K. (1976b) Philosophical eminence, beliefs, and zeitgeist: An individual-generational analysis. *Journal of Personality and Social Psychology*, **34**, 630–40.
Simonton, D. K. (1976c) Do Sorokin's data support this theory? A study of generational fluctuations in philosphical beliefs. *Journal for the Scientific Study of Religion*, **15**, 187–98.
Simonton, D. K. (1977a) Eminence, creativity and geographic marginality: A recursial structural equation model. *Journal of Personality and Social Psychology*, **35**, 805–16.
Simonton, D. K. (1977b) Creative productivity, age, and stress: A biographical time-series analysis of 10 classical composers. *Journal of Personality and Social Psychology*, **35**, 791–04.
Simonton, D. K. (1978) The eminent genius in history: the critical role of creative development. *The Gifted Child Quarterly*, **22**, 187–200.
Simonton, D. K. (1980a) Intuition and analysis: A predictive and explanatory model. *Genetic Psychology Monograph*, **102**, 3–60.
Simonton, D. K. (1980b) Techno-scientific activity and war: A yearly time-series analysis, 1500–1903 A.D. *Scientometrics*, **2**, 251–5.
Simonton, D. K. (1981) Presidential greatness and performance: Can we predict leadership in the White House? *Journal of Personality*, **48**, 306–23.
Simonton, D. K. (1983a) Quality, quantity, and age: The careers of 10 distinguished psychologists. *International Journal of Ageing and Human Development*, **15**, 231–8.
Simonton, D. K. (1983b) Creative productivity and age: A mathematical model based on a two-step cognitive process. *Developmental Review*, **4**, 77–111.
Simonton, D. K. (1984a) Leaders as eponyms: Individual and situational determinants of monarchal eminence. *Journal of Personality*, **52**, 1–21.
Simonton, D. K. (1984b) *Genius, Creativity and Leadership*. Cambridge: M.A.L. Harvard University Press.
Simonton, D. L. (1987) Creativity, leadership, and chance. In: R. J. Sternberg (Ed.), *The Nature of Creativity*. Cambridge: Cambridge University Press.
Simonton, D. K. (1988a) *Psychology, Science, and History: An Introduction to Historiometry*. London: Yale University Press.
Simonton, D. K. (1988b) *Scientific Genius*. Cambridge: Cambridge University Press.
Simonton, D. K. (1989) Shakespeare's sonnets: A case of and for single-case historiometry. *Journal of Personality*, **57**, 1–26.
Simonton, D. K. (1990) *Psychology, Science and History*. New Haven: Yale University Press.
Simonton, D. K. (1991a) Latent-variable models of posthumous reputation: a quest for Galton's *G*. *Journal of Personality and Social Psychology*, **60**, 607–19.
Simonton, D. K. (1991b) Career endmarks in science: Individual differences and interdisciplinary contrasts. *Developmental Psychology*, **27**, 119–30.
Simonton, D. K. (1992) Emergence and realization of genius: The lives and works of 120 classical composers. *Journal of Personality and Social Psychology*, **61**, 829–46.
Simson, G. (1982) Genie und Irrsinn. In: E. Dreher (Ed.), *Der Mensch zwischen Anschaung und Abstraktion*, pp. 34–52. Basel: Krager.
Slade, P. D. (1976) An investigation of psychological factors involved in the predisposition to auditory hallucinations. *Psychological Medicine*, **6**, 123–32.

Slater, E. (1936) Inheritance of manic-depressive insanity. *The Lancet*, **1**, 429–31.
Slater, E. (1953) *Psychotic and neurotic illnesses in twins. Medical Research Council Report No. 278*. London: Her Majesty's Stationery Office.
Slaughter, F. G. (1950) *Immortal Magyar: Semmelweis, Conqueror of Childbed Fever*. New York: Henry Schuman.
Slominsky, N. (Ed) (1956–8) *Baker's Biographical Dictionary of Musicians* (5th edition). New York: Schirmer.
Smetslund, J. (1984) What is necessarily true in psychology. In: J. R. Royce & L. P. Mos (Eds.), *Annals of Theoretical Psychology*, **2**, 241–72.
Smith, R. (1990) *Collins Dictionary of Artificial Intelligence*. London: Collins.
Smith, S. (1992) Genetic vulnerability to drug abuse. *Archives of General Psychiatry*, **49**, 723–7.
Snapper, A. (1956) Mediating verbal responses in transfer of training. *Unpublished A.B. honors thesis, Harvard University. Quoted by Simonton, 1975.*
Snyderman, M. & Rothman, S. (1987) Survey of expert opinion on intelligence and aptitude testing. *American Psychologist*, **42**, 137–44.
Sobel, R. S. & Rothenberg, A. (1980) Artistic creation as stimulated by superimposed versus separated visual images. *Journal of Personality and Social Psychology*, **39**, 953–61.
Solomon, P. R., Kiney, C. A. & Scott, D. R. (1978) Disruption of latent inhibition following systemic administration of parachlor-orophenylalanine (PCPA). *Physiology and Behaviour*, **20**, 265–71.
Solomon, P. R., Nichols, G. W., Kierman, J. & Kamer, R. S. (1980) Differential effects of lesions in medical and dorsal raphe of the rat: Latent inhibition and septo hippocampal serotonin levels. *Journal of Comparative and Physiological Psychology*, **94**, 145–54.
Sorokin, P. A. (1925) Monarchs and rulers: A comparative statistical study. *Social Forces*, **4**, 22–35.
Sorokin, P. A. (1926) Monarchs and rulers: A comparative statistical study. II. *Social Forces*, **4**, 523–33.
Sorokin, P. A. (1937) *Social and Cultural Dynamics*, Vol. 2. New York: American Books.
Sorokin, P. A. (1937–41) *Social and Cultural Dynamics*, 4 vols. New York: American Books.
Sorokin, P. A. (1951) *Social Philosophies of an Age of Crisis*. Boston: Beacon Press.
Southern, M. & Plant, W. (1968) Personality characteristics of very bright adults. *Journal of Social Psychology*, **75**, 119–26.
Spearman, C. (1904) 'General intelligence', objectively observed and measured. *American Journal of Psychology*, **15**, 201–93.
Spearman, C. (1923) *The Nature of 'Intelligence' and the Principles of Cognition*. London: Macmillan.
Spearman, C. (1927) *The Abilities of Man*. London: Macmillan.
Spearman, C. (1931) *Creative Mind*. London: Macmillan.
Spearman, C. & Jones, L. (1950) *Human Ability*. London: Macmillan.
Spence, D. P., Klein, L. & Fernandez, R. J. (1984) Size and shape of the subliminal window. *Unpublished MS. Rutgers University Medical School. Quoted by Balay & Shevrin, 1988.*
Spence, J. (1984) Masculinity, femininity, and gender-related traits: A conceptual analysis and critique of current research. In: B. A. Maher & W. Maber (Eds.), *Progress in Experimental Research*, **13**, 2–97.
Spence, J. (1993) Gender-related traits and gender ideology: Evidence for a multifactorial theory. *Journal of Personality and Social Psychology*, **64**, 624–35.
Spence, J. & Helmreich, R. L. (1978) *Masculinity and Femininity: Their Psychological Dimensions, Correlates and Antecedents*. Austin: University of Texas Press.
Spence, J. D. (1984) *The Memory Palace of Matteo Ricci*. New York: Viking Penguin.
Spence, K. W. (1956) *Behaviour Theory and Conditioning*. New Haven, Conn.: Yale University Press.
Sperry, R. W. (1974) Lateral specialization in the surgically separated hemispheres. In: F. O. Schmitt & F. G. Worden (Eds.), *The Neurosciences: Third Study Programs*, pp. 5–19.

Cambridge: MIT Press.
Spitz, H. H. (1986) *The Raising of Intelligence*. London: Erlbaum.
Springer, S. & Deutsch, G. (1989) *Left Brain, Right Brain*, (3rd Edition). New York: Freeman.
Stamp, J. (1937) *The Science of Social Adjustment*. London: Macmillan.
Staszewski, J. (1989) Exceptional memory: The influence of practice and knowledge and the development of elaborative encoding strategies. In: W. Schneider and F. Weinert (Eds.), *Interaction among Aptitudes, Strategies, and Knowledge in Cognitive Performance*. New York: Springer Verlag.
Stavridou, A. & Furnham, A. (1994) The relationship between psychoticism, trait-creativity and the attention mechanism of cognitive inhibition. *Personality and Individual Differences* (in press).
Stayte, S. (1977) Incidental learning, creativity and schizophrenia. *London: Unpublished Dissertation, Institute of Psychiatry*.
Stein, L. (1978) Reward transmitters: Catecholamines and opioid peptides. In: M. A. Lipton, A., Di Marcio and K. F. Killam (Eds.), *Psychopharmacology: A Generation of Progress*. New York: Raven Press.
Stein, M. I. (1962) Creativity and the scientist. In: B. Barber and W. Hirsch (Eds.), *The Sociology of Science*, pp. 329–43. New York: Free Press.
Stein, M. I. (1975) *Stimulating Creativity*. New York: Academic Press.
Stelmack, R. M., Houlihan, M. & McGarry-Roberts, P. (1993) Personality, reaction-time, and event-related potentials. *Journal of Personality and Social Psychology*, **65**, 399–409.
Sternberg, C. (1956) Interests and tendencies toward maladjustment in a normal population. *Personnel and Guidance Journal*, **35**, 94–9.
Sternberg, R. J. (1985) *Beyond IQ*. Cambridge: Cambridge University Press.
Sternberg, R. J. (Ed.) (1988). *The Nature of Creativity*. Cambridge: Cambridge University Press.
Sternberg, R. J. & Detterman, D. H. (Eds.) (1986) *What is Intelligence*? Norwood, N. J.: Ablex.
Sternberg, R. J. & Wagner, R. K. (1986) *Practical Intelligence: Nature and Origins of Competence in the Everyday World*. New York: Cambridge University Press.
Stigler, S. M.l (1986) *The History of Statistics*. Cambridge: Belknap Press.
Storfer, M. D. (1990) *Intelligence and Giftedness*. Oxford: Jossey-Bass.
Storr, A. (1983) *The Dynamics of Creation*. Harmondsworth: Penguin.
Storr, A. (1989) Genius and psychoanalysis: Freud, Jung and the conceptual personality. In: P. Murray (Ed.), *Genius: The History of an Idea*, pp. 213–30. Oxford: Blackwell.
Strelau, J. & Eysenck, H. J. (Eds.) (1987) *Personality Dimensions and Arousal*. New York: Plenum Press.
Suedfeld, P., Tetlock, P. E. & Strenert, S. (1992) Conceptual/integrative complexity. In: C. P. Smith (Ed.), *Handbook of Thematic Analysis*. Cambridge: Cambridge University Press.
Sullwold, L. & Huber, G. (1986) *Schizophrene Banisstorungen*. Berlin: Springer.
Suppe, F. (Ed.) (1974) *The Structure of Scientific Theories*. Chicago: University of Illinois Press.
Sutton-Smith, B. & Rosenberg, B. (1970) *The Sibling*. New York: Holt, Rinehardt & Winston.
Swartberg, M. & Stiles, T. C. (1991) Comparative effects of short-term psychodynamic psychotherapy: A meta-analysis. *Journal of Counselling and Clinical Psychology*, **59**, 704–14.
Swerdlow, N. R. & Koob, G. F. (1987) Dopamine, schizophrenia, mania, and depression.: Toward a unified hypothesis of cortico–striato–pallidothalamic function. *Behavioural and Brain Sciences*, **10**, 197–245.
Swerdlow, N. R., Koob, G. F., Geyer, M. A., Mansbach, R. & Braff, D. I. (1988) A cross-species model of psychosis. In: P. Simon, P. Soubrie & D. Widlocher (Eds.), *Animal Models of Psychiatric Disorders*. Zurich: Karger.
Swingle, P. G. (1992) *Subliminal Treatment Procedures*. Sarasota: Professional Resource Press.
Szelenberger, W. (1979) Visual evoked response modified recovery cycle and personality

dimensions in healthy and schizophrenic subjects. *Biological Psychiatry*, **14**, 141–53.
Taggart, W. M. & Torrance, E. P. (1984) *Administrator's Manual: Human Information Processing Survey*. Bassenville, Il.: Scholastic Testing Service.
Taggart, W. M., Kroeck, K. & Escoffier, M. R. (1991) Validity evidence for the Myers–Briggs Type Indicator as a measure of hemisphere dominance: Another view. *Educational & Psychological Measurement*, **51**, 775–83.
Tanner, W. P. & Swets, J. A. (1954) A decision-making theory of visual detection. *Psychological Review*, **61**, 401–9.
Taylor, C. W. (1947) A factorial study of fluency in writing. *Psychometrika*, **12**, 239–62.
Taylor, C. W. (1988) Various approaches to and definitions of creativity. In: R. J. Sternberg (Ed.), *The Nature of Creativity*, pp. 99–121. Cambridge: University Press.
Taylor, P. J. (1980) ECT for schizophrenia. *Lancet*, **1**, 1380–2.
Taylor, R. E. & Eisenman, R. (1968) Birth order and sex differences in complexity–simplicity. Color–Form preference and personality. *Journal of Projective Techniques and Personality Assessment*, **32**, 383–7.
Tendler, A. D. (1945) Signified features of disturbance in free association. *Journal of Psychology*, **120**, 65–89.
Terman, L. M. (1917) The Intelligence quotient of Francis Galton in childhood. *American Journal of Psychology*, **28**, 209–15.
Terman, L. M. (1925) *Genetic Studies of Genius: Vol. I. Mental and Physical Traits of a Thousand Gifted Children*. Stanford: Stanford University Press.
Terman, L. M. & Oden, M. H. (1947) *Genetic Studies of Genius: Vol. IV. The Gifted Child Grows Up*. Stanford: Stanford University Press.
Terman, L. M. & Oden, M. H. (1959) *The Gifted Child at Mid-life*. Stanford: Stanford University Press.
Terry, L. S. (1929) *Johann Sebastian Bach*. Leipzig: Insel Verlag.
Thalbourne, M. (1994) Belief in the paranormal and its relationship to schizophrenic-relevant measures: A continuity study. *British Journal of Clinical Psychology*, **33**, 78–80.
Theios, J. (1973) Reaction-time measurements in the study of memory processes: Theory and data. In: G. H. Bower (Ed.), *The Psychology of Learning and Motivation*, Vol. 7, pp. 43–85. New York: Academic Press.
Thoren, V. E. (1990) *The Lord of Uraniborg*. Cambridge: Cambridge University Press.
Thorndike, E. L. (1950) Traits of personality and their intercorrelation as shown in biography. *Journal of Educational Psychology*, **41**, 193–216.
Thorne, A. & Gough, H. (1991) *Portraits of Type*. Palo Alto, CA: Consulting Psychologists Press.
Thurstone, L. L. (1927) A law of comparative judgment. *Psychological Review*, **34**, 273–86.
Thurstone, L. L. (1938) *Primary Mental Abilities*. Chicago: University of Chicago Press.
Thurstone, L. L. & Thurstone, T. G. (1941) *Factorial Studies of Intelligence*. Chicago: University of Chicago Press.
Tipper, S. P. (1985) The negative priming effect: Inhibitory priming by ignored subjects. *Quarterly Journal of Experimental Psychology*, **37a**, 571–90.
Tipper, S. P. & Bayliss, G. C. (1987) Individual differences in selective attention and priming. *Personality and Individual Differences*, **8**, 667–75.
Tipper, S. P. & Cranston, M. (1985) Selective attention and priming: Inhibitory and facilitatory effects of ignored primes. *Quarterly Journal of Experimental Psychology*, **37a**, 591–611.
Torrance, E. P. (1962) *Guiding Creative Talent*. Englewood Cliffs, N. J.: Prentice-Hall.
Torrance, E. P. (1972a) Career patterns and peak creative experiences of creative high school students 12 years later. *Gifted Children Quarterly*, **16**, 75–88.
Torrance, E. P. (1972b) Predicting validity of the Torrance Test of Creative Thinking. *Journal of Creative Behavior*, **6**, 236–52.
Torrance, E. P. (1974) *Torrance Tests of Creative Thinking: Norms – technical manual.*

Princeton: Personnel Press/Ginn.
Torrance, E. P. (1981) Predicting the creativity of elementary school children (1958–1980) – and the teachers who made a 'difference'. *Gifted Child Quarterly*, **25**, 55–62.
Torrance, E. P. (1984) Sounds and images production of elementary school pupils as predictors of the creative achievements of young adults. *Creative Child and Adult Quarterly*, **7**, 8–14.
Torrance, E. P. (1988) The nature of creativity as manifest in 15 testings. In: R. J. Sternberg (Ed.), *The Nature of Creativity*, pp. 43–75. Cambridge: Cambridge University Press.
Torrance, E. P., Taggart, B. A. & Taggart, W. M. (1984) *Human Information Processing Survey*. Bassenville, Il.: Scholastic Testing Service.
Toulouse, E. (1910) *Henri Poincaré*. Paris: Flammarion.
Treisman, A. (1964) Verbal cues, language and meaning in selective attention. *American Journal of Psychology*, **77**, 205–19.
Trethowan, W. H. (1977) Music and mental disorders. In: M. Critchley & R. E. Herson (Eds.), *Music and the Brain*, pp. 398–442. London: Heinemann.
Tsanoff, R. A. (1949) *The Ways of Genius*. New York: Harper and Bros.
Tsuang, M. T. (1979) Schizo-affective disorders dead or alive? *Archives of General Psychiatry*, **36**, 633–4.
Tsuang, M. T., Bucher, K. D. & Fleming, J. A. (1983) A search for 'schizophrenia spectrum disorders': An application of a multiple threshold model to blind family study data. *British Journal of Psychiatry*, **143**, 572–7.
Tsuang, M. T., Dempsey, G. M., Dvoredsky, N. & Strauss, A. (1977) A family history study of schizo-affective disorder. *Biological Psychiatry*, **12**, 331–8.
Tsuang, M. T., Faraone, S. V. & Fleming, J. A. (1985) Familial transmission of major affective disorders. *British Journal of Psychiatry*, **146**, 268–71.
Tsuang, M. T., Winokur, G. & Crowe, R. R. (1980) Mortality risks of schizophrenia and affective disorders among first-degree relatives of patientls with schizophrenia, mania, depression and surgical conditions. *British Journal of Psychiatry*, **137**, 497–504.
Uhl, C. R., Blum, K., Noble, E. P. & Smith, S. (1993) Substance abuse vulnerability and D_2 receptor genes. *Trends in Neuroscience*, **16**, 83.
Underwood, B. J. & Schulz, R. W. (1960 *Meaningfulness and Verbal Learning*. Chicago: Hippincott.
Upmanyu, V. V. & Kaur, K. (1986) Diagnostic ability of word association emotional indicators. *Psychological Studies*, **32**, 71–8.
Upmanyu, V. V. & Upmanyu, S. (1988) Utility of word association emotional indicators for predicting pathological characteristics. *Personality Study and Group Behavior*, **8**, 13–22.
Usher, A. (1954) *The History of Mechanical Inventions*. Cambridge: Harvard University Press.
Uttley, A. M. (1956) Conditioned probability-machines and conditioned reflexes. In: C. E. Sharman & M. McCarthy (Eds.), *Autom. Studies*. Princeton, N. J.: Princeton University Press.
Vaillant, G. E. (1962) The predicting of recovery in schizophrenia. *Journal of Nervous and Mental Disease*, **135**, 534–49.
Vallery-Radot, R. (1937) *The Life of Pasteur*. New York: Sun Dial Press.
Venables, P. H. (1963–4) The relationship between level of skin potential and fusion of paired light flashes in schizophrenics and normal subjects. *Journal of Psychiatric Research*, **1**, 279–87.
Venables, P. H. (1980) Primary dysfunction and cortical lateralization in schizophrenia. In: M. Kourkham, D. Lehmnann & J. Aupt (Eds.), *Function and States of the Brain: Their Determinants*. Amsterdam: Elsevier.
Venables, P. H. & O'Connor, N. (1959) A short scale for rating paranoid schizophrenics. *Journal of Mental Science*, **105**, 815–18.
Venables, P. H. & Wing, J. H. (1962) Level of arousal and the sub-classification of schizophrenia. *Archives of General Psychiatry*, **7**, 114–19.

Verma, R. M. & Eysenck, H. J. (1973) Severity and type of psychotic illness as a function of personality. *British Journal of Psychiatry*, **122**, 573–85.
Vernon, P. E. (1950) *The Structure of Human Abilities*. London: Methuen.
Vernon, P. E. (1979) *Intelligence: Heredity and Environment*. San Francisco: Freeman.
Vernon, P. E. (1989) The nature–nurture problem in creativity. In: J. A. Glover, R. R. Ronning & C. R. Reynolds. (Eds.) *Handbook of Creativity*, pp. 53–110. New York: Plenum Press.
Vernon, P. E. & Parry, J. B. (1949) *Personnel Selection in the British Forces*. London: University of London Press.
Veroff, J., Atkinson, J. W., Feld, S. C. & Gurin, G. (1960) The use of thematic apperception to assess motivation in a nationwide interview study. *Psychological Monograph*, **74** (Whole No. 499).
Vigotsky, L. S.k (1934) Thought in schizophrenia. *Archives of Neurology and Psychiatry*, **31**, 1063–77.
Visher, S. S. (1948) Environmental background of leading American scientists. *American Sociological Review*, **13**, 65–72.
Volkman, L. (1928) Ars memorativa. *Jahrbuch der Kunsthistorischen Sammlung in Wien*, **3**, 191–200.
Wagner, A. R. (1978) Expectancies and the priming of STM. In: S. H. Hubie, H. Fowler & W. M. Honig (Eds.), *Cognitive Processes in Animal Behavior*, pp. 177–209. Hillsdale, N. J.: Lawrence Erlbaum.
Walberg, H., Rasher, S. & Parkerson, J. (1979) Childhood of eminence. *Journal of Creative Behavior*, **13**, 225–31.
Walberg, H., Strykowski, B. F., Ronai, E. & Hung, S. S. (1984) Exceptional performance. *Review of Educational Research*, **54**, 87–112.
Wallace, A. (1986) *The Prodigy*. London: Macmillan.
Wallace, D. B. & Gruber, H. E. (Eds.) (1989) *Creative People at Work: Twelve Cognitive Case Studies*. New York: Oxford University Press.
Wallach, M. A. (1971) *The Intelligence/Creativity Distinction*. New York: General Learning Press.
Wallach, M. A. (1985) Creativity testing and giftedness. In: F. D. Horowitz and M. O'Brien (Eds.), *The Gifted and Talented*. Washington, D. C.: American Psychological Association.
Wallach, M. A. & Kogan, N. (1965) *Modes of Thinking in Young Children: A Study of the Creativity and Intelligence Distinction*. New York: Holt, Rinehart & Winston.
Wallach, M. A. & Wing, L. W. (1969) *The Talented Student*. New York: Holt, Rinehart & Winston.
Wallas, G. (1926) *The Art of Thought*. London: Watts.
Waller, N. G., Bouchard, T. J., Lykken, D. T., Tellegen, A. & Blacker, D. M. (1993) Creativity, heritability, familiality: Which word does not belong? *Psychological Inquiry* (in press).
Waller, N. G., Bouchard, T. J., Lykken, D. T., Tellegen, A. & Blacker, D. M. (1993) Why creativity does not run in families: A study of twins reared apart. *Psychological Inquiry*, **4**, 235–7.
Wankowski, J. A. (1973) *Temperament, Motivation and Academic Achievement*. Birmingham: University of Birmingham Educational Survey and Counselling Unit.
Ward, P. B., McConaghy, N. & Catts, S. V. (1991) Word association and measures of psychosis-proneness in university students. *Personality and Individual Differences*, **12**, 473–80.
Watkins, M. J. & Gardiner, J. M. (1979) An appreciation of generate-recognize theory of recall. *Journal of Verbal Learning and Verbal Behavior*, **18**, 687–704.
Watson, D. L. (1938) *Scientists are Human*. London: Watts.
Watts, F. (1990) Special issue on evaluative conditioning: Editorial. *Cognition and Emotion*, **4**, 1–2.

Webb, E. (1915) Character and Intelligence. *British Journal of Psychology*, Monograph Supplement, Vol. **1**, No. III.
Weber, R. & Perkins, D. (1992) *Inventive Minds: Creativity in Technology*. Oxford: Oxford University Press.
Wegener, A. (1915) *Die Entstehung der Kontinente und Ozeane*. Braunschweig: Vieweg.
Weinberg, I. & Lobstein, J. (1943) Inheritance in schizophrenia. *Acta Psychiatrica Scandinavica*, **18**, 93–104.
Weinberger, J. (1989) Response to Balay & Shevrin: Constructive critique or misguided attack? *American Psychologist*, **74**, 1417–19.
Weiner, I. (1990) Neural substrates of latent inhibition: The switching model. *Psychological Bulletin*, **108**, 442–61.
Weisberg, P. S. & Springer, K. J. (1961) Environmental factors in creative function. *Archives of General Psychiatry*, **5**, 64–74.
Weiss, V. (1977) Die Heritabilitaten sportlicher Tests. *Arzllicke Jugenkunde*, **68**, 167–72.
Weiss, V. (1982) *Psychogenetik*. Jena: Gustav Fischer.
Weiss, V. (1993) Leistungsstufen der Begabung und dreigliedriges Schulsystem. *Zeitschrift fuer Padagogische Psychologie*, **7**, 171–200.
Welsh, G. (1949) *Baron–Welsh Art Scale*. Palo Alto, Cal.: Consulting Psychologists Press.
Welsh, G. (1975) *Creativity of Intelligence: A Personality Approach*. Chapel Hill, N. C.: University of North Carolina Press.
Westcott, M. (1961) On the measurement of intuitive leaps. *Psychological Reports*, **9**, 267–74.
Westcott, M. (1964) Empirical studies of intuition. In: C. Taylor (Ed.), *Widening Horizons in Creativity*, pp. 34–53. New York: Wiley.
Westcott, M. (1966) A note on the stability of intuitive thinking. *Psychological Reports*, **19**, 194.
Westcott, M. (1968) *Toward a Contemporary Psychology of Intuition*. New York: Holt, Rinehart and Winston.
Westcott, M. & Ranzoni, J. (1963) Correlates of intuitive thinking. *Psychological Reports*, **12**, 595–613. (Monograph Suppl. 5–V12).
Westfall, R. S. (1973) Newton and the fudge factor. *Science*, **179**, 751–8.
Westfall, R. S. (1980) *Never at Rest: A Biography of Isaac Newton*. Cambridge: Cambridge University Press.
Wexler, B. E. (1979) The density of neuroses in the human hippocampus. *Neuropathology and Applied Neurobiology*, **4**, 249–64.
White, M. & Gribbin, J. (1992) *Stephen Hawking: A Life in Science*. London: Viking.
Whorf, B. (1960) *Language, Thought and Reality*. Cambridge, Mass.: Technology Press.
Whyte, L. C. (1962) *The Unconscious Before Freud*. New York: Anchor Books.
Willerman, L., Schultz, R., Rutledge, J. N. & Bigler, E. D. (1991) *In vivo* brain size and intelligence. *Intelligence*, **15**, 223–8.
Williams, L. P. (1965) *Michael Faraday*. London: Chapman & Hall.
Willis, S. L. (1985) Towards an educational psychology of the adult learner. In: J. E. Birren & K. W. Schaier (Eds.), *Handbook of the Psychology of Aging*. New York: Vankostrend Reinhold.
Wilson, C. (1968) Kepler's derivation of the elliptical path. *Iris*, **59**, 75–90.
Wilson, G. D. (1990) Personality, time of day and arousal. *Personality and Individual Differences*, **11**, 153–68.
Wise, R. A. (1987) Intravenous drug self-administration: A special case for positive reinforcement. In: M. Bozarth (Ed.), *Methods for Assessing the Reinforcing Properties of Abuse of Drugs*, pp. 117–42. New York: Springer Verlag.
Wolpert, L. (1992) *The Unnatural Nature of Science*. London: Faber & Faber.
Woodman, R. W. & Schoenfeldt, L. F. (1989) Individual differences in creativity. In: J. A. Glover, R. R. Ronning and C. R. Reynolds (Eds.), *Handbook of Creativity*, pp. 77–91. New York: Plenum Press.

Woodworth, R. S. (1941) *Heredity and Environment*. New York: Social Science Research Council, Report 47.
Woody, E. & Claridge, G. (1977) Psychoticism and thinking. *British Journal of Social and Clinical Psychology*, **16**, 241–8.
Wundt, W. (1896) *Lectures on Human and Animal Psychology*. New York.
Wyatt, R. J., Murphy, D. L., Belmaker, R., Cohen, S., Donnelly, C. H. & Pollin, W. (1974) Reduced monoamine oxidase activity in platelets: a possible genetic marker for vulnerability to schizophrenia. *Science*, **173**, 916–18.
Yasamy, M. T. (1987) Schizoaffective disorder: A dimensional approach. *Acta Psychiatrica Scandinavica*, **76**, 609–18.
Yates, F. A. (1966) *The Art of Memory*. London: Routledge & Kegan Paul.
Zahn, T. P. (1968) Word association in adoptive and biological parents of schizophrenics. *Archives of General Psychiatry*, **19**, 501–3.
Zeller, E. (1837) Bericht uber die Wirksamkeit der Heilanstalt Winnenthel. *Allgemeine Zeitschrift fur Psychiatrie*, **1**, 1–39.
Zenhausern, R. & Kraemer, M. (1991) The dual nature of lateral eye movements. *International Journal of Neuroscience*, **56**, 169–75.
Zhao, H. & Jiang, C. (1985) Shifting to world's scientific center and scientists' social age. *Scientometrics*, **8**, 59–80.
Zubin, J., Eron, L. D. & Schumer, F. (1965) *An Experimental Approach to Projective Techniques*. New York: Wiley.
Zuckerman, H. (1977) *Scientific Elite: Nobel Laureates in the United States*. New York: The Free Press.
Zuckerman, M. (1979) *Sensation Seeking: Beyond the Optimal Level of Arousal*. New York: Erlbaum.
Zuckerman, M. (1989a) Sensation seeking: A comparative approach to a human trait. *Behavioral and Brain Sciences*, **7**, 413–471.
Zuckerman, M. (1989b) Personality in the third dimension: A psychological approach. *Personality and Individual Differences*, **10**, 391–418.
Zuckerman, M. (1991) *Psychobiology of Personality*. Cambridge: Cambridge University Press.
Zuckerman, M., Kuhlman, D. M. & Camac, L. (1988) What lies beyond E and N? Factor analyses of scales believed to measure basic dimensions of personality. *Journal of Personality and Social Psychology*, **54**, 96–107.
Zuriff, G. E. (1985) *Behaviorism: A Conceptual Reconstruction*. New York: Columbia University Press.
Zusne, L. (1976) Age and achievement in psychology: The harmonic mean as a model. *American Psychologist*, **31**, 805–7.

Index

Page numbers in italic type refer to figures